The Canadian Niagara Power Company Story

The Canadian Niagara Power Company Story

NORMAN R. BALL

The BOSTON MILLS PRESS

In association with

FortisOntario Inc.

A BOSTON MILLS PRESS BOOK

Published by Boston Mills Press, 2005
132 Main Street, Erin, Ontario N0B 1T0
Tel: 519-833-2407 Fax: 519-833-2195
e-mail: books@bostonmillspress.com
www.bostonmillspress.com

In Canada:
Distributed by Firefly Books Ltd.
66 Leek Crescent
Richmond Hill, Ontario, Canada L4B 1H1

In the United States:
Distributed by Firefly Books (U.S.) Inc.
P.O. Box 1338, Ellicott Station
Buffalo, New York 14205

Library and Archives Canada Cataloguing in Publication

Ball, Norman R., 1944-
The Canadian Niagara Power Company story / Norman R. Ball.

Includes bibliographical references and index.
ISBN-13: 978-1-55046-462-7
ISBN-10: 1-55046-462-0

1. Canadian Niagara Power Company — History. 2. Electric power production — Ontario — Niagara River (N.Y. and Ont.) — History. I. Title.

HD9685.C34C35 2006 333.79'32'06071338 C2005-906417-X

Publisher Cataloging-in-Publication Data (U.S.)

Ball, Norman R., 1944-
The Canadian Niagara Power Company story/ Norman R. Ball.

[320] p.: photos. ; cm.

Includes bibliographical references and index.

Summary: A history of the Canadian Niagara Power Company, which has been generating power continuously at its station upstream from Niagara Falls since 1905, and a look at the impact of harnessing that electricity.

ISBN-13: 978-1-55046-462-7
ISBN-10: 1-55046-462-0

1. Canadian Niagara Power Company — History. 2. Electric power production — Ontario — Niagara River (N.Y. and Ont.) — History. I. Title.

333.79' 32 dc22 HD9685.C4C342B35 2006

The publisher acknowledges for their financial support of our publishing program, the Canada Council, the Ontario Arts Council and the Government of Canada through the Book Publishing Industry Development Program (BPIDP).

Design by Gillian Stead
Indexed by Philippa Campsie
Jacket photography by George Fischer
Printed in Canada by Friesens Printers

Contents

This striking crest, sometimes referred to as the company seal, was adopted by the Niagara Falls [NY] Power Company and its affiliated companies, namely Cataract Construction Company, Canadian Niagara Power Company, Niagara Junction Railway Company, and the Niagara Development Company.

The crest was modelled in 1895 by Frederick MacMonnies, a distinguished American sculptor living in Paris. It was featured most prominently in the solid granite carvings for the Edward Dean Adams Generating Station in Niagara Falls, New York, and the William Birch Rankine Generating Station in Niagara Falls, Ontario. The former station has been demolished, but the granite crest survives on Goat Island, where it was moved along with the reconstructed doorway and a bust of Nikola Tesla. In addition to letterhead and other official purposes, the crest also appeared on medals struck from a steel die engraved by Paulin Tasset, also a Parisian. The medals were given to key officials of the company, the French government, and five cultural institutions including the Metropolitan Museum of Art, New York, The Art Institute of Chicago, and the Buffalo [and Erie County] Historical Society.

(Reproduction by permission of FortisOntario)

Foreword

In 1892, when the Canadian Niagara Power Company came into being, revolutionary technological change was in the air. Some argued that not even the beauty of Niagara Falls should stand in the way of generating electric power for industry. Others believed that there was no place for industry of any kind at Niagara Falls. The private investors who created Canadian Niagara sought a middle ground. They planned to build a powerhouse in the Queen Victoria Niagara Falls Park that would contribute to the beauty of the area and the financial well-being of the park which the government of Ontario had created to preserve the beauty of the Falls. And so, from its earliest days, Canadian Niagara was tied to and reflected community values.

The company's community involvement took many forms, including responding to emergencies caused by major storms, donations to local charities and events, and helping people weather hard times such as the Depression of the 1930s or during the 1980s when jobs in many other companies were lost to automation, and in more recent years with the scaling back in production at Rankine.

Employees also had a sense of community involvement. United by hard work, pride in their achievements, and in many cases being the second or third generation to work for Canadian Niagara, they thought of themselves as the Canadian Niagara family. But the family is changing. Canadian Niagara, once American-owned, is now a Canadian-owned Fortis company. In 2009 Canadian Niagara's historic water rights on the Niagara River will end, and stewardship of the Rankine Generating Station will pass into the hands of the Niagara Parks Commission.

Canadian Niagara will continue as FortisOntario, but as one era ends and another begins, it is time to reflect upon all that the company and its employees have achieved and to capture something of the history and feelings of what it has meant to be a part of the Canadian Niagara family, whether as customer or employee. That process of discovery has taught us that there is not one single Canadian Niagara story, but many Canadian Niagara stories, far more than we could ever put in one book. We thought the research, writing, and publication of this book would tell the Canadian Niagara story, but now we know it is just a beginning, a challenge for others to tell more Canadian Niagara stories. FortisOntario will proudly continue the tradition of making and telling many new stories in the future. Our stories will extend beyond the Niagara Frontier, but wherever we go, our stories will continue to reflect the values of the communities and families that define who we are as a company. That, after all, is the heart of Canadian Niagara.

Bill Daley, President and CEO
FortisOntario

Introduction

When the Ontario Legislature gave a group of canny, far-sighted American investors monopoly rights to generate hydroelectric power on the Canadian side of the Niagara River, there were smiles all round. The investors, led by William Birch Rankine, Francis Lynde Stetson, and Albert D. Shaw, had bought water rights to go along with their access to funds and their confidence in the future of large-scale hydroelectric power. Both political parties in the legislature were so happy they pushed the necessary bill through without debate, lest the "lunatic" Americans change their minds. There was another reason for broad smiles — $50,000, the amount the newly formed Canadian Niagara Power Company had paid in advance to the Queen Victoria Niagara Falls Park for two years of water rights rental. The Park Commissioners desperately needed the money, because both political parties had vowed the government would never give even one cent to the parks for upkeep and operations, a vow from which they never wavered.

The company's promise of hydroelectric power took a little longer to fulfil than expected. Canadian Niagara was supposed to be generating power by November 1, 1898, but it took until 1905 before the generators were turned on. In the interval, Canadian Niagara had lost its monopoly rights and two other companies were moving ahead to generate power on the Canadian side of the Niagara River. Both of these companies soon sold out to the Hydro-Electric Power Commission of Ontario, later Ontario Hydro, but Canadian Niagara remained a small bastion of private power in a beautiful park on the Niagara River.

What in 1905 had been the world's largest-capacity turbines and generators were soon surpassed, and today there are single generators with more output than the eleven in the Rankine Generating Station. Nonetheless, the company has supplied power to both sides of the Niagara River for a century.

In the early years, it was difficult to foresee Canadian demand for all of the power generated at Niagara Falls, Ontario. Most of it went to New York State, notably Niagara Falls and Buffalo. However, the company eventually expanded its Canadian customer service area from Niagara Falls, Ontario, to Fort Erie, Bridgeburg, Amigari, Ridgeway, Stevensville, Crystal Beach, and Point Abino. And throughout Canadian Niagara's history, it has always provided power cheaper than its neighbouring competitors.

The Canadian Niagara Power generating station, now called the William Birch Rankine Generating Station, quickly became an important tourist attraction. More important, Canadian Niagara helped attract industry to the region. And through services such as an appliance showroom and sales centre, cooking classes, full-time staff nutritionists, and time-payment plans, the company ushered part of the Niagara Frontier into the electrical age. The Service Centre and Showroom, opened in 1927, was Fort Erie's most distinctive commercial building and a source of considerable municipal pride.

It takes many skills and jobs to run a company that generates electricity and maintains a complete system to deliver power to individual customers, along with metering, billing, and service, in spite of occasional lightning, snow, or ice storms. As a company that offered good wages and steady work, Canadian Niagara soon became a preferred employer in the area. Today second- and third-generation employees are not unusual.

Naturally the jobs have changed over time. The First and the Second World Wars, the Depression of the 1930s, the postwar boom, and the effects of industry restructuring and automation have changed the company, as they have all North American electrical utilities. But today, Canadian Niagara, now FortisOntario, is clearly established as part of the new electrical utility industry for the 21st century.

Despite the many changes, there are constants in the company's history. First, there is a strong commitment to employees, shown in the tradition of on-the-job training and company-encouraged education, along with a preference for promoting from within the ranks of current employees and an unwillingness to lay off employees when times are bad. Second, there is a high level of community involvement, including donating parkland, sponsoring sporting and cultural events, financial contributions for local educational institutions, and support for employee events. Above all, there is an intense feeling of employee loyalty and a sense of *family* — a word that comes up repeatedly when employees, past and present, talk about working for Canadian Niagara.

The loyalty and sense of family works two ways. On the one hand, the company benefits from employees' loyalty and commitment and the desire to serve the communities that they live in. One employee explained that when they are in the midst of cleaning up and getting power back on after a fierce winter storm, they are doing it for their parents, brothers, sisters, friends, and neighbours, for the people on their sports teams or the members of their church or service club. On the other hand, the sense of family means that more is expected of the company. When the company fails to measure up to employees' expectations, the judgments are harsher and the sense of disappointment felt more keenly than it might be if Canadian Niagara were just another company.

Canadian Niagara Power Company is an unusual company. In trying to capture both the essence and some of the details of its history, I have written a story that has many strands, but is not, I hope, too fragmented. *The Canadian Niagara Power Company Story* is partly a history of technology, partly a social and corporate history, and partly the story of adapting to change. I had to leave out more than I was able to fit in — stories, details, background, connections. So this book is just an introduction to the company, with its buildings, technology, distribution network, employees, and customers, each of which merits its own book.

Norman R. Ball
Toronto, 16 August 2005

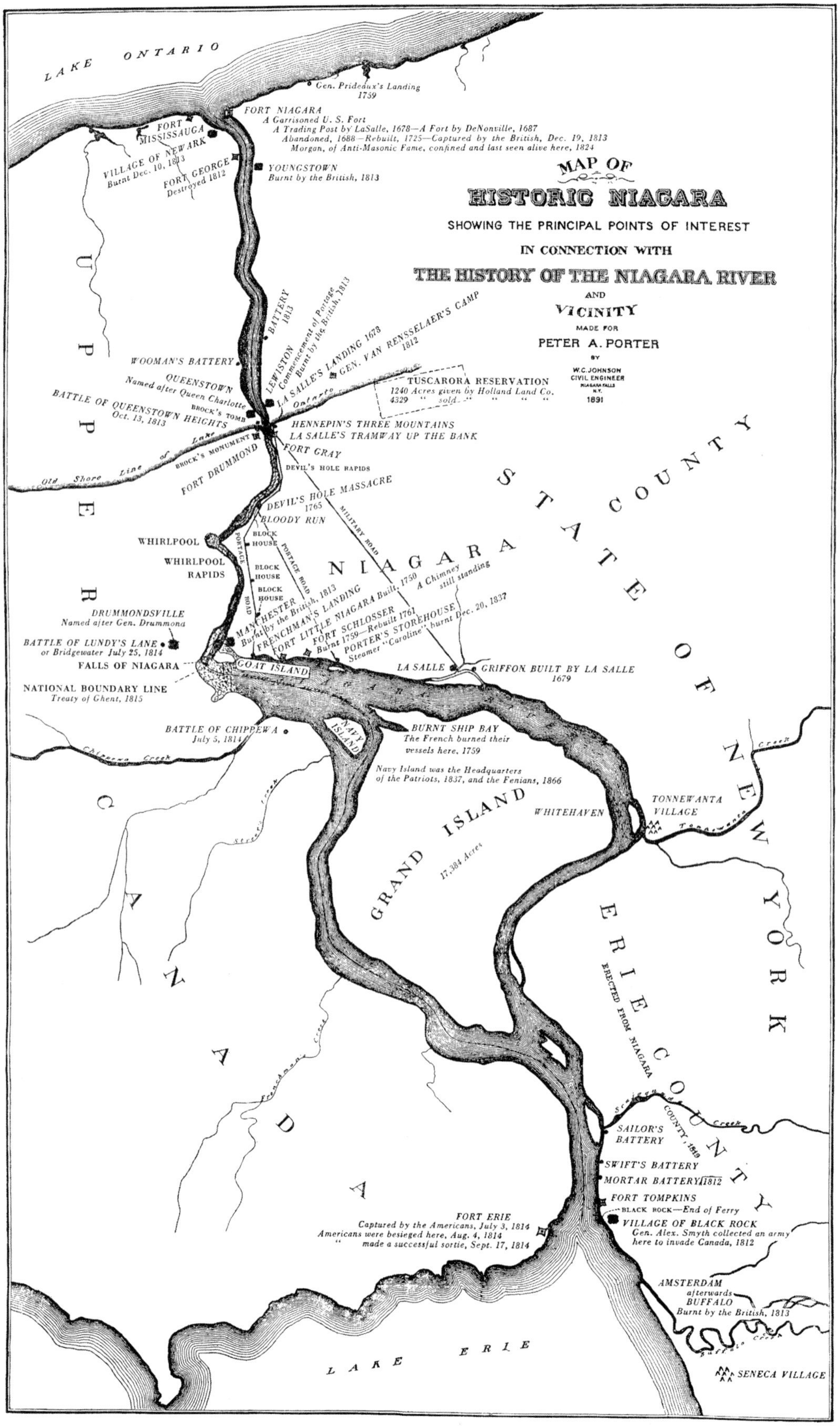
MAP OF
HISTORIC NIAGARA
SHOWING THE PRINCIPAL POINTS OF INTEREST
IN CONNECTION WITH
THE HISTORY OF THE NIAGARA RIVER
AND
VICINITY
MADE FOR
PETER A. PORTER
BY
W.C. JOHNSON
CIVIL ENGINEER
NIAGARA FALLS
N.Y.
1891
LAKE ONTARIO
Gen. Prideaux's Landing 1759
FORT NIAGARA
A Garrisoned U. S. Fort
A Trading Post by LaSalle, 1678—A Fort by DeNonville, 1687
Abandoned, 1688—Rebuilt, 1725—Captured by the British, Dec. 19, 1813
Morgan, of Anti-Masonic Fame, confined and last seen alive here, 1824
FORT MISSISSAUGA
VILLAGE OF NEWARK Burnt Dec. 10, 1813
FORT GEORGE Destroyed 1812
YOUNGSTOWN Burnt by the British, 1813
BATTERY 1813
LEWISTON Commencement of Portage Burnt by the British, 1813
LA SALLE'S LANDING 1678
GEN. VAN RENSSELAER'S CAMP 1812
TUSCARORA RESERVATION
1240 Acres given by Holland Land Co.
4329 " sold " " " "
WOOMAN'S BATTERY
QUEENSTOWN Named after Queen Charlotte
BROCK'S TOMB
BATTLE OF QUEENSTOWN HEIGHTS Oct. 13, 1813
HENNEPIN'S THREE MOUNTAINS
LA SALLE'S TRAMWAY UP THE BANK
FORT GRAY
BROCK'S MONUMENT
FORT DRUMMOND
Old Shore Line of Lake Ontario
DEVIL'S HOLE RAPIDS
DEVIL'S HOLE MASSACRE 1765
BLOODY RUN
MILITARY ROAD
WHIRLPOOL
WHIRLPOOL RAPIDS
BLOCK HOUSE
PORTAGE ROAD
UPPER CANADA
NIAGARA COUNTY
STATE OF NEW YORK
ERIE COUNTY
ERECTED FROM NIAGARA COUNTY, 1819
DRUMMONDSVILLE Named after Gen. Drummond
BATTLE OF LUNDY'S LANE or Bridgewater July 25, 1814
FALLS OF NIAGARA
MANCHESTER Burnt by the British, 1813
FRENCHMAN'S LANDING
FORT LITTLE NIAGARA Built, 1750
A Chimney still standing
FORT SCHLOSSER Burnt 1759—Rebuilt 1761
PORTER'S STOREHOUSE
Steamer "Caroline" burnt Dec. 20, 1837
LA SALLE
GRIFFON BUILT BY LA SALLE 1679
GOAT ISLAND
NATIONAL BOUNDARY LINE Treaty of Ghent, 1815
BATTLE OF CHIPPEWA July 5, 1814
Chippewa Creek
NAVY ISLAND
BURNT SHIP BAY The French burned their vessels here, 1759
Navy Island was the Headquarters of the Patriots, 1837, and the Fenians, 1866
GRAND ISLAND 17,384 Acres
WHITEHAVEN
TONNEWANTA VILLAGE
Tonnewanta Creek
Street's Creek
Frenchman Creek
Scajaquada Creek
SAILOR'S BATTERY
SWIFT'S BATTERY
MORTAR BATTERY 1812
FORT TOMPKINS
BLACK ROCK—End of Ferry
VILLAGE OF BLACK ROCK
Gen. Alex. Smyth collected an army here to invade Canada, 1812
FORT ERIE
Captured by the Americans, July 3, 1814
Americans were besieged here, Aug. 4, 1814
" made a successful sortie, Sept. 17, 1814
AMSTERDAM afterwards BUFFALO Burnt by the British, 1813
Buffalo Creek
SENECA VILLAGE
LAKE ERIE

Niagara Falls, Where All Is Not as It Seems

CHAPTER 1

On January 2, 1905, the Canadian Niagara Power Company formally opened its powerhouse on the grounds of the Queen Victoria Niagara Falls Park. It was the most sophisticated hydroelectric generating station in the world at the time: something worth celebrating, even if it was not quite finished. Only two of eleven generating units were installed. The remainder would take another 19 years to complete. The powerhouse itself, which had been built to less than half its final length, sat in the middle of a construction site. Blasting and excavating would continue for some time to come. But by January 2, enough had been accomplished to launch the station in a spirit of international cooperation and corporate pride.

William H. Beatty, a Canadian and president of the wholly American-owned company, turned on the first of what was then one of the largest-capacity generators in the world. Francis Lynde Stetson, a New York corporate lawyer and company shareholder, turned on the second 10,000-horsepower unit. Cheers rang out for King Edward (on behalf of the Canadians and British present), President Theodore Roosevelt (for the Americans), and William B. Rankine, who represented the parent Niagara Falls Power Company on the other side of the river.

Rankine had helped bring the Canadian Niagara Power Company into being and had kept it alive during a protracted and difficult infancy. Sadly, he would be dead before the end of the year. In 1927 the station was renamed the Rankine Generating Station. Although there is only a small plaque to tell the passerby that this is his station, that is neither insult nor oversight, merely the habit of a company so deliberately low-profile that it was only in 1992, one hundred years after the company had been founded, that the mayor of Niagara Falls convinced Canadian Niagara to put its name prominently on the building. Such modesty seems odd in a city renowned as a magnet for publicity-seekers. But all is not as it seems in Niagara Falls. No power was produced on January 2, and apparently no journalists noticed. And on July 27 that year, when power really did start to flow, nobody noticed either.

Today Canadian Niagara's quiet, almost self-effacing, cut-stone power station, with its green tile roof, seems to fit naturally into its park setting. There seems a certain inevitability to its placement, design, and very existence there. Yet there is nothing "natural" or "inevitable" about the building or even about Niagara's electrical industry. The history of the company and of the creation of the powerhouse is one of surprises and twists. It very nearly did not come into being, and the city around it developed in an unpredictable, sometimes startling, way.

In a way, the powerhouse is one answer to the questions that arose in the 19th century. What do you do with the precipitous edge of the Niagara escarpment and the mighty waters of the Niagara River? What do you do at the site of a waterfall second to none in terms of the volume of water flowing over it — What do you do with a waterfall with the world's largest millpond behind it? Lakes Erie, Huron, Michigan, and Superior? How do you sort out the competing interests that vie to use, enjoy, or protect one of the wonders of the world?

At left:
Edward Dean Adams included this 1891 map in his 1927 book *Niagara Power*. In addition to showing the sites of many historic events, it shows the falls of Niagara far upriver from the escarpment at Queenston Heights. It also clearly indicates the sharp bends in the River that favoured the shorter portage route on the east or American side of the Niagara River. This in turn spurred earlier economic activity on the same side. (Reproduction by permission of John Burtniak)

This is the story of a company that found a way to answer those questions, and did so inside a park. Its opening signalled that natural beauty and industry could co-exist. Before it opened, most people would have thought such co-existence impossible. Niagara Falls was a place where interests collided, where people lost money to tourist traps and get-rich-quick schemes, where everyone was out to grab a piece of the action. It often brought out the worst in people. Niagara Falls was, from the very beginning, a problem to be solved.

An Inconvenient Impediment

Those who saw the Falls before the 19th century saw something very different from today's Falls. The first European to arrive was Father Louis Hennepin, a Recollet priest who arrived on December 7, 1678. There was nothing in the Old World to compare to it. Hennepin was looking at a volume of water far greater than the amount that tourists see today. The noise was probably the loudest thing he had ever heard in his life. His reaction? He was scared to death. He called it "the most Beautiful and at the same time most Frightful Cascade in the World." This "prodigious quantity of water" produced "a noise, that is more deafening than the loudest Thunder."[1]

Hennepin was the first in a long tradition of writers who had strong reactions to the Falls. American novelist Harriet Beecher Stowe was so moved by the sight, she wanted to die. "I felt as if I could have *gone over* with the waters; it would be so beautiful a death."[2] Standing on Table Rock, Charles Dickens found that "the first effect, and the enduring one — instant and lasting — of the tremendous spectacle was Peace. Peace of Mind: Tranquillity: Calm recollections of the Dead: Great Thoughts of Eternal Rest and Happiness; nothing of Gloom or Terror."[3] Some people found the rapids above the Falls as impressive as the Falls themselves. About two miles above the Falls the river narrows, the waters flow even faster, and drop more than 55 feet in less than a mile before the final drop of 165 feet into the gorge.

But practical, hard-headed types saw the Falls as an inconvenient impediment to orderly settlement, shipping, travel, commerce, the growth of industry and nearly everything associated with progress. The Falls were in the way. Something had to be done about them.

The Falls were also unlike other waterfalls that Europeans had encountered, such as the Chaudière waterfalls on the Ottawa River or the Rideau Falls, where the Rideau River joins the Ottawa River. At most waterfalls, the change in elevation is at roughly the same location on both land and water. Those travelling by boat leave the water, climb up or down the change in elevation on land, and rejoin the river. But the Niagara River isn't like that. The edge of the escarpment is near Lewiston and Queenston. So were the Falls themselves around 12,000 years ago. But as a result of erosion from the enormous volume of water, the Falls has clawed its way seven miles upstream from the land face of the escarpment.[4] This made the portage route longer than it would have been if the Falls had coincided with the escarpment on land. People portaging upstream had first to scale the escarpment on land, then, because the river is in a deep gorge at that point, continue the land journey for miles to the point at which the river is roughly the same height as the top of the escarpment, where they could continue the journey by water.

In 1678, Father Hennepin became the first European to see Niagara Falls. From that time on, the Falls seemed to be shrouded in confusion and controversy. This illustration, taken from a map by Herman Moll and published in London, England, in 1715, gives estimates of the height of the Falls ranging from 100 fathoms (600 feet) to half a league (7,920 feet). Clearly the artist had never seen a real beaver. However, lack of knowledge or experience rarely got in the way of early writers on Niagara Falls.

(Reproduction by permission of the Metropolitan Toronto Reference Library)

The first portage route was on what is now the American (east) side in the 18th century. Portaging was so labour-intensive that an entire village of Senecas took on the job of helping travellers. "Women and children carried the bundles up the steep bluff for almost 2 km [over a mile]. The climb was so difficult that they called it Duj-jih-heh, or 'The Crawl-on-all-Fours.' At the top, men and ponies took over, carrying the goods along the portage to the bateau landing at Fort Schlosser"[5] upstream from the Falls.

After the American Revolution, when the Niagara River became an international boundary, the British needed an alternative portage on the Canadian (west) side. They created one from Queenston to Chippawa. It was longer and led inland, away from the river, bypassing the Falls. And since commerce tended to follow transportation routes, the new portage directed economic activity upstream to Fort Chippawa, the first good harbour above the Falls.[6]

Then came the War of 1812. From the vantage point of the 21st century, the War of 1812 between Britain (Canada) and the United States sometimes seems like a tiff between otherwise friendly allies. In fact, it was a brutal and bloody affair with heavy loss of life and property, not easily forgotten by those who lived through it. The peace brought new problems. Newark (now Niagara-on-the-Lake) lost any chance of becoming a capital city, because it was too close to the border and therefore open to enemy attack. And settlement slowed on the Canadian side of the Niagara River. New settlers favoured safer places such as the Talbot settlement near London.

The Americans, too, wanted water routes that led away from the border. Ships using the Great Lakes and the St. Lawrence had to go through an area that was British on both sides of the river. That was an unacceptable security risk. Moreover, that route didn't give access to a year-round harbour. Therefore the Americans built the Erie Canal, which opened in 1825, with Buffalo as its Western terminus. The Erie Canal took ships, cargo, and barges out of the Great Lakes system at the eastern end of Lake Erie and straight to the American eastern seaboard. The impact was immediate. "Transportation costs fell dramatically, and travel time between Buffalo and New York City was reduced from twenty to six days.... Farms bordering the route doubled and quadrupled in value; undeveloped areas of either side blossomed...and towns and villages sprang up along the whole line of the canal."[7]

The beneficiaries of the new route included Lockport, Pendleton, the Tonawandas, Black Rock, and Buffalo — especially Buffalo. By 1900, Buffalo's population of over 352,000 made it the eighth largest city in the United States.[8] But Niagara Falls, Chippawa, Queenston, and Lewiston languished. They were too far from the new route to prosperity. The canal that had catapulted Buffalo into prominence as a shipping port, as well as a centre of commerce, finance, industry, railroads, milling, and manufacturing, had bypassed Niagara Falls completely.

Canada responded to the Erie Canal with the privately financed Welland Canal, which opened in 1829. Unlike its American counterpart, the Welland Canal stayed within the Great Lakes system, connecting Lakes Erie and Ontario. But like the Erie Canal, it avoided the Falls and continued to reroute prosperity away from the Niagara River. It spurred growth in Port Colborne, Welland, Port Robinson, Allanburg, Thorold, Merritton, St. Catharines, and Port Dalhousie. The water flowing through the canal could also be tapped for power. The settlements on the new transportation corridor became part of an emerging energy and industrial corridor.

In 1841, the Legislature of Upper Canada bought the struggling Welland Canal Company and began enlarging the canal. The Second Welland Canal was completed in 1845. The Third Welland Canal, further enlarged, opened in 1887. In each case, the increased capacity of the canal spurred further industrial activity along the route. Clearly, advanced transportation technology was not helping Niagara Falls develop as either an energy or an industrial centre. Rather, it had become a tourist centre, and not a very appealing one at that.

A Tourist Trap

Tourism had started early. In 1791, Queen Victoria's father, the Duke of Kent, visited the Falls and stopped for refreshments at a log cabin inn on the Canadian side. He left no record of what he thought about the place. Four years later, the Duc de Rochefoucault-Liancourt rode on horseback from Chippawa to the Falls. It was a difficult ride through thickets and swamps and in the end, he didn't even get a good view of the Falls. His complaints identified a continuing theme in the history of the Falls: government unwillingness to spend money on improvements.[9]

During the 1820s, Niagara Falls became more of a tourist destination. As transportation improved, starting with the Erie Canal in 1825 and continuing with railway lines and bridges that crossed the Niagara River, it became easier to get to the Falls. But once they arrived, tourists needed something to do in addition to just star-

ing at the water. Even a strong supporter of the movement to preserve the beauty of the Falls had to admit that "Young people cannot sit in silence gazing at the Falls, through all of a long summer day, thinking of aesthetic sublimities, or communing with the Absolute and Infinite."[10] People wanted other attractions. And given the laws of supply and demand, they got them.

By 1827, Niagara Falls had become the site of publicity stunts, some of them appalling. That year an estimated eight to ten thousand people flocked to the Falls to see a condemned Great Lakes merchant vessel called the *Michigan*, which had been filled with animals (including a lion), set on fire and let loose to tip over the Falls. It didn't work as planned. The ship broke up before reaching the Falls and some of the animals escaped. However, as the *Colonial Advocate* reported, "the tavern keepers cleared a great deal of money."[11]

Clearing a great deal of money seemed to be the point of most local businesses. Tourists were there to be fleeced. Conditions, however, differed somewhat on the two sides. On the American side, the river's edge was privately owned. On the Canadian side, however, a 66-foot wide strip along the river's edge was reserved as crown property. That distance of 66 feet, the length of a surveyor's chain, was called the Chain Reserve, or sometimes the Crown Reserve or Military Reserve.

The Horseshoe or Canadian Falls are larger and more scenic than the American Falls, so tourism developed more on the Canadian side. Table Rock, the best viewing point on the Canadian side, was the place where tourists were parted from their money. In 1866, the United States consul at Clifton reported, "Upon arrival at the [Table Rock] house...parties are immediately waited upon by some one of the employees, politely invited to enter the building, proceed to the observatory, below the rock, or 'wherever they please,' and in most cases assured that it is all '*quite free*' except what they choose to give to the guide, while the keeper of the house receives his pay in profits on such fancy articles as they desire to purchase. The [photographic] artist also solicits patronage, and offering to perform his work at low prices — which "When the visitors are ready to depart they are reminded they have bills to settle, and

Early surveyors, engineers, promoters, visionaries and a few investors dreamed of the riches to be won by harnessing the seemingly endless waters of Niagara. The late 19th-century projects on the American side brought surface water from a diversion canal to individual industrial sites, where water dropped down shafts or penstocks to turbines or waterwheels and then discharged into the Niagara River below the Falls. The results, as seen in this photograph, were not pretty. Niagara Falls won global fame for industrial ugliness and the ruination of an area of natural beauty. The Save Niagara Movement led to the creation of parks on both sides of the river, but it did not end concerns over the future of Niagara. In this undated postcard, tourist boats and a scenic bridge seem at odds with the factories and water discharging down the Niagara Gorge and into the river. (Reproduction by permission of the author) exorbitant prices are charged for use of oil cloth clothing, pictures, etc. In case of refusal to comply with these demands, the proprietors and employees of the house take the law into their own hands, the visitors are falsely imprisoned — not allowed to depart without paying every cent that it is thought fit to demand — insulted, assaulted and quite frequently knocked down, dragged out and otherwise maltreated."

The situation got so bad that in the 1870s a Royal Commission was appointed to look into the town's tourist trade. It deplored "the difficulty of escape from the organized band of cabmen, fancy and variety store keepers, guides, sight showers, picture takers, oil clothes furnishers, conductors under what is alleged to be the sheet of water, hotel keepers and runners, all working to plunder." E.B. Wood, the head of the Royal Commission, wanted the Ontario Government to take control of the Chain Reserve and the path to the foot of the Falls, but as often happens with Royal Commission recommendations, the government was slow to respond.[12] Meanwhile, the tourist operators weren't the only ones trying to turn the Falls into money.

The Monstrous Mill-dam

Not everyone was filled with sublime thoughts when they looked at the Falls. One visitor declared, "Nature having prepared everything, there remained little else to do but build [mills]."[13] And indeed, that is exactly what people did.

As early as 1725, the Niagara River was driving a modest sawmill. Other mills followed. In 1824, Niagara Falls, New York, mapped out its future by calling itself Manchester. Americans might have broken away politically from Britain, but when it came to industrial development, they wanted to be like Britain. "By the early part of the nineteenth century, 'Manchester' was a common name for any American township that showed promise for industrial development."[14] But it would take more than a name change to harness the power of the Falls.

For one thing, there were cheaper alternative sources of power. Cheap coal and improved steam engines let manufacturers and mill owners install their own steam engines on their own property. In the early to mid-19th century, no technology existed to harness Niagara's powers at a competitive cost. And there wasn't much demand for power in a small tourist town. The nearest potential customers were in Buffalo, and conveying power over long distances was not yet feasible. The power of Niagara Falls had to be made portable if it was to be exploited on a large scale, and if it could not be exploited on a large scale, it could not be exploited at all. Finally, according to some observers, "The scenic beauty of the great cataract made the situation still more difficult for those who wished to put the river to practical use. Sooner or later, the public was certain to resist any methods of utilization of the falls that encroached upon their natural grandeur."[15]

This last consideration seemed to carry little weight in the early years of development. Visitors to the American side wrote about disfigurement by "ugly stone dams, gristmills, outdoor clotheslines, heaps of sawdust, stables, advertising placards, shanties, lumberyards, a pulp mill and a gas works."[16] Things were a little better on the Canadian side, where the Chain Reserve kept mills away from the river's edge and the Welland Canal had drawn away traffic and business. On the American side, industry prevailed, and on the Canadian, exploitive tourism.

The First Industrialists

One of the first to try to exploit the Falls' potential for power was Augustus Porter, who first visited the area in the mid-1790s.[17] Porter later returned to live in Niagara Falls, New York, where he and his family became industrial, business, and community leaders. He started by rebuilding an earlier sawmill and adding a gristmill to a site first developed in 1757. In 1806, when the State of New York was auctioning off what was called the One Mile Strip along the river above the Falls, Porter and his partner Benjamin Barton bought 119 acres of it. He later bought Goat Island and an area on the nearby shore, and turned the channel between Goat Island and the mainland into a millrace, but he kept Goat Island as a tourist site, building a toll bridge to it and charging visitors for this view of the Falls.

Porter had great hopes for the area. By 1825 the local industries included two woollen cloth factories, two clothier's shops, several carding and spinning machines, a forge, and a papermill. But given the potential of the area, that wasn't much. These businesses used short millraces or loop canals which diverted the water from the river to a waterwheel that generated power. The water flowed back into the river a short way downstream. The volume of water was relatively small and so was the head under which it generated power. Because this type of power required that the mills be located right at the river's edge, industry and power generation looked like the enemy of beautiful landscapes.

In 1825, Porter issued an "Invitation to Eastern Capitalists & Manufacturers" to develop hydraulic operations on the upper rapids. Nothing came of this proposal. Twenty-two years later, in January 1847, he tried again, offering a right of way to anyone willing to undertake the construction of a canal.[18]

At this point one needs to examine a map. From a Canadian power development perspective, the River Bent the Wrong Way. As it flows downstream, the Niagara River turns sharply east. Porter wanted a canal cut overland from the river above the Falls to a point half a mile below the Falls, so that the river above and below the falls and the canal would form a triangle. That would provide about 3,000 feet of frontage along the top of the bank where factories could be located. These businesses could draw water from the canal and let it drop into pits to drive waterwheels to generate mechanical power. The deeper the pit, the more power one could extract from the water. It was easy to get rid of the water after it had been used: simply cut a short tunnel to the gorge and the water would come out and run down the cliff face. It was an interesting idea, but it never came to pass. Before anyone could take him up on the proposal, Porter died in 1849 at age 80.

The First Attempts at Generating Power

Still, Porter's idea got others thinking. In 1853, Caleb W. Woodhull, former mayor of New York City, and a group of Boston investors incorporated the Niagara Falls Hydraulic Company to build a canal for transportation and power generation. The Porter family, now regarded as one of the founding families of the area, donated land for the canal right of way. The company planned to build a canal 70 feet wide and 10 feet deep, with wharves along the company's land on the upper river. Alas, within

Above and opposite:
The massive ice build-up and hoar frost on the building in this photograph taken in February 1904 might suggest a picturesque scene. However, as the ice melted and exposed layers of dirt and soot, as in the photo taken April 1904, it is hard to see this as an improvement on Niagara's natural beauty. This annual cycle provided powerful ammunition for the Save Niagara Movement.
(Reproduction by permission of the estate of Daniel Dumych)

16 months the company was bankrupt, done in by a combination of geology (tough Niagara limestone) and inadequate blasting technology (the workmen used gunpowder, which was too weak shatter the limestone).

The Niagara Falls Hydraulic Company was reorganized in 1858 as the Niagara Falls Water Power Company. More money was spent, more work done, but the results still fell short of the expectations. On September 1, 1860, Horace H. Day took control of the company. Day had what earlier money could not buy: dynamite, a new, much more powerful explosive. By July 1, 1861, he had built a canal 36 feet wide and about eight feet deep.[19]

Then one of the great tragedies in American history struck Day's canal, along with the lives, families, and businesses of countless Americans. The Civil War sucked dry the investment capital needed to take advantage of a power canal system. The canal languished until 1875, when Charles B. Gaskill built a flour mill on the canal basin beside the gorge. His mill used a fraction of the power available: the water fell only 25 feet before it flowed through a short tunnel and was discharged from the face of the cliff into the gorge nearly 200 feet below.[20] Gaskill's mill lasted about a year. He and the other tenants on the canal could not raise enough money to pay the mortgage and interest on bonds. Creditors foreclosed on the mill.

Between 1853 and 1876, attempts to harness the powers of Niagara consumed more than $800,000. Jacob Schoellkopf, a German-born miller, bought the canal at auction on May 1, 1877, for $71,000 plus an additional $8,000 for Horace Day, whose entire fortune had gone into the canal.[21] From an investment point of view,

Schoellkopf was buying at ten cents on the dollar. Mrs. Schoellkopf probably thought all her husband had bought at the auction was a ticket to ruin. Mr. Schoellkopf disagreed. Along with other investors, he issued yet another prospectus, this one for 100 mill and factory sites on the canal. He created a new incorporated company, the Niagara Falls Hydraulic Power and Manufacturing Company.[22] So far, he was following in the steps of his unsuccessful predecessor.

But things worked out differently this time. Schoellkopf didn't see himself as a power developer or entrepreneur; he was just a businessman looking for cheaper power. This set him apart from the companies that were simply trying to sell power without understanding the need for that power. The Niagara Falls Hydraulic Power and Manufacturing Company prospered as a source of traditional mechanical power derived from water power. By 1882, the hydraulic company was "supplying water power to seven mills located on the canal basin."[23]

Unfortunately, a technological limitation turned the public against the canal, not for what it did, but for its effects on the Niagara River gorge. At that time, it was not technically possible to build waterwheels that could stand up under a very high head of water. The shafts that were sunk to use the water ranged from 25 to 75 feet deep, but the gorge was much deeper than that. The tunnels from these pits emerged well above the water level, and discharged water from the cliffs above the river.[24] The holes spewing out their individual waterfalls disfigured the Niagara Gorge and generated opposition to industrial development in the area.

The Beginning of Electrical Generation and Electricity as a Commodity

The coming of electricity to the area did not excite much interest at first. The *Niagara Falls [N.Y.] Gazette* devoted little space to reporting that on July 2, 1879, "the Brush Electric Light apparatus in Prospect Park became so far completed that the machine was put in motion for the first time" and the arc lights yielded "a most beautiful and powerful light." There was no rhapsodizing about the future possibilities of electricity; the arc lights were just one more attraction to entertain the jaded tourists.[25]

The display was the work of Arthur Schoellkopf, Jacob's son, and some associates. They used the mechanical power generated by the waters of Niagara moving through the hydraulic canal to drive a generator in Quigley's Mill on the canal basin. The direct-current Brush dynamo supplied enough power for sixteen 2,000-candle-power arc lamps in several stores, two mills, and two office buildings.[26] This development represents an important milestone in the history of electricity at Niagara Falls. At first, companies had generated power for their own use. Now power was being generated by a company for sale to others. It was the beginning of electricity as a commodity.

Schoellkopf and his partners proved it was possible to generate electricity on a small scale. But soon a proposal came forward for a much bigger project. In 1886, Thomas Evershed, an engineer working for the State of New York, proposed a radically different plan for generating electricity, or other types of power, at the Falls.

The hydraulic canal system from which Schoellkopf and his partners made money was a surface canal system. Evershed took a completely different approach. Instead of an open canal, the water would flow through an enormous tunnel that would begin above the Falls, run all the way underneath the town to mills or generating sites, and discharge water back into the river near the Clifton Suspension Bridge. The Niagara River Hydraulic Tunnel, Power, and Sewer Company picked up Evershed's proposal and secured a charter from New York State to divert river water into a tunnel. Over the next few years, the proposal was altered, almost beyond recognition, as engineers and managers wrestled with questions of construction and logistics. But the biggest questions of all were: What could be done with all that power? Who would use it? Since there wasn't much industry in Niagara Falls, how could power be transmitted to industries elsewhere? From the vantage point of the 21st century, the answers to these questions seem obvious. In the 1880s, they were anything but.

Meanwhile, other developments changed the environment in which industry would operate in Niagara Falls.

Protecting the Beauty of Niagara

Niagara Falls wasn't a pretty sight any more, what with its smoky factories, tacky tourist attractions, and the holes in the cliff. Visitors were complaining — and spending less time (and money) in the area. Worse, they were starting to avoid the Falls in favour of pristine environments such as the Grand Canyon or Yellowstone Park. People rallied to the cry of "Save the Falls!" They pointed to places where parks had been created to preserve the beauties of natural areas such as Yosemite, or to create a more natural-looking environment in the city, as with New York's Central Park. Parks, it appeared, were the answer.

In New York State, the Niagara Falls park would be called the State Reservation; in Ontario, it would be the Queen Victoria Niagara Falls Park. To decide what lands needed to be acquired and determine the general layout of the park, both governments brought in the top landscape architect of the day, Frederick Law Olmsted, the designer of Central Park in New York City, the grounds of the Parliament Buildings in Ottawa, and, closer to Niagara, the Buffalo parks system.

A VIEW OF NIAGARA AS IT MAY BE A FEW YEARS HENCE.

Editorial cartoonists had a field day with fears over the future of the Falls. In 1883, nine years before Canadian Niagara received its charter, *Harper's Weekly* envisioned the Niagara gorge on both sides of the river filled with the smoke, fumes, noise and stench of industry. (Reproduction by permission of the Buffalo and Erie County Public Library, Buffalo, New York.)

Despite this joint effort, New York State and the Government of Ontario approached the creation of the park in very different ways. In New York State, the lands along the river front had been auctioned off to private landowners in the early 19th century, whereas Ontario had its Chain Reserve. Creating a park in Ontario just meant adding more land to this strip, but it had a starting point. In New York State, everything had to be bought back from the private sector.

In New York, after several failed attempts, legislation was finally passed and approved on April 30, 1883. The government of New York State bought the land and gave the newly created State Reservation an annual operating budget. On July 15, 1885, the American park officially opened.

Things took longer in Ontario. The government passed *An Act for the Preservation of the Natural Scenery about Niagara Falls* on March 30, 1885, and appointed Commissioners to undertake the work of preservation. In 1887 the government passed *The Queen Victoria Niagara Falls Park Act*. However, there was a catch. Because of intense political pressure and the perceived political dangers of — horrors! — spending money on something as frivolous as a *park*, Ontario started with and never deviated from the idea that the province would not give the park any money. Not a penny. The park had to be self-sufficient financially or there would be no park.

In assigning responsibility without financial resources, the government gave the Park Commissioners the right to enter into money-making ventures within the boundaries of the Park. These included renting franchises to develop power from the waters flowing through it. The Park could also issue debentures to acquire lands or for other purposes. If ever a group had an incentive to encourage electric power creation, this was it.

H.M.Pettit.

The Birth of the Canadian Niagara Power Company

CHAPTER 2

The newly formed Park Commission needed money. In 1889 it welcomed more than 250,000 visitors and received very little in revenues. There was no admission charge to the park itself, and the money from concessions granted for such things as the right to charge for taking photographs of tourists were paltry. But the Commissioners had more than money on their minds. They took seriously their duty to protect the Falls and improve the environment. Through a financially crippling bond issue, they had been able to buy up properties and knock down or close some of the more objectionable of the so-called attractions. They wanted to ensure that visitors had a pleasant experience when they came to the Falls. One of the big problems the Commissioners grappled with was transportation.

Most visitors depended on hackmen or coachmen to take them from the train station or boat dock to the park. Many hackmen collected money from their passengers and then took them, not to the park, but to other destinations, the proprietors of which paid them handsomely for delivering tourists for the fleecing. The Park Commissioners were powerless to control the throngs of independent hackmen. But what if they could bring in an alternative transportation system and put the hackmen out of business? What if they could create an electric railway? And if a railway offered the possibility of revenues for the park, so much the better.

Early in 1889, a group of businessmen from the United States, led by one Colonel Albert D. Shaw, offered to lease the water power of the Canadian Falls to generate electricity, which could be used to run a railway. The Commissioners were interested, but they knew that Shaw and his colleagues were asking permission to venture into the unknown. Long negotiations followed. The government and the Park Commissioners agreed that a deal could be struck, under certain conditions.

The first stipulated "that all constructions required for electric works should be placed under the bank of the river in a manner not to detract from the park design." This is a peculiar requirement, since there was little or no room to build *under* the bank of the river, except downstream of the Falls. Furthermore, any construction needed "approval of the Commissioners both as to position and character." The second part was the money: two years' rent in advance at $25,000 a year for a total of $50,000. It was too much for the investors.[1]

The Commissioners had spent a lot of time and had come up empty-handed. But they had learned something. Next time, they would demand $10,000 payment up front as a show of good faith.

Col. Shaw came back "on behalf of an eminent English electrician and well-known capitalists in London for privileges of a similar character" to those offered the Americans. The "eminent electrician" was Sebastian Ziani de Ferranti, one of the world's greatest pioneers in the use of alternating current (AC). While still legally a minor, the young Ferranti had patented an AC alternator in 1882 and, with his father's consent, formed a company to develop and market it. In October 1890,

Opposite:
How much water should be diverted for power? Surely not all. But Lord Kelvin, an eminent 19th-century scientist and power consultant, had suggested just that. To do anything else would be a sinful waste of a god-given resource — water to generate electric power. And the tourists? Plant brightly coloured vegetation on the faces of the American and Canadian falls, he suggested. The 1906 illustration from *Leslie's Weekly* clearly expresses the fear that no water will be left for scenic beauty after industry has taken it all. (Reproduction by permission of Queen's University, Kingston, Ontario)

under his direction, the Deptford Station had begun transmitting AC electrical energy to London at 5,000 volts — an important event in electrical history. He was not the first to demonstrate high-voltage transmission of AC, but he was the first to design and construct a full-scale central station and distribution system.[2]

Sometime in September 1889, Ferranti and Shaw came to an understanding. If Shaw could negotiate reasonable terms for the right to generate hydroelectric power at Niagara Falls from the Queen Victoria Niagara Falls Park Commission, Ferranti would form a syndicate of British capitalists to finance the undertaking. On December 16, 1889, Shaw wrote to Park Commission Chairman Casimir Gzowski (later Sir Casimir) to ask about terms for the right to use the water power, either on the Canadian or American side of Niagara Falls.

In their annual report for 1889, the Commissioners expressed their confidence that the group would buy a franchise for the right to construct and operate an electric railway from Queenston to the park, thereby banishing the "curse" of the hackmen.

With $10,000 in the bank, the negotiations continued. In the meantime, two other groups came forward with proposals for electric railways. The Commissioners could not, however, entertain these other proposals while they were offering the English group a combined generating and electric railway package. The package involved an annual payment of $5,000 plus 5 percent of the gross earnings of the railway. The English group agreed to the terms and Gzowski made a research trip to England to investigate how electric technology was developing.

Below is the last page of the agreement signed on April 7, 1892, by Casimir S. Gzowski, chairman of the Niagara Falls Park Commission, Arthur D. Shaw, Francis Lynde Stetson and William B. Rankine. The handwritten final clause states that living up to the terms of the agreement would not be the personal responsibility of the three investors, but of the entity they would create — the Canadian Niagara Power Company. This agreement became the basis of a Bill proposed by Provincial Parliament. (Reproduction by permission of FortisOntario)

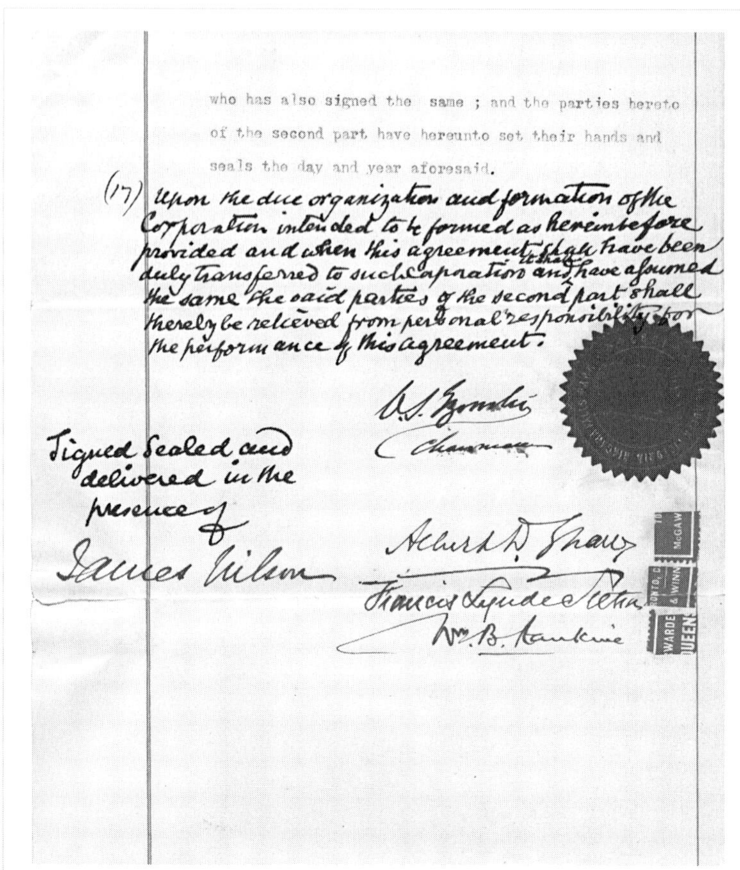
who has also signed the same ; and the parties hereto of the second part have hereunto set their hands and seals the day and year aforesaid.

(17) Upon the due organization and formation of the Corporation intended to be formed as hereinbefore provided and when this agreement shall have been duly transferred to such Corporation and have assumed the same the said parties of the second part shall thereby be relieved from personal responsibility for the performance of this agreement.

Signed Sealed and delivered in the presence of James Wilson

C. S. Gzowski, Chairman

Arthur D. Shaw

Francis Lynde Stetson

Wm. B. Rankine

Alas, the March 1, 1891, deadline came and went with no further money and no signed contract. It was a heavy blow. The Commissioners knew that every failed proposal weakened their bargaining position. They were in an unenviable position — caught between a government that wouldn't offer a nickel because it wouldn't look right, and investors who smelled blood each time they failed. Furthermore, during the course of negotiations, the franchises for the electric railway and for generating power had become separated and could not be put back together.

An Electric Railway at Last

The Commissioners were running out of time. They signed a contract on December 4, 1891, with the Niagara Falls Park and River Railway to run a railway between Chippawa and Queenston. This time, the proponents had money and proven railway experience. The four signatories were front men for the real power behind the throne — the railway builders William Mackenzie and Donald Mann, who later went on to build railways in South America, China, and across Canada. And

The all-electric Niagara Falls Park and River Railway delighted tourists and Park Commissioners alike. It provided a scenic trip from Queenston to Chippawa at a fair price. The railway power generating station also supplied power to Canadian Niagara Power Company customers before the Rankine Generating Station opened. (Reproduction by permission of the author)

unlike all who had come before, and some who would come later, Mackenzie and Mann did what their front men said they would do. They actually built the railway.

The Commissioners had asked for an electric railway. Arc lights had lit the streets of Niagara Falls, Ontario, since 1885, powered first by electricity from Niagara Falls, New York, and then, beginning in 1891, by locally generated power from steam-driven Edison General Electric equipment owned by the Niagara Falls Light and Power Company. But the lighting company had no electricity to spare. The railway company would need its own generating station. At first, the company petitioned to make only the centre portion from the Whirlpool to Dufferin Islands electric and rely on steam at either end. The Commissioners refused. Gzowski wisely understood that if you want an emerging technology to serve you, you have to push it.

Construction started in 1892 on the first electric generating station on the Canadian side of the Falls. The single-track railway opened May 24, 1893, and carried 354,000 passengers in its first year.[3] The Park Commission received its $10,000 franchise fee; no one knows how much money the railway made. It still had to pay off investments for the track, power lines, generating station, and generating equipment.

A Contract for Generating Power

Although the English syndicate was out of the running for the railway, the never-say-no Shaw was still interested in the power generation rights. He found other American investors and on April 7, 1892, the Commissioners, with the approval of the Government of the Province of Ontario, signed an agreement with Albert D. Shaw, of Watertown, New York, and Francis Lynde Stetson and William B. Rankine, both of New York City.[4]

The agreement is an interesting document both for what it contains and what it omits.[5] It contained provisions for protecting the look of the park, and generous financial terms for the park. For ten years starting November 1, 1892, the company would pay $25,000 a year; the first two years of rent were to be paid before signing the contract. There were provisions in case the company didn't pay its rent on time. There was also a timetable. By November 1, 1898, the company was to be supplying at least 10,000 horsepower in the form of electricity or pneumatic power, and construction was to start by May 1, 1897. But the Commissioners did not put a maximum figure on the amount of power the company could generate or the amount of water that could be withdrawn from the river.

The Niagara Falls Park and River Railway kept the right to generate power within the park, but only for its own needs. Otherwise, the company had a clear monopoly.

Although no one foresaw significant Canadian demand for electricity, Canadian interests were protected to some extent. "The company whenever required shall...supply...in Canada to the extent of any quantity not less than one-half the quantity generated, at prices not to exceed the prices charged to cities, towns, and consumers in the United States at similar distances from the Falls of Niagara for equal amounts of power and for similar uses." Moreover, the company had to show that the prices they charged were in accord with this clause. This gave the American investors what they wanted, namely the right to export power to the United States to supplement what they were already generating there.

What was missing? There was no penalty clause for failing to generate electricity on time. There was only Clause 13: "If the company should at any time or times continuously neglect for the space of one year effectually to generate electricity or pneumatic power as hereby agreed by the company, unless hindered by unavoidable accident," the contract would be considered void. At the time, no one could have imagined that clause coming into play.

The Canadian Niagara Power Company Is Formed

This agreement became the basis of hurriedly passed legislation, which transferred all the rights and obligations in the agreement between Shaw, Stetson, and Rankine, and the Park's Commissioners to the Canadian Niagara Power Company. Among other things, the enabling legislation allowed the company to issue up to 30,000 shares, each valued at $100 for a total of $3 million for fees, plans, and, most important, construction work. The legislation also gave the company the power to issue bonds up to a total of $5 million.[6]

Introduced on April 6, 1892, by the Attorney General as an "Act to Confirm a Certain Agreement Made Between the Queen Victoria Niagara Falls Park and the Canadian Niagara Power Company, and to Enable the Said Company to Carry the Agreement into Practical Effect," the matter attracted little attention. A week later, on April 14, the day the session was to be prorogued, the Clerk Assistant read out a list of numbered bills that had passed "severally," that is to say in a group. Item number 105 was the legislation involving the Canadian Niagara Power Company. Apparently, "The Government was so afraid that 'the lunatics'...might change their minds, that an arrangement was speedily concluded with the leader of the Opposition...to permit the passage through the Legislature of the necessary Act without discussion and without delay."[7]

The incorporators of the new company, who were also the shareholders, met on July 16, 1892. Alfred D. Shaw was named President, Francis Lynde Stetson Vice President, and William B. Rankine Secretary and Treasurer. (Stetson and Rankine were also officials of the Niagara Falls [New York] Power Company and the Cataract Construction Company.) The founders were rewarded in the form of certificates entitling them to paid-up capital stock in the company as well as debenture bonds bearing 6 percent interest. The company issued these three men 5,000 shares of the 30,000 it was entitled to issue, for a total of $575,000 — of that, $500,000 was in shares and $75,000 bonds paying 6 percent interest.[8]

The three founders, along with provisional directors John D. Irwin and Charles J. Holman, hurriedly agreed to these arrangements before noon at a meeting in the Clifton House on July 16, 1892. They had to hurry, because at twelve noon the shareholders had to hold a second meeting to approve what they had done in the previous meeting. They then adjourned briefly, to reconvene half an hour later to approve the minutes of the noon meeting. They delegated Secretary Rankine to send a copy of the Agreement of Transfer to the Commissioner of the Queen Victoria Niagara Falls Park. And President Shaw was to confer with a Mr. James Wilson about becoming Assistant Secretary and Resident Agent at $100 a year.[9]

Hurry Up and Wait

Given all the haste surrounding the signing of the contract, the passing of the legislation, and the three meetings held in a single day, one might reasonably assume that the company intended to build a hydroelectric generating plant immediately. One would, however, be wrong.

On the day the new company legally came into being, a Niagara Falls, New York, newspaper stated that "The company will await the completion of the great tunnel upon this side [American] of the river, and its success will greatly determine what will be done by them in Canada."[10] The *Toronto Mail* reported that "Mr. Rankine, when asked if there was any probability of work commencing this fall in the new tunnel, replied that there was not."[11] *The Globe* in Toronto cited Shaw, who stated that "they will await the completion of the tunnel on the United States side before commencing operations in Canada."[12] Clearly no one expected instant action, although the land speculators got busy. Even before the agreement was final, the *Globe* had reported, "There has been a sudden rise in real estate on the Canadian side of the river at Niagara Falls. The creation of the Canadian Niagara Power Company has caused the boom."[13]

But after all that furious activity in 1892, matters came to a complete halt. By law there was to be a meeting of the company on the first Thursday in June every year at three o'clock in the afternoon. Yet in the bound collection of corporate minutes, there is no record of the June 1893 general meeting of the company. After those three rapid-succession meetings on July 16, 1892, no meetings took place until February 1894, when the directors met in C.J. Holman's law office on Bay Street in Toronto. Stetson was absent. They were there to review three plans showing proposed locations for powerhouses, water channels, and tunnels. It was agreed that Shaw and Rankine would submit these plans to the Commissioners of the Queen Victoria Niagara Falls Park for approval.

It wasn't much to show for the 18 months that had elapsed. What had happened in the meantime? The records suggest financial troubles, rocky relations with the Park Commissioners, or both. Moreover, the American parent company had run into financial and scheduling problems with their work in Buffalo and Niagara Falls, New York. But the delay was partly intentional: the longer Canadian construction could be postponed, the more it would benefit from American experience.

The next recorded directors' meeting took place on April 6, 1895, in the office of James Wilson, Park Superintendent. This was presumably the same James Wilson whom Shaw was supposed to hire as agent for the company at $100 a year. It was an important meeting, because the company had to make a rent payment. The first two years' rent had been paid in 1892 and now the first semi-annual payment was coming due. The minutes mention a letter sent by the Niagara Falls Power Company in the United States, which had offered to help Canadian Niagara pay its rent to the Government of Ontario. The letter suggests that shares had been transferred from Rankine and Stetson to the Niagara Falls Power Company, although such a transfer is not explicitly mentioned. Whatever the truth of the matter, the officers of the company decided to issue 6 percent debenture bonds payable by January 1, 1900. On April 30, 1895, the Niagara Falls Power Company lent Canadian Niagara the $12,500 it needed to pay the semi-annual rent due on May 1, 1895. The Niagara Falls Power Company took $25,000 of the bonds as collateral.[14] It was the beginning of a pattern.

The company was also at pains to distinguish between corporate and personal responsibility for signed agreements. At the meeting of April 6, 1895, the directors agreed to send documentation to the Queen Victoria Niagara Falls Park Commission to remind the Commissioners that it was the *company*, not the individuals, who had signed the agreement. This was an important point, because neither the company nor the individuals were doing much, and relations were getting strained. And the directors had a new plan.

How to Make Money from Electricity Without Generating, Transmitting, or Selling It

Canadian Niagara had the right to generate electric power within the park boundaries and sell it outside the park boundaries. But the company had no power to sell. On the other hand, the Niagara Falls Park and River Railway Company could generate more power than it could use. But without the permission of the Canadian Niagara Power Company, it had no right to sell power outside the park. The opportunity for a profitable exchange did not escape the owners of Canadian Niagara.[15]

The two companies started negotiations in July 1895.[16] On August 17, 1895, Canadian Niagara proposed to allow "the Railway Co. to sell its surplus power [outside] the Park," in return for 25 percent of the gross proceeds.[17] Only three years after receiving a charter to generate and sell power, Canadian Niagara Power Company had found a way to make money from electricity without generating, transmitting, or selling it. But it was another case of hurry up and wait.

By this time, the Park Commissioners and the Ontario government were getting impatient. Canadian Niagara was not going to get money for doing nothing. It would pay the Railway Company $10 a year for each unit of horsepower generated using its equipment. In return, the railway was to provide what was called "power upon the shaft." A rotating shaft in the powerhouse would power generators that Canadian Niagara was to install to generate electricity. Furthermore, within 15 days of the signing date of November 27, 1897, Canadian Niagara had to have signed "contracts for the supply of the machinery necessary to the generation of power" and within five months of signing, "be prepared to provide electric power."

Prodded by this requirement, the directors stopped dragging their feet. They bought the needed equipment and installed it in the railway powerhouse. The railway generated direct current power to meet its own needs. To meet local industrial demand for alternating current, Canadian Niagara installed Westinghouse AC generating equipment — three-phase, 25-cycle, 2400-volt, 10-pole alternators. This combination gave the now-demolished railway generating station an important place in Canadian history. "Not only was it the first hydraulic power station to provide revenue to the Queen Victoria Niagara Falls Park Commission, it also produced direct current for the railway, park lighting for Queen Victoria Park and alternating current for local industries."[18]

Given its importance, *The Electrical World* sent a reporter to take a look. He dismissed the electric railway's equipment as "primitive…such as might have been found in service in the average town of the United States about six years ago." It was not a very thoughtful judgement. What he was looking at was a small-scale, technologically forgiving, low-cost installation that did its job effectively and reliably. Instead of having the water run down the penstock and then through an elbow to the turbine, the penstock was located directly over the turbine. The rotating shaft from the turbine went straight up through the centre of the penstock to the powerhouse floor. Since the generators on the powerhouse floor had horizontal shafts, the two had to be coupled, and for this, the designers had chosen traditional milling technology.

The teeth of one set of bevel gears were of wood, and the other of cast iron. This wood-and-cast-iron combination meant that if something jammed or if things got seriously out of alignment, the wooden teeth would simply break, and power would stop being transmitted. New teeth could be fitted once the problem had been resolved. It also meant that if there were any settling or misalignment, the wooden teeth would conform. The gear wheels drove horizontal line shafts with pulleys mounted on them and joined to the generator pulleys by wide leather belts, probably made of buffalo hide.

There were two turbines and therefore two shafts rising up above the powerhouse floor. Originally they had been coupled, but by uncoupling them, one turbine was able to turn the shaft to generate DC power for the railway load, and the other to generate AC power for Canadian Niagara. In the event of a breakdown of one turbine, the shafts could be recoupled and a portion of each company's needs met.

There were no governors on the turbines and no overhead crane. The water inlet was controlled by hand. "Enough generators are at any time running to handle any probable peak of the load, with a gate opening sufficient for the same, and enough of the controllers in operation to take up any probable difference between the external load and the output of the machines with that gate opening. As the current taken by the [railway] cars rises and falls, slight variation in voltage affects the controller relays so that the current taken by the controllers falls and rises, maintaining the machine load practically constant." It was an ingenious, inexpensive feedback system, which controlled both the 600-volt DC equipment for the railway and the 2400-volt AC equipment for Canadian Niagara.

The equipment also showed a flair for simple, carefully thought-out design that would be displayed more fully when the company got around to building its own generating station. For example, the reporter noted that "the alternating switchboard…is remarkably simple and neat, considering the amount of apparatus which it controls."[19] The rear of the switchboard panels were a complex maze of wires and switches, but the front panels had simple switches and dials, intended to be easy to operate during an emergency. It was a small start, but it was a start.

The Park Commission Defends Canadian Niagara

As far as the public was concerned, it was too small a start. By rights, the company should have been constructing its own independent facility, not putting equipment into someone else's building. Worse still, in the same year, 1897, the company asked the government for an extension on its contractual obligation to generate power.[20]

The Park Commissioners did their best to keep the peace. They had to. The $25,000 that came in from the company annually was their largest source of income and they needed it desperately. And if nothing else, the company always met its financial obligations. After all, there was a clearly worded forfeiture clause covering failure to pay the rent.

The Park Commissioners apparently took the position that the contract with Canadian Niagara was first and foremost a revenue source and that although the contract appeared to contain a fixed schedule, it was really a starting point for further negotiations as the necessary technology was developed — or not. The Commissioners also supported the company's arrangement with the Niagara Falls Park and River Railway Company as an acceptable interim solution. Unfortunately, neither the company nor the Commissioners took time to explain this perspective to local residents or to the politicians and bureaucrats in Queen's Park.

The Park Commissioners supported the Company's request for an 18-month extension. But there were conditions. The rent had to be guaranteed. And if Canadian Niagara and the Niagara Falls Park and River Railway agreed to arrange for temporary use of surplus power, the power could not exceed the capacity of the railway generating station's wheelpits and buildings.[21] In other words, the railway generating station could not be expanded, thereby allowing Canadian Niagara to delay building its own generating station even further.

Stalemate

The government turned down the request for an extension. For the first time, the Park Commissioners did not get what they had recommended for Canadian Niagara. The government wanted to see electricity generated as soon as possible, but accepted the Commissioners' recommendation that Canadian Niagara and the Niagara Falls Park and River Railway cooperate to use the surplus power from the railway to supply local users for the time being. But this would in no way relieve the Canadian Niagara Power Company from its requirement to generate power.

The Commissioners tried to put up a good front by reiterating their faith in the long-term prospects of the Canadian Niagara Power Company. But they also called it "a matter of regret…that the carrying out of its undertaking should have been so long delayed by the Power Company."[22]

To get things moving, the Commissioners instructed James Wilson, Park Superintendent and a civil engineer, to report upon the options for developing power for commercial purposes along the west shore of the Niagara river between the Cantilever Bridge and Queenston. He submitted a detailed report on June 20, 1897, which became part of the Commissioners' annual report for 1897.[23] The report dealt with land covered by Canadian Niagara's monopoly clause. It suggested that the idea of having a single company generate power within the park for sale outside the park was losing support.

At the same time, the Commissioners found themselves in a battle over the interpretation of the original contract. To them, it had seemed perfectly clear. "The failure of the power company to have the quantity of power specified in the agreement ready at the time stipulated and also their failure to carry on the extensive construction works necessary for the development of the power during the year prior to November 1st, 1898…voided the franchise of the Company."[24] The company had missed the deadline; therefore the company had lost its rights under the contract.

Canadian Niagara, however, claimed it had until November 1, 1899. This claim may seem unreasonable, but the company had nothing to lose by trying and the contract was unclear. It included penalties for non-payment of rent, but it did not state the consequences of noncompliance with the projected completion date. The wording of Clause 13 was particularly ambiguous: "If the company should at any time continuously neglect for the space of one year to effectually generate electricity or pneumatic power as hereby agreed by the company, unless hindered by unavoidable accident, the Lieutenant-Governor in Council may then and from thenceforth declare this agreement, the liberties, powers and authorities thereby granted and every of them to be forfeited and thenceforth the same shall cease and determine and be utterly void and of no effect whatever." A smart lawyer could argue that up until November 1, 1898, the company had only neglected to get on with construction, and now, after failing to do so, it was entitled by Clause 13 to a year of neglecting to generate electricity. The case went to court.

It was heard by the High Court of Justice for Ontario at Osgoode Hall, Chief Justice Sir W. R. Meredith and Justices Ross and McMahon presiding. The judges had to answer the following questions:

(1) Was the agreement void because of the failure of the Company to complete, since the Company had not been hindered by any "unavoidable accident"?

(2) Could the Park Commissioners declare the rights listed in the contract to be forfeited by the Company?

(3) If the contract was not completely void, were the Government and the Commissioners relieved from their commitment not to grant or confer upon any other Company or person the right to take or use the waters of the river?

The company's smart lawyers prevailed. The Court answered "no" to each question. The contract remained in force, and Canadian Niagara hung on to its monopoly. It had until November 1, 1899, to carry out the works specified in the agreement.[25]

The ruling, which came down on December 20, 1898, bought the company time but didn't settle anything. The company could not meet its contractual obligations by November 1899 any more than it could by November 1898.

Meanwhile, the Park Commissioners, who had not wanted to make any decisions until the ruling came down, had not responded to the revised set of specifications and detailed drawings submitted by Canadian Niagara for approval in October 1897. After December 20, 1898, the Commissioners told Canadian Niagara that they were "awaiting the convenience of the Company to proceed with the consideration of their plans as finally revised." But they were starting to distance themselves from the company they had so warmly defended earlier.

The Monopoly Ends

Public and official sentiment had turned against the Canadian Niagara Power Company. The public was largely in favour of cancelling the monopoly clause and letting other companies have an equal chance to generate power. Someone had to break the stalemate. The Legislature did so with an Act[26] that authorized the Commissioners not only to negotiate with the Canadian Niagara Power Company for the surrender of the exclusive right to use the waters of the Niagara River within the Park, but also "to enter into agreements with any other persons or companies for using the waters of the Niagara or Welland Rivers to generate electric or other power at certain points within or without the park."[27]

Things started to happen quickly. On July 15, 1899, the Park Commissioners and Canadian Niagara Power Company signed a new agreement.[28] The Commissioners had learned a thing or two since 1892. The new contract contained tougher wording and fewer loopholes. The first contract had expressed trust in the company. The new one expressed mistrust and sought to impose greater control on the company.

To be fair, Canadian Niagara was not the only electric company to be regarded with suspicion. By 1899, very few people in Canada or the United States had much trust in electrical or street railway utilities. There had been too many promises broken, too many interlocking directorates, and too many mergers as one utility bought up another supposedly competing utility to eliminate competition. These mergers were a favourite means of thwarting municipalities and other agencies that wanted to

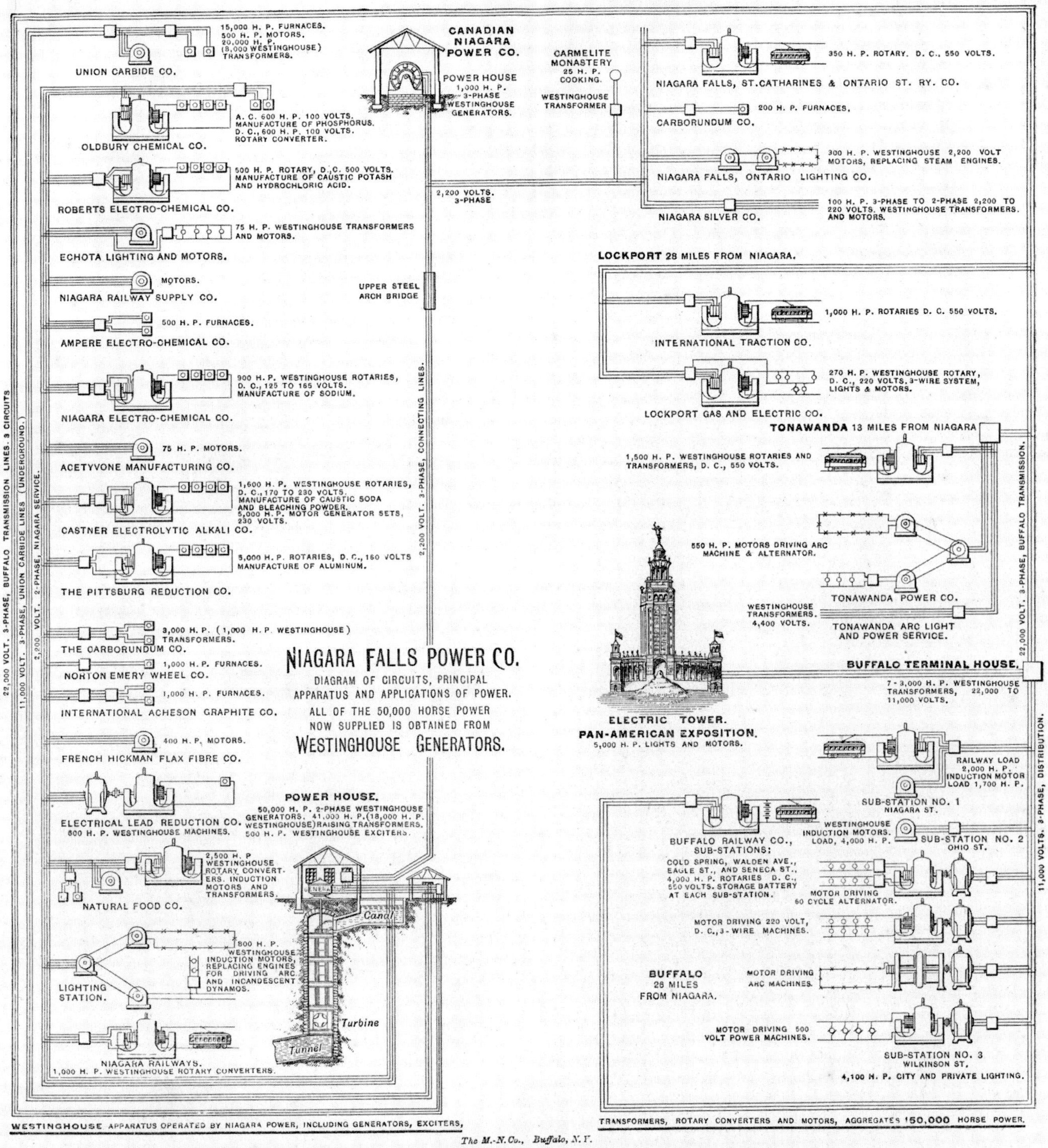

FIGURE 17.

The Pan-American Exposition in Buffalo in 1901 was a dazzling electrical showcase. When Philip P. Barton of the Niagara Falls Power Company addressed delegates from the New England Cotton Manufacturers Association, the association published his speech, which included this schematic of the entire Niagara Falls Power Company system, including Canadian Niagara. The fact that industrial customers are identified by name shows just how small the electrical industry was at the time. The Natural Food Company made shredded wheat. The Pittsburgh Reduction Company had moved to Niagara Falls to get cheap, reliable power. It later became Alcoa (Aluminum Company of America).

(Reproduction by permission of the Buffalo and Erie County Public Library, Buffalo, New York)

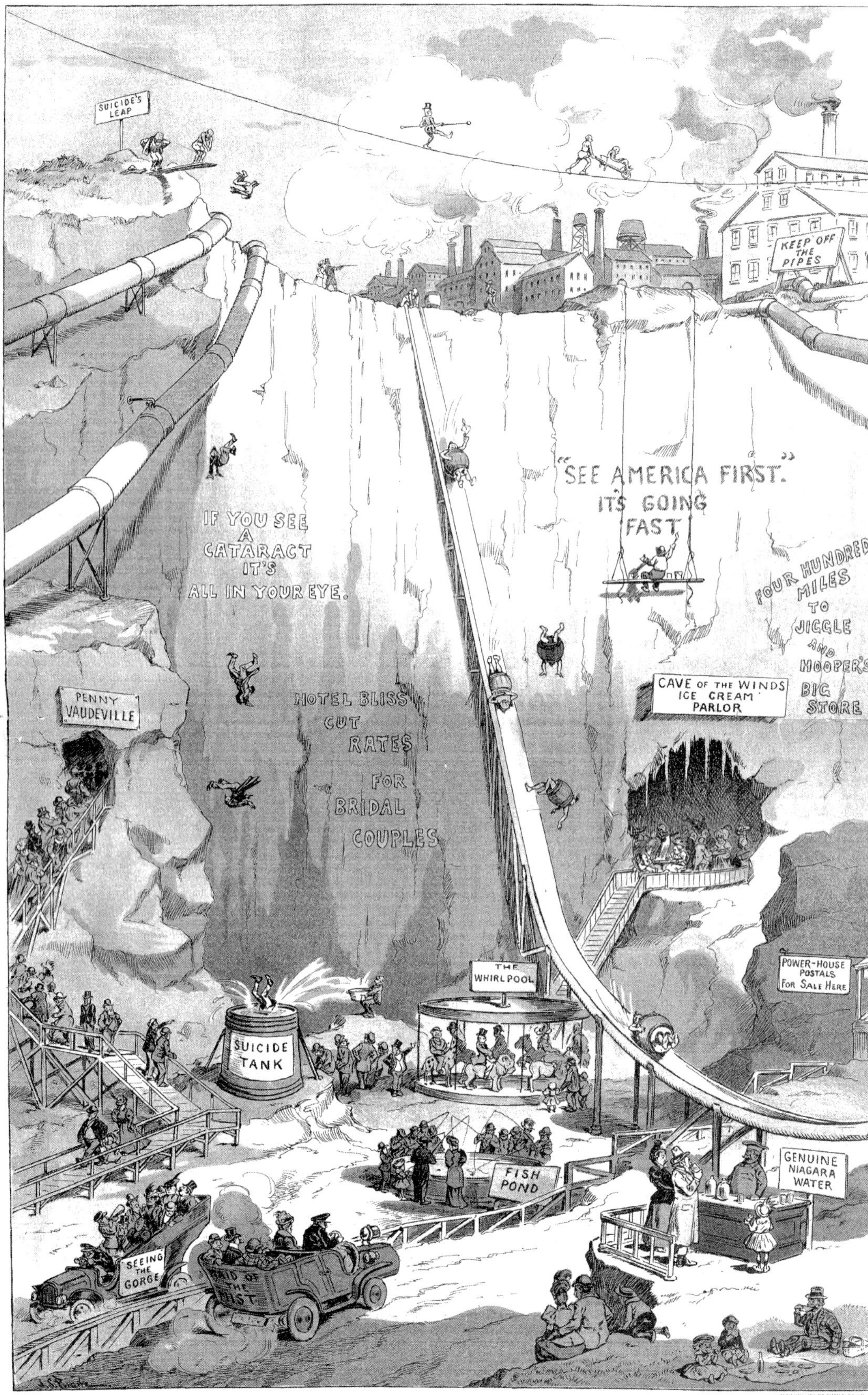

SAVE NIAGARA FALLS——FROM THIS.

see rapid development and price competition. The Park Commissioners were careful to introduce an anti-amalgamation and anti-price-fixing clause in the 1899 contract.

The new agreement contained plans, specifications, location details, and a map. All excavated material was to be placed where the Commissioners wanted it. There were restrictions on where steam power and compressed air could be used, and stipulations about maintaining public traffic flow through the park. Then were even provisions for cleaning up the site after construction, "by the first day of July, 1904, in every respect in perfect condition."

There were stiff provisions relating to timing, forfeiture, and fines, limitations on the right to sue, and a new rental schedule. The date for "completion of water connections for the development of 25,000 horse power and having actually for use, supply and transmission 10,000 developed horse power" was set at July 1, 1903." If that date was not met, the company would suffer "forfeiture of all licenses and powers." There were even prohibitions against lawsuits. For example, now that more than one company would be taking water from the river, Canadian Niagara's intake might be reduced. In these circumstances, the company had "no claim or right of action against the Commissioners," but it could "deepen such point of intake to such extent as to restore the supply of water to the volume or quantity necessary for the purposes of the company."

The Park Commissioners made some concessions in the amount of rent money owing — at least "until a market for a reasonable amount of the product could be found and the business of the company placed upon a secure basis." The annual rent was reduced, but once the company started generating a significant amount of power, a sliding scale for rental charges based on output would give the Park Commission more income than under the old agreement.[29] In addition, the Park Commissioners were free to make income-producing agreements with other companies wanting to generate electric power within the park.

Rent was payable twice a year, on May 1 and November 1, and applied to "power supplied or used either in Canada or the United States." These rates were to hold until May 1, 1949, after which the government could change them. The new contract also specified how arbitration was to be carried out if the two parties failed to come to an agreement.

In the same annual report that described the new terms for the Canadian Niagara Power Company, the Park Commissioners announced that under the authority of the Act that had led to Niagara Power losing its monopoly, the Commissioners had concluded an agreement with the Ontario Power Company, "a company holding a charter under the Dominion Government, to utilize the waters of the Welland River for power purposes."[30]

At left: In 1906, *Puck*, an irreverent and humorous American journal, saw the commercial possibilities in a Niagara Falls no longer occupied by all that water. It would be a fine place to continue the Niagara Falls tradition of fun, games, rides, signs plastered everywhere, and who knows what else. (Reproduction by permission of the Buffalo and Erie County Public Library, Buffalo, New York.)

What Had Been Lost or Gained?

The Queen Victoria Niagara Falls Park Commission had lost very little and gained a great deal. First of all, it had earned back some credibility, having got rid of a contract that turned out to be an embarrassment. There was a short-term income loss of $5,000 a year, because the yearly $35,000 from Canadian Niagara had been replaced by $15,000 a year each from Canadian Niagara and the Ontario Power Company. But as soon as either company was generating more than 10,000 electrical horsepower a year, that income would rise and would soon be much higher than the Commission could ever have earned from Canadian Niagara under the contract of 1892. And the power companies had to do things such as landscaping, filling in watercourses, and building up river banks, as well as road and bridge building, at no cost to the Commission.

What had the people of Niagara Falls gained? The new agreement seemed to herald the beginning of a new age of prosperity. There was far more water flowing over the Falls on the Canadian side than there was on the American side, but most industrial development had so far been on the American side. The Welland Canal

Cartoonists and other critics insisted that the choice was between the beauty of the Falls or the ugliness of development. However, the Queen Victoria Niagara Falls Park and the Canadian Niagara Power Company worked together to have both beauty and hydroelectric power at the same time. This early architectural rendering, which was later modified, captures some of the essential features of the Rankine Generating Station. Many windows add to a sense of openness in a building that sits relatively close to the ground. The lower set of windows are on a wall that projects outwards, so as to protect the inner forebay, a lesson learned from the experiences of the parent company in Niagara Falls, New York. The absence of any weirs or diversions in front of the low arched bridge expresses the optimism that enough water would flow into the forebay. It is a beautiful building and one is grateful that an earlier design, which A. Monro Grier likened to "a cross between a mediaeval castle and a modern ice cream parlour," was never built. (Reproduction by permission of FortisOntario)

had also drawn traffic and business away from the Niagara River. It was hoped that under the new contract, electricity would bring industry to the area and build on the economic potential many felt had been lost when the Welland Canal did not follow the Niagara River.

What about the company? It had lost a lot. In 1892, it was the darling of the Park Commissioners, the Ontario Legislature, and many citizens. By 1899, it had lost credibility and political support. And it had lost a lot of money, particularly the fees it had spent on high-priced Toronto lawyers, who had failed to save the 1892 agreement. One of them, Wallace Nesbitt, was about the most expensive lawyer you could hire in Canada, and he had a reputation for seldom failing. But Canadian Niagara had done so many things wrong that not even the Nesbitt magic could prevail.

There was some merit in Canadian Niagara's claim that it had to wait until the technology had been proven in the United States before it could proceed. But Canadian Niagara's image wasn't helped when entrepreneurs from Hamilton started building a generating plant at DeCew Falls on October 5, 1897, with equipment from the Royal Electric Company of Montreal. On August 25, 1898, the DeCew Falls plant started transmitting power to Hamilton at 22,400 volts. And Hamilton was farther from DeCew Falls than Buffalo was from Niagara Falls, Ontario. Canadian Niagara's public reputation was at an all-time low. And that was saying something in an industry so widely criticized for unfairness that it provided a major impetus in many cities for political reform movements. Some construction work could have been started even before the details of the final installation had been settled. Others had taken this route, but Canadian Niagara did not. The company also apparently chose not to keep the community, the Commissioners, or politicians informed about its plans. It could have built some sympathy and understanding for its plight. Instead, through its failure to communicate and its reliance on confrontation and legal battles, the company lost any public support it had ever had.

Had the company gained anything, other than perhaps humility? In the long run, the loss of the monopoly rights probably saved the company from being absorbed by Ontario Hydro. The two companies that followed Canadian Niagara on the Canadian side of the river had licenses to generate more power and eventually sold out to Ontario Hydro. With the growing public power movement and one company holding a stranglehold on the development of Niagara power on the Canadian side, Canadian Niagara might not have survived as a private company. But that is only speculation.

What is not speculation is that the new agreement, with its timetable, penalty clauses, and approved plans, gave the company a chance to redeem itself. It had squandered its opportunity to make a good first impression, but it could make a good second impression.

Disaster Strikes

The owners of the Niagara Falls Park and River Railway Company had sold out to the Buffalo Railway Company in April 1899, and this group in turn merged with an electric railway line on the United States side and changed its name to the International Railway Company. When all the ink had dried on the agreements, the one Canadian-owned Niagara Falls Park and River Railway Company emerged as the Park and River Division of the International Railway Company. These companies were part of an increasingly complicated network of interlocking companies. Many of them had the same shareholders. The names Edward Dean Adams, Francis Lynde Stetson, and William B. Rankine kept cropping up.

But a few months after the acquisition, disaster struck. On September 4, 1899, about 8 p.m., the lights went out and the electric railway cars stopped. In the Niagara Falls Park and River Railway powerhouse, Mr. Parker, the electrician on duty, noticed a short circuit on the alternating current line owned by Canadian Niagara. When he went to the basement to investigate, he found the place on fire.

The quick-thinking Parker shut off the water gates to the penstocks and called the fire department. He tried to put out the flames with two barrels of water, but it was too late. Within minutes, the whole building was on fire. All the equipment was destroyed or heavily damaged. Only the stone walls remained. The *Thorold Post* reported that "With the shafting and other apparatus the total loss is $75,000, damage to building $10,000." The Canadian side needed power, and quickly. In response, the town of Niagara Falls started up the steam engines in its own electric light plant and had street and house lights back on by 11:15.[31]

On the night of the fire, Burt Van Horn, General Manager of the International Railway Company, which owned the railway and powerhouse, told a reporter from *The People's Press* in Welland that "Traffic will not be delayed.... As soon as we learned of the fire, cars loaded with wire were sent to Niagara Falls, and men will work all night putting the wire across the upper arch bridge, connecting with the line across the river. Cars will be running as usual there tomorrow, from power furnished by the generators on the American side of the river."[32] The electric railway cars were back running, the lights were back on, but all was not back to normal.

In their annual report for 1903, the Queen Victoria Niagara Falls Park Commissioners included this handsome rendering of a Canadian Niagara Power Company generating station that would attract well-dressed ladies and gentlemen. It would prove to be an important tourist attraction in its early years. (Reproduction by permission of the Buffalo and Erie County Public Library, Buffalo, New York.)

By the end of the month, workers had started to clear away the wreckage. "It is a pitiful sight to see the great dynamos, once so perfect and valuable, now broken, silent and covered with rust. It will be some time yet before the wreckage will be all cleared away," lamented *The People's Press*.[33] The railway company was unwilling to rebuild until it had settled with the insurance company,[34] but local industry needed more power than was coming across the border with the single extra cable installed.

On October 3, 1899, *The People's Press* reported that "The carborundum works have been entirely crippled ever since the burning of the powerhouses. Orders keep pouring in, and nothing can be done to fill them. It is rumoured that another cable will be laid across the upper bridge to supply the factory and the electric light plant."[35] Finally, in November word came that the powerhouse would be rebuilt.[36]

The Park Commissioners responded to the need for electrical power by allowing the Canadian Niagara Power Company to import power from the American side, "sufficient to supply the Niagara Falls Park and River Railway's needs and the needs of the industries which were under contract with the Company at the time of the fire."[37] In other words, Canadian Niagara Power Company was still not generating its own power. Now it was importing power from the United States.

A Beginning Long Deferred

After signing the new contract in July 1899, the company had begun test borings to get a better understanding of the underlying geology of the powerhouse site and appeared to be poised for action. But after the fire at the powerhouse in September of that year, the company seemed to fall back into inertia. Little happened for a year and a half. The turn of the century came and went. Finally, on May 23, 1901, construction started on the company's own powerhouse.

The next day the *Welland Tribune* wrote that "it was not generally believed that the company was about to proceed immediately to construct works, as previous announcements of a similar character have proven groundless."[38] Canadian Niagara had to do more than build a generating plant and transmission lines.

Even before construction was finished, visitors to Niagara Falls were sending postcards of the Canadian Niagara Company powerhouse to friends. This one was postmarked in Ottawa on December 17, 1907.
(Reproduction by permission of John Burtniak)

Building the Powerhouse

CHAPTER 3

Who were the people behind the Canadian Niagara Power Company? The principals were men who were deeply involved in power schemes on the American side, preoccupied with the affairs of the Niagara Falls Power Company. They were certainly deeply committed to generating power at Niagara, but they already had their hands full.

The busiest of all was the man for whom the Canadian Niagara Power plant is named — William Birch Rankine. He was a lawyer, a native of New York State, and a tireless promoter of hydroelectric power. Above all, as secretary-treasurer of the Niagara Falls Power Company and of Canadian Niagara, he was the man who Made Things Happen.

Rankine had been born in 1858, the son of a clergyman. He attended Canandaigua Academy, a well-respected prep school, before going on to Hobart College and then Union College, where he was a member of the Alpha Delta Phi fraternity. He graduated in 1877.

His first job after graduating was in Niagara Falls, New York, in the law office of none other than A. Augustus Porter, grandson of the Augustus Porter who had bought extensive river frontage in the early 1800s and had tried to encourage entrepreneurs to take advantage of Niagara's potential. While living in Niagara, he became interested — or, if one reads between the lines of his biographical notices, *obsessed* — with the idea of generating power from the Falls. But this was the late 1870s and he was barely into his twenties. He was too young and too far ahead of his time.

Four of the key players in the early years of the Canadian Niagara Power Company, shown here on September 1, 1904, at the generating station construction site: the long-suffering Resident Engineer Cecil B. Smith, the dapper Corporate Secretary A. Monro Grier, the workaholic Vice President William Birch Rankine, and Edward Adams from the American parent company. (Reproduction by permission of FortisOntario)

Rankine took his bar examinations in 1880, and moved to New York City to practise law. After three years at a large firm, he formed his own partnership, Hawksworth & Rankine, where he earned a reputation as a successful litigator. But the prospect of power from Niagara kept nagging at him, and he returned to the Falls from time to time to talk to the Porters and other local industrialists. He followed the progress of the Evershed Plan and its proponents, and the financial ups and downs of the Niagara Falls Power Company, which was trying to carry out the Evershed Plan.

He looked around New York City for potential investors in power generation, using any and all connections available to him. One was his fraternity membership. It so happened that Francis Lynde Stetson, legal counsel to J. Pierpont Morgan, was also an Alpha Delta Phi fraternity man. Rankine approached him with his idea. Stetson, twelve years his senior, was at first sceptical, but gradually he became caught up in the younger man's enthusiasm and agreed to talk to Morgan. Whatever he said clearly infected Morgan with the same enthusiasm.

The first joint venture of Stetson and Rankine, with Morgan's backing, was the Cataract Construction Company, formed in 1889. (By this time, Stetson had formed his own law firm with Grover Cleveland, who had just been defeated in his candidacy for a second term as U.S. President.) Stetson brought in Edward A. Wickes as a third partner in the Cataract Company; Wickes was said to have connections to Vanderbilt money. Some sources describe Cataract as "the construction arm of the Niagara Falls Power Company," others as a holding company, created to "own the entire capital stock of the Niagara Falls Power Company and act as its financial agent."[1] The company's backers included many of the big New York financiers. Rankine was Secretary and Treasurer of the company, but in many senses, he *was* the company.

For the next few years, Rankine shuttled back and forth between New York City and Niagara Falls, New York. Although the word "workaholic" had not been coined in the 1890s, it is clear that this is what he was. He wasn't a particularly robust man (apparently his heart was so bad that at one point, a doctor told him he probably had only 24 hours to live[2]), but he juggled dozens of commitments, and rarely took time off. He was also something of a micromanager who involved himself in every last detail of the projects he undertook. When Rankine put his name on the papers creating the Canadian Niagara Power Company in 1892, he was a mere 34, and it was just one of his many projects relating to power at the Falls.

In 1897, he moved back to Niagara Falls, New York, to be closer to the work that was closest to his heart. He immediately overcommitted himself to just about every local cause and organization in the town. In addition to being secretary and treasurer for at least five companies — Canadian Niagara Power Company, the Niagara Falls Power Company, the Niagara Development Company, the Niagara Junction Railway Company, and the Cataract Power and Conduit Company — he was president of the Niagara Falls Water Works Company and of the Tonawanda Electric Light and Power Company, and vice-president of the Niagara Falls Suspension Bridge Railroad Company.

That wasn't all. He had directorships and trusteeships galore, everything from the Francis Hook & Eye Fastener Company and the Shredded Wheat Company to the Niagara Tachometer and Instrument Company and his alma mater, Union College. He raised funds for the library, the hospital, and the YMCA, and served on the vestry of his church. In 1902, his friend, Nikola Tesla, gave him a photograph, signed on the side: "This is to my friend, W.B. Rankine, who thinks while others sleep, and works while others think and does while others try, who is in many enterprises and many hearts. . . ." Rankine was everywhere, doing everything. He wasn't married at the time, which was probably just as well, since he was probably never home. When he died before reaching fifty, W. Caryl Ely of the International Railway Company said of him, "He practically devoted his life to the great work at Niagara Falls, and I think that he worked too hard. It's too bad."

When construction began on the Canadian side, he was at his peak, overseeing every little detail. His point man in Niagara Falls, Ontario, was A. Monro Grier.

Grier had been born in Australia in 1860 and educated in England, and had done legal work for the Canadian Pacific Railway. His correspondence, which displays a fondness for witticisms, suggests that he was an interesting contrast to the work-obsessed Rankine. In photographs, his round face appears jovial, whereas Rankine looks serious and exhausted, with perceptible bags under his eyes. (Grier later became president of the company from 1919 until his retirement in 1933.)

It was Grier's job to keep Rankine up to date on everything to do with the construction of the Canadian plant. Unfortunately, for the first few years, the news was seldom encouraging. Grier's letters, as well as those of Cecil B. Smith, the Resident Engineer, are full of reports of delays, cost overruns, unpaid bills, labour problems, and complaints about noise and construction debris. In other words, they were probably fairly typical of most construction projects. Yet what was being built was not, at the time, typical. It was new technology, with a new design.

Redesigning the Generating Station of the Future

When it opened in 1905, the Canadian Niagara Power Company generating station, or powerhouse, was the most advanced hydroelectric station in the world, with the largest capacity generators and turbines. The company managers had delayed planning and building until emerging technology caught up with their vision of the future.[3]

Given the uncertainty of the times over how far one could transmit electric power, the original contract had allowed Canadian Niagara to produce and generate electric or pneumatic (compressed air) power. The original plans would have put the generating plant on land across from Cedar Island — an island which no longer exists — and drawn water from what was known as The Gap, between the island and the shore. The Gap needed to be widened to let more water in and the channel needed to be deepened. The plan called for three powerhouses, which would have allowed for sequential building. Each powerhouse would have had 10 generators of 5,000 horsepower each. This sounds like a larger version of Niagara Falls Power Company on the American side, where 10 generators, each of 5,000 horsepower capacity, stood in two powerhouses on opposite sides of the same forebay canal.

However, it was not clear whether the inlet would let in enough water during periods of extreme cold and ice. And the prospect of digging wheelpits for 30 turbines must have been intimidating. Three powerhouses would have crowded the limited space allotted to the company and probably been visually overpowering.

So that plan was abandoned. One of the drawings accompanying the 1899 contract shows an arrangement similar to what was eventually built, but with several important exceptions. While there was to be only one powerhouse, it was shown as 925 feet long, over 50 percent longer than the 570 foot powerhouse that stands today. Second, the powerhouse in the drawing has 20 turbines in the wheelpit and 20 generators on the floor. The plan indicated that the canal, forebay, and powerhouse would be built to partial capacity and then enlarged, to allow the company to start generating income before final capacity was reached. These plans also changed.

Between signing in 1899 and finalizing plans in 1901, the technology of turbines and electric generators had advanced so much that manufacturers could design machinery for units of 10,000 horsepower. Doubling the capacity of the generating units was an extraordinary development with far-reaching implications. It meant that as the cost of hydraulic and electrical machinery dropped, the canal, powerhouse, and wheelpit could all be much shorter (and therefore cheaper), the opportunities for increased development in the park (and the potential for profits) would increase, and the whole operation could be run more conveniently and cheaply.

But, of course, these changes meant that the company had to rethink its plans and that meant yet more delays. The new plan, which involved building the powerhouse in stages, offered the opportunity to meet the terms of the contract and start earning money earlier, with a smaller initial investment. But there were significant disadvantages to it, too. First, in the long term, it is invariably more expensive to do the job in stages than it is to do it all at once. Second, extending a working wheelpit is very hazardous — not just to workmen, but to existing equipment. The wheelpit would have to be excavated by drilling, blasting, and removing rock. Each operating turbine had to be connected to its generator by a long, vertical shaft, which had to be kept in perfect alignment. Blasting could ruin the alignment. Then there was the danger to existing equipment from flying rock. And it would have been very hard to keep the fine rock dust out of electrical parts. The whole idea made short-term accounting sense, but not construction and engineering sense. Nor did it make long-term financial sense, with the additional costs and the high risk of power disruptions during second-phase construction.

In the end, the company decided to go ahead with all excavation work in one phase. At that point, construction moved ahead on a number of fronts simultaneously.

The Canadian Niagara Power Generating Site: An Overview

To understand how the site was envisioned, planned, and built, one needs to think in terms of all the parts that come together to form the working site. The first need was to draw a large quantity of water in from the river and direct it towards the forebay, at which point it would drop down through the penstocks to the turbines in the wheelpit. The wheelpit houses equipment, most notably the turbines, one for each generator. While the turbines are deep in the ground to take advantage of the height the water falls through the penstocks, the generators are at ground level on the main floor of the powerhouse, which covers the wheelpit and equipment. Each generator is joined to its turbine by a long vertical shaft.

After the water has passed through the penstocks and turbines, it has to rejoin the river. To do this it flows through a long underground tunnel and falls into the river below the Horseshoe Falls. But not all of the water drawn from the river goes through the penstocks, turbines, and tunnel. Station operators do not want ice in the turbines. So as much as possible, any water containing ice that enters the canal into the forebay must be drawn off and floated away through a separate sluiceway back to the river.

The canal, the forebay, the powerhouse, the wheelpit, the generators and turbines, the underground tunnel, and the sluiceway — it all made for an ambitious construction project. And everything had to be started at once.

Under the direction of Clemens Herschel, consulting engineer, the day-to-day work was overseen by Cecil B. Smith, who arrived on the job as Resident Engineer in August 1901 and remained until December 1904, when construction was nearly finished. He brought extensive railway and construction experience to the job as well as a master's degree in engineering, at a time when engineering degrees were rare. When he left Canadian Niagara, he served on the Hydro-Electric Power Commission of Ontario, where he recommended that generating electricity be left to the private sector and transmission to the public sector (his advice was ignored). Smith later formed an influential consulting engineering company and went on to design and build several other major railway and electric generating projects. At Canadian Niagara, Smith certainly had his hands full — and in some ways, as we will see later, he also had his hands tied.

The Cofferdam

The company would have liked to put the powerhouse on the edge of the river. It would have saved construction costs, but for a powerhouse in a park, aesthetics and river views overrode engineering considerations. The river bank location would have blocked the public's view of the upper rapids, so the powerhouse had to be set back, separated from the river bank by a scenic walking area as well as a vehicle and horse carriage road and the double tracks of the Niagara Falls Park and River Railway.

In order to draw enough water from the river, the site had to be modified. The canal which carried water to the forebay had to be made deeper, with a wide entryway from the river. And out in the river itself, the company had to build something to guide as much water as possible to the canal entrance.

January 9, 1902. The cofferdam was the prelude to all other major construction. This specially constructed dam kept the Niagara River's raging waters and ice out of the construction site so the company could build the road bridge, forebay, and generating station. Here, looking north towards the crest of the Falls, the south end of the cribwork extends out from the river and then turns to run parallel to the shore. The inner and outer compartments have been filled with rocks to hold everything in place against the swift current. Workmen drove planks or sheet piling down against the inner walls of the crib to begin making the crib waterproof. Hard hat divers plugged leaks that defeated surface workers. (Reproduction by permission of FortisOntario)

May 20, 1902. With the sheet piling secure and not much water leaking in, workmen filled the inner compartment with a traditional mixture known as "puddle" and tamped, or rammed, it down as they proceeded. The puddle is visible on the lower right. Now the cofferdam was waterproof, but it had water on either side of it, so it was time to install steam engines and pumps on the cofferdam and begin exposing dry land to work on.

(Reproductions by permission of FortisOntario)

June 19, 1902. With the water pumped out, workmen put sheeting on the inside of the cofferdam as a further precaution against water leaking in. Now, safe from the incursions of the Niagara River, they could build the forebay and the bridge running over it.

In a river as rapid as the Niagara, construction could not take place right in the water. So the first challenge was to build a three-sided cofferdam, which, with the shore as the fourth side, would block off an area of the river where water could be pumped out and construction started. The cofferdam had to be sturdy, with walls wide enough for construction cranes and other vehicles to travel along. It had to fit the irregular river bed almost perfectly so that water could not get into the spaces between the inner and outer cofferdam walls and leak into the construction area.

The cofferdam, which consisted of two rows of rock-filled cribs, was built out from the shore in sections. First, engineers had to take soundings to get an accurate profile of the ups and downs of the river floor. Each crib, something like a huge box with an open top, had to be made upside down to make an accurate profile that would match the river bottom. The crib was then flipped over, floated into position, and sunk by filling it with rocks (cross-pieces were added to keep the rocks in the crib). The current was so fast — 10 to 12 feet per second — that the cribs would have been smashed to pieces if they hadn't been held in position with wire rope held by steam-powered hoisting engines.

A rock-filled crib alone did little to stop water coming in. The third, innermost wall and a filling of "puddle" made the structure watertight. Puddling was a technique familiar to canal builders, who had to make the stone walls of excavated locks waterproof. Originally, canal builders took local clay and mixed it with water to form a thick paste or slurry which was then rammed into place between the back of the stone wall and the ground behind it. There was no clay available nearby, so Smith instructed his men to mix about two-thirds gravel and one-third clay to make the cofferdam watertight.

The exposed walls of the cofferdam were further covered with tongue-and-groove sheeting driven down by mauls (heavy wooden hammers similar to sledgehammers) to the river bottom. In places, water still came through, and divers had to go into the river, locate the leaks, and make them tight by ramming bags of cement into the openings. Then, to seal the bottom, the surface of the rock was coated with concrete, and a heavy line of timber, about a foot above the bottom, attached to the sheeting. Eventually the cofferdam was pumped out by centrifugal pumps that could move about 10,000 gallons a minute. All this work was needed to keep the cofferdam in place for two years while other work was going on.

Building the cofferdam was a difficult and dangerous task. In 1901, Canadian Niagara hired the firm of James Barry as principal contractor for the cofferdam, and he seems to have done his job well. But that didn't mean there weren't problems. Barry hired French-Canadian labourers — an unpopular move. Word got around that Barry "was importing cheap French-Canadian labor to supplant local men." There were also lay-offs. Once again, the Canadian Niagara Company had a public relations battle on its hands. Barry explained that he had hired the workers from Quebec because they were experts in this kind of work. And he explained that the lay-offs had been necessary because the workers from Quebec had arrived a week later than expected; the work on shore had got so far ahead that he had to lay off some of the framers who were building the sections that would have to be placed in the rapids by the French workers. Barry also ordered derricks with steam-powered hoisting engines to assist the Quebec workers. Once the machinery and the French workers had arrived, there would be work again for the laid-off locals. And the workers from Quebec were not cheaper than local labour. Barry was paying their fares both ways and "paying them higher wages than the men who work on shore" because he was asking them "to work in places where he would not ask local men to go."[4]

Labour unrest and rumour mills were common on most large construction sites at the time. The work was hard, the pay generally no more than barely adequate, and many workers spoke little English and found it hard to separate fact from rumour. Moreover, there were no social safety nets for those who lost their jobs; construction workers lived day to day and fought hard to keep them.

Canal and Forebay

Once the area enclosed by the cofferdam had been pumped out and secured from incoming water, contractors could start on the canal and forebay.

In January 1902, the St. Catharines firm of Dawson & Riley won the contract for canal, forebay, and wheelpit for a cost of just under $500,000, with a deadline of July 1, 1903, as well as a potential $50,000 bonus for completion on time and the threat of penalties for delays. It was not the best choice for the work.

Water had to flow from the river under a bridge, which would carry traffic going through the park. The waterway widened out to the forebay, which stretched along the 570-foot front of the powerhouse. The river bottom exposed by the cofferdam had to be deepened by drilling, blasting, and hauling rock away. To get the canal and forebay area to the required depth and shape, workers started clearing away sand, clay, gravel, and boulders to a depth of about 14 feet. Although the work could have been done by steam shovels, Dawson & Riley's men worked with picks and shovels,

March 14, 1902. The boiler and compressor plant was an essential part of the construction job, which would disappear upon completion. Here, four of the five steam boilers are in place, three chimney stacks are up, and the black smoke marks on the brickwork indicate that two are in service; the one on the left longer than the one on the right. Compressed air was particularly crucial for machinery used underground to make the tunnel that would ultimately take spent turbine water to the face of the Niagara gorge. (Reproduction by permission of FortisOntario)

October 27, 1902. Once the forebay had been dug deep enough, work started on the bridge piers. The pony, which has worn down a circular track by running round and round, is the power source for the gin-pole crane lifting the blocks of stone. The pony powers the geared mechanism that winds or unwinds the cable that then runs underground to the base of the crane, up the diagonal arm, and down to the massive pieces of stone. Later photographs show steam hoisting engines. (Reproduction by permission of FortisOntario)

and hauled the material to the dump in one-yard wagons. This was not a recipe for rapid progress. And after the loose overburden had been removed, the contractors had to remove one to three feet of bedrock. This work also proceeded slowly.

This lackadaisical approach seems to have been typical of Dawson & Riley's work. Almost from the start, the project started to fall behind schedule. Both Smith and Grier insisted that Dawson needed to put more men on the job, but apparently he never did. Smith and Grier implored, they threatened, they cajoled, they expostulated, but nothing seemed to make things go faster. Dawson seemed immune. At one point, in a letter to Rankine, Grier described Dawson as a "decent, stubborn, honest, illiterate man."[5] He wrote that in August 1903, after the contract deadline had come and gone, and the work was still far from complete. Oddly enough, the one thing he did not do was take the work away from Dawson, who plodded along at his own pace.

Finally, in July 1903, Grier asked Smith to take over as superintendent to see if he could move things along any faster. Grier was still optimistic in January 1904, when he wrote to Rankine, "I am not without some hope that the combined efforts of Mr. Brackenridge [chief engineer of the Cataract Construction Co.] and Mr. Smith, if put forth daily, may result in compelling Mr. Dawson to do what he evidently dislikes doing, that is to say, putting more men when the conditions are not favourable for procuring a hundred cents in the dollar from the labourers."[6] Even so, at the official opening in 1905, the work was far from finished.

Because the Canadian Niagara Power Company site was so close to the lip of the Falls, the generating station had to be built back from the river so as not to spoil the tourists' view of the Falls. This meant that a bridge was needed to carry rail, motor vehicle, carriage, bicycle, and pedestrian traffic over the forebay entrance. In this photo, taken June 16, 1903, the bridge piers have been completed, and workmen have erected the temporary "falsework" to support the steel ribs that would be covered first with concrete, then a waterproof asphalt coating, before putting in the final roadbed and street railway tracks.

(Reproductions by permission of FortisOntario)

May 3, 1904. The completed bridge is one of the most striking and least appreciated bridges in the region. Without the concealed steel ribs, the arches could not have been so flat or low. It represents an important compromise. The Queen Victoria Niagara Falls Park needed a handsome bridge that blended in well with the generating station and was not so high as to seem out of place. Canadian Niagara needed clearance under the bridge for water and, unfortunately, for ice. The street railway needed a level track bed. All three got what they needed. The cofferdam in the background is holding out the Niagara River. The dark stains near the bottom of the cofferdam are water stains.

The Forebay and the Bridge

Meanwhile, the plans were still changing, even as work proceeded. When the company decided to build the canal and forebay to their final dimensions from the beginning, it had to rethink the design. Also, with a generating station intended to last at least 100 years, the long-term performance of building materials was extremely important. At the time, concrete made from Portland cement was becoming popular for routine construction work. But not enough was known about its performance under adverse or unusual conditions.

To match the walls of the powerhouse, the exposed walls of the canal and forebay were to be cut stone, backed with concrete. But what about below water level, where it would not be visible? Plans were prepared showing solid concrete walls below low water, but consulting engineer Clemens Herschel rejected them. The canal would be faced with stone from top to bottom, at considerable cost.

Next, the company had to design a larger, permanent bridge. The original plan was to build a steel bridge across the entrance canal, to carry the tracks of the International Railway Company (formerly the Niagara Falls Park and River Railway), and provide room for a driveway and a sidewalk. But that would have been a temporary bridge, removed when the canal was widened and given its final form.

Given the change in plans, the company opted for a long-term solution that would stand the test of time. The engineers designed a stone and concrete bridge 55 feet wide, made of 5 spans of 50 feet each. It cost nearly twice as much as the steel bridge would have, but it stands to this day. It is a graceful structure with very low arches, built with stone from the Queenston Quarry Company. A more conventional design with higher arches would have meant raising the level of the roadway, which would have interfered with the view of the station from the river's edge and could have caused difficulties for the street railway.

Today, motorists and bicyclists crossing the span do not get a side view and cannot see how beautiful it is. We do not know who designed the bridge, only that it is extraordinarily sensitive to its context and far exceeds the minimum required.

The Tailrace Tunnel

Spent water from the turbines flows through a tunnel beneath the park and part of the city of Niagara Falls to rejoin the river below the falls. It is one of the most amazing parts of the entire project, but almost completely invisible.

The 2,200-foot-long tunnel and the portal, where water is discharged into the river, were built by the firm of Anthony C. Douglass, a Niagara Falls, New York, mining engineer and contractor who later became mayor of that city. The first design called for two tunnels to come out of the powerhouse and then join. This was how the tailrace was later built for the generating station of the Electrical Development Company of Ontario, just upstream from the Canadian Niagara site. Twin tunnels allowed operators to keep part of the generating station running while part of the tunnel was being worked on or inspected. The disadvantage was greater cost. And if constructed properly, the single tunnel should not need to be closed for maintenance. It was a bold move that paid off.

January 12, 1902. With the photography equipment available at the time, shots of tunnel construction are very rare. This one of work on the north heading or working face shows the heavy timbering needed to prevent collapse. The photographer would have used a long exposure, measured in minutes, not seconds, and the jagged white tracks in the lower left show the trail of the candle or lantern used to light the scene. (Reproduction by permission of FortisOntario)

January 29, 1902. Above ground at the head frame for one of the shafts leading to the tunnel, excavated rock is drawn up. The same machinery and bucket was used to lower explosives, timbering, bricks, and other equipment. The building site had its own stone crushing plants where excavated stone could be crushed and used for the concrete in the tunnel. Much of the excavated rock was used to re-landscape parts of the park. The Loretto Academy sits atop the hill in the background. It lost its dome in the fire of January 1938. (Reproduction by permission of FortisOntario)

A number of factors determined the route. The beginning of the tunnel had to be an extension of the tailrace in the wheelpit. The portal, where the water emptied into the river, had to be at right angles to the cliff face. In between, to keep the water flowing quickly, reduce turbulence, and avoid wear and tear on the tunnel lining, curves had to be as gradual as possible. Also, Canadian Niagara could not tunnel under the wheelpit of the Niagara Falls Park and River Railway powerhouse, nor under its discharge tunnel.

For most of its length, the tunnel sloped 7 feet for every 1,000 feet. But if this slope were maintained for the entire tunnel, the water would have emerged out of the face of the gorge above the water level. Water was still being discharged like this on the American side, disfiguring the beauty of the Falls. The problem could be avoided if the water was discharged below the water level. This meant the end of the tunnel had to drop sharply through a "reverse vertical" or ogee curve. The water would enter nearly horizontally at the top of the portal, drop on a sharp slope, and then take a horizontal path to exit under the surface of the river.

The final design called for a single horseshoe-shaped tunnel 25 feet tall and 18.8 feet wide at the widest point on the inside. The original design had been only 21 feet high, but it was enlarged. The tailrace water had to move, but the faster it went, the more likely it would wear the inside of the tunnel too quickly. Increasing the cross-section of the tunnel allowed a given volume of water to pass at lower velocity, but increased construction costs.

Construction started with the digging of a vertical shaft on the line of the tunnel 1,470 feet from the north end (downriver) of the wheelpit. From the base of this shaft, tunnellers moved in two directions, one towards the wheelpit, the other towards the portal into the Niagara River below the falls. Except for the digging at the portal end, everything that went in or out of the tunnel during excavation and construction had to go through this vertical shaft: workmen, equipment, building materials, explosives, and excavated rock.

This work seems to have proceeded largely on schedule. The shaft was sunk during the summer of 1901 and all excavation was complete by December 1902 (while Dawson's crew were still flailing away at the forebay and canal — a much less demanding job). Douglass clearly knew what he was doing.

The one problem that occurred was unexpected, and completely mysterious. April 25, 1902, began like any other day in the tunnel. The night shift had just come on and the workmen were to start blasting on the working face to loosen the rock for excavation. As a newspaper reported the next day, John Clark, "a colored employee" who worked as a "nipper" — that is, he prepared the dynamite for blasting — made the discovery. For safety reasons, dynamite was usually kept above ground. Sometimes there were exceptions, and that night there was some in the tunnel about 100 feet north of the shaft. When Clark went to get it, he saw that someone had tampered with it. Dynamite was detonated electrically. Someone had changed the wiring so that when the workmen thought they were detonating the dynamite at the north face, they would have set off two cases of dynamite further back in the tunnel. The blast would probably have caused the tunnel to collapse, at least in the north arm, where 17 men were working.

The portal was the end of the tunnel, where water from the generating station flowed out of the face of the gorge and rejoined the river. When photographed from a distance on September 10, 1902, the portal work site looked rather scenic. The men on the job site knew that it was about the most difficult and dangerous site of all. Equipment and building materials, including massive blocks of stone, had to be lowered by derrick from the top of the gorge. Much of the excavated rock had to be loaded into flat-bottomed boxes or skips and hauled up. (Reproduction by permission of FortisOntario)

June 23, 1902. The portal work site was a dangerous rock pile. Above, rock that could slide down at any time had to be held back by a growing array of crib work. And all day long, most of the workmen handled wet rocks and stood in water on slippery rocks. The mist from the Falls was always there in greater or lesser degrees, depending on the wind. (Reproduction by permission of FortisOntario)

Police investigations led nowhere. Some thought it might have been a hoax, others speculated that it was a racial incident, because the whites worked days and the blacks worked nights. The mystery is still unsolved.[7]

Despite this disturbing incident, work went on. After the blasting cleared the space, the next step was to shore up the rock. Some types of rock require very little shoring up after tunneling. But much of the Canadian Niagara tailrace tunnel went through shale, which, when disturbed by air and water, becomes unstable. Therefore, as the tunnel proceeded, the roof had to be shored up as fast as the headings were driven. The workmen excavated the top part of the 25-foot-high tunnel first, reinforced it, then dug down the remaining depth in two steps known as benches. As the benches were cleared, wall plates were inserted for further reinforcement.

Construction of this kind looks chaotic to outsiders. The smell of dynamite lingers in dimly lit work areas. The sound of drilling echoes throughout the tunnel and everything blasted out has to be dragged to the ever-receding vertical shaft and hauled up. But within the apparent disorder, a fine precision must rule. Each arm of the tunnel consisted of straight and curved sections, and each arm had to be guided by underground surveying to arrive at an exactly predetermined destination.

On land, surveyors could see the target, the point the tunnel was to arrive at, but underground, everything had to be done blind. Only when the workmen got to the destination could the surveyors find out how accurately they had surveyed underground. How did they do it? They suspended two piano wires 12 feet apart in the shaft. These provided the reference point from which all underground alignment was fixed. As the tunnel proceeded, the surveyors put in new reference points which were related to the original two piano wires. Deeper holes had to be dug periodically into the limestone and "plumb posts" inserted to act as the new reference points. This was exacting work. Any small error in one plumb post would be carried on to the next and the error would be magnified as the job proceeded.

Amazingly, the underground surveying was a success. Resident Engineer Cecil B. Smith declared the results achieved by the contractor A.C. Douglass as "almost perfect," the "result of extreme care in the use of plumbobs, and taking the average of a large number of observations in this [underground] and surface instrument work."

By December 1902, the excavation was finished. But the tunnel was far from complete. The interior had to be as smooth as possible. With water flowing by at 30 feet per second, or just over 20 miles per hour, any irregularities would cause turbulence that would erode the tunnel prematurely. And repairing a tunnel would be exceedingly costly, because the only way to shut off the water flow for repairs would be to shut the generating station down completely. What was needed was a smooth, long-lasting surface.

The tunnel of the Niagara Falls Power Company in Niagara Falls, New York, had given good service for seven years. It was lined with brick. But had the company paid too much for success? Had the tunnel been over-engineered? Would a concrete lining do as well? Certainly it would be cheaper. But concrete then was not the highly engineered building material that it is today, even though it was being used successfully in increasingly demanding applications.

The original plans called for Canadian Niagara's tunnel to be lined completely by four layers — or rings, as they were called — of brick. But the engineers chose to line the walls and floor with concrete covered with a better brick than their counterparts had used across the river. It was only a single layer of 4-inch thick brick, but

November 25, 1903. The workmen, who seem dwarfed by the portal site, provide a sense of scale to the massive tunnel that ran from the generating station. (Reproduction by permission of FortisOntario)

with a concrete backing, the thickness of the brick was far less important than its durability. If the water wore through a layer of brick, that spot would be so turbulent that other layers would soon be damaged. Better to have a single layer of superior brick, particularly one that would "slime" well — that is, it would easily acquire a thin layer of smooth vegetation which would offer even less resistance to the water flowing by.

On the roof of the tunnel, the part known as the arch or invert, the difficulty of pouring concrete had to be considered. There were no concrete pumps; everything had to be placed by hand. Moreover, with the somewhat unstable shale roof, it would have been difficult and hazardous to build formwork and pack concrete. Therefore the roof of the tunnel was made of four layers of brick backed by something Smith called "dry packing" — probably sand, or fine stone gravel. Because the tunnel was not designed to have water running through up at roof level, the upper bricks were not quite as smooth, or expensive, as the vitrified bricks in the walls and floor, which had an almost glasslike surface. The four layers of brick in the arch was more for stability than protection from wear.

The process of lining the tunnel started with laying the concrete floor. Vitrified bricks were laid on the floor and then in stages the workmen created formwork for the concrete walls, poured concrete, and continued laying the four-inch vitrified brick up to the point at which the arch started and the four-layer brick construction started. The bricks for the arched ceiling had to be supported by formwork called arch centres while being put in place, a standard piece of construction at the time.

Most underground tunnels would have been finished once the side forms and arch centres had been removed. But not this one. The whole tunnel was scraped clear of all remaining mortar to make it smooth.

At the portal end, where the water picked up speed, dropped more steeply, and turned through a smaller radius, even more reinforcement was needed. This part was lined with granite two feet thick. The work went on in winter, and must have been very hard on the workers, who would have been cold, wet, and constantly in danger of slipping on ice or wet rocks, timber, and equipment.

Lining the tunnel and building the headwall at the portal end took until May 1904. Later that summer, the shaft, which had served as entry and exit for the tunnel workers, was bricked up and the outlet excavation completed. This opened the way to removing the cofferdam between the river and the portal. At this point, the engineers discovered that the water channel in the river had to be deepened. This would have been easier behind a cofferdam, but in the summer of 1905, the men did the job with dynamite and a bucket attached to a derrick.

Improving the Park and Dealing with the Public

During construction, the contractors for the tunnel, canal, forebay, and wheelpit had removed about 310,000 cubic yards of earth and rock. Under the terms of the 1899 contract between Canadian Niagara and the Park Commissioners, the latter had the right to direct where excavated material was to be placed or specify if it had to be removed from the park. As usual, aesthetic considerations came first, not contractor convenience or price. Shale from the tunnel was used to fill in the lagoon between Cedar Island and the shore. Filling in this area added more usable space to the park from which visitors could enjoy the views of river, rapids and water racing to the falls. But that project didn't take all of the rock excavated from the tunnel.

The rest of the material excavated from the tunnel, canal and wheel-pit allowed the Park Commissioners to implement a plan they had formed years before and had partly carried out in 1895: the reclamation of a large area of rock surface at the west end of the Horseshoe Falls left bare as the falls receded and the central portion of the channel deepened. Filling this area made it possible to construct additional walks and driveways and improve the look of the park. The fill let visitors stand closer to the water's edge safely and get a better view. Nonetheless, rumours inevitably began to circulate that the crest of the Canadian Falls had been shortened by several hundred feet by the power companies for their own selfish reasons.

Sometimes, it must have seemed that Canadian Niagara could do nothing right. The stream of complaints poured in from all sides. The chairman of the executive committee of the State Reservation on the American side, Alexander J. Porter, wrote to complain about the noise of blasting. The sisters at the Loretto Convent claimed that the blasting had damaged their building. Caryl Ely of the International Railway Company complained about dumping of rock by the contractors.[8] Suppliers wrote dozens of letters about late payments, or non-payments. And in the middle of everything, with work proceeding underground, above ground, and in the river, the company had to deal with a strike.

A Brief but Worrisome Strike

In the early 1900s, Niagara Falls was a construction boom town as well as a tourist centre. By 1903, three major electrical generating plants were under construction on the Canadian side for the Canadian Niagara Power Company, the Ontario Power Company, and the Electrical Development Company. Jobs were plentiful. Then some contractors decided they were paying too much for labour. On November 2, 1903, they reduced the hourly rate for labourers from $1.75 to $1.50 for those who worked in dry places, but not for those working in wet or hazardous places.

When they heard the news, about 20 men downed tools and walked to other job sites, where they picked up 100 or so followers. As the number of strikers grew, they became less tolerant of non-strikers. At the Canadian Niagara wheelpit excavation, strikers throwing rocks and earth down the pit convinced Dawson & Riley's workers this was not a good day to work. Work continued, however, on the tunnel.

By late in the day, 700 to 800 workers were marching around the Park to make sure that no-one was working. The police ordered all drinking establishments near the construction sites closed, but strikers got hold of 25 kegs of beer. There were soon many drunks on the bluff above the park near the Loretto Convent. The Park Commissioners called in the militia, which arrived at about midnight. About 50 soldiers from the No. 1 Company of the 44th Battalion and the Lundy's Lane Rifle Corps guarded each construction site, and the moonlit night passed fairly peacefully.

On the second day, workers gathered on the Michigan Central Railroad tracks in front of the Loretto Convent. They could see men working on derricks and around the boiler house on the A.C. Douglass excavation site. The strikers yelled and threw rocks at the workers. Some men with revolvers started shooting, but no-one was hurt. At that point, the militia closed in with bayonets drawn. The strikers were ordered to disperse, and fortunately, they did so.

From that point on, the strike ran out of steam. The next morning, November 4, about 20 men were in jail, charged with inciting a riot. Another 75 to 100 were missing from work. Everyone else was back on the job. The most dangerous events of the strike had been the attacks against men at Canadian Niagara work sites, particularly the wheelpit and at the tunnel excavation sites. Their attackers had come from other sites.[9]

In that era, wildcat strikes over reduced or missed wages were not uncommon. This strike stood out from so many others because it was short-lived and there were no casualties. As strikes go, it was mercifully uneventful.

The Wheelpit

Turbines and generators could be linked by either horizontal or vertical shafts. Horizontal shafts between turbine and generator meant that the generators would have to be placed at the bottom of the wheelpit, at the same level as the turbines. But no-one was sure that the bottom could be kept dry enough for generators, so the engineers decided to place the generators at the top, vertically linked to the turbines at the bottom.

Digging the deep wheelpit was a quarrying or open pit mining operation that relied on the most up-to-date equipment. In this photo, taken on March 14, 1902, portable steam engines and drilling machines — both mounted on temporary movable rails — made the holes for explosives and the endless rounds of drill, blast and muck. Mucking is a mining term for removing blasted rock or rubble. (Reproduction by permission of FortisOntario)

Difficult as building the tunnel was, excavating the wheelpit was the most troublesome and demanding part of the project. This work also fell to Dawson & Riley, probably not the best choice of contractors.

The wheelpit is exactly 566.84 feet long. In order to have it extend down to the tunnel 171.1 feet below the powerhouse floor, the contractors had to remove the overburden and excavate through 153 feet of rock at a width of 20.5 feet. It was a complex task. The wheelpit is where water goes down from the inner forebay, through the penstocks to the turbines, and out the draft tubes to the tailrace and tunnel. Long rotating shafts run up to the generators on the powerhouse floor, but in between, a host of auxiliary equipment has to be housed in recesses or rooms running off the main excavation. In effect, the wheelpit was a multi-storey subterranean workplace, only 18 feet wide after lining, where almost everything had to be fastened to, or in some way supported by, the walls.

Blasting for a wheelpit is not like blasting in a quarry to get rock for crushing. The walls of the wheelpit must remain structurally sound. The less they are shattered, the less they will leak water into the wheelpit, and the less likely it is that they will buckle, deform, or break apart.

After stripping off 15 feet of earth, boulders, and gravel overburden, workmen had to excavate 100 feet of limestone and 50 feet of shale. To blast rock out to precise dimensions and prevent the remaining rock from shattering, they used a proce-

As the wheelpit deepened, the same sequences of operations continued, but the walls had to be channelled carefully so that the explosives fractured only the rock to be removed and not the rock remaining as part of the wall. (Reproduction by permission of FortisOntario)

At Niagara Falls, ice is just one of the things that have to be accepted as a natural force. Water seeped through the exposed walls and froze. During storms, snow or freezing rain and ice fell into the pit. The ice delayed construction and added to the danger. A piece of falling ice could be every bit as deadly as a piece of falling rock. (Reproduction by permission of FortisOntario)

dure known as channelling. It required two distinct steps. First, they made a vertical channel about 6 feet deep behind the exposed face of the rock section, or bench, to be blasted out. This smooth saw cut separated the piece to be blasted from what would be exposed as the new edge or face of the wall. Then, using compressed-air drills, they drilled horizontal holes, extending in to meet the channel cut. This isolated the "bench." Finally, they put dynamite into the holes, detonated it, and blasted off the face of the bench down to the bottom of the channel cuts.

As the wheelpit became deeper, the excavation became more complicated. The side walls have to bear the weight of floor beams and girders supporting the generating station floor, as well as the heavy generators and other equipment. Floor beams and girders rested on castings built into the side walls.

In areas where parts or castings had to go into the wheelpit wall, precisely shaped recesses or holes had to be created without damaging the sides of the wheelpit. To do this, the workers used a mining technique known as gadding and broaching. The workmen drilled a series of horizontal holes around the area to be removed. A small charge of dynamite could blow out these detached masses of rock without disturbing the surrounding rock. If all went well, the adjoining walls remained intact.

Near the bottom of the wheelpit, three chambers, each about 40 feet by 16 feet by 30 feet high, were built at right angles to the wheelpit. These chambers were created in a way similar to tunnelling, but unlike the tailrace tunnel, which was cut through shale and had to be shored up with timber, the wheelpit went through limestone and needed no timbering. (The three chambers were in the first 270 feet of the wheelpit. Two more were built under a later contract, which brought the wheelpit to its full length.)

Recesses had to be cut into the side walls for a wide range of castings, some as heavy as 25 tons. All of the recesses had to be made using the careful and laborious technique of gadding and broaching. Each opening had to be excavated accurately and the castings placed in position equally accurately with the help of reels of piano wire carrying heavy plumbobs immersed in water.

Finally, the entire wheelpit, including the chambers, was lined with 24 inches of solid brick backed by 4 inches of hollow brick to make it watertight. The hollow bricks allowed for seepage from the surrounding rock to be carried safely to the bottom of the pit, without harming the equipment.

One construction crew worked at excavating to create the wheelpit, while deeper down, underneath the area being excavated, tunnellers were driving the tunnel forward. Working at two levels went well until the tunnel had gone about two-thirds of the length of the wheelpit above it. Then came a tunneller's nightmare; the blasting above and below had weakened the shale. Water began to leak into the tunnel. The tunnel roof was close to collapse.

Yellow pine is traditionally used for the vertical timbers to shore up the roof. However, Douglass, the tunnel contractor, switched from yellow pine timber to hemlock, which is stronger. But as the tunnellers extended further into the area beneath the wheelpit, water continued to flow down through the shale, weakening it. The hemlock was compressed by the immense weight it was bearing. Although the tunnel roof did not collapse, the damage had been done. The sides of the wheelpit were shattered, and large masses of rock broke away. Before the brick lining could be put in, workers had to fill up the cavities with concrete.

Then another problem surfaced. As early as 1897, William Rankine had told the Canadian Power League about a potentially disastrous problem in the wheelpit on the American side: the sides of the wheelpit were moving inwards. By 1902, it was clear that the same thing was happening on the Canadian side. Workmen measured the total width of the wheelpit to see if it changed as the excavation proceeded down to its full depth. They found that as the pit grew deeper, it became narrower. This was serious. Too much movement could lead to collapse. But even if the walls remained intact, the movement could ruin the alignment of the long shafts and other pieces of machinery that the wheelpit had to house. The wheelpit walls had to bear a great deal of weight, because they would support the heavy arches that held up the floor and eventually the full complement of 11 generators and other equipment.

The engineers decided to put in heavy cast-iron struts across the pit at units No. 4 and No. 5, to thrust against cast-iron strut bases, which had been put in at more than 20 places in anticipation of trouble. Putting in the strut bases was a brilliant piece of engineering foresight that undoubtedly drew on the experience of wheelpit deformation at Niagara Falls, New York. Here was one of the benefits of delaying construction past the original target dates. The company paid for the delay, certainly, but saved itself trouble in the long run.

There was still plenty of work to do after the wheelpit reached its final depth. The exposed rock walls had to be covered in poured concrete, followed by hollow weeping tile and then a brick facing. The weeping tile allowed water that seeped through the rock and concrete to drain down harmlessly. If the weeping tile had not been there, or if it were plugged, the hydrostatic pressure could become so strong that it could blow the brick off the wall or crack sections of the wall. The brick floor, or invert, shown here formed the bottom of the tailrace that would take spent water from the turbines, through the tunnel, and back to the river below the Falls. In this photo of September 19, 1903, the draft tubes for units #1 and #2 are being readied for installation. (Reproduction by permission of FortisOntario)

The struts largely stopped the inward motion. Given the natural changes in the rock and the varying forces on it, one cannot completely stop all movement. But when Smith found monthly variation at the thrust deck between 12 and 24 one-thousandths of an inch, he and everyone else could proudly proclaim that they had built a stable wheelpit.

Why had the walls started to move inwards? In 1909 engineer A.H. Van Cleve, who had worked on both sides of the river, wrote, "When the inward movement of the walls of wheel-pit No.1 of The Niagara Falls Power Company first attracted attention in 1897 the theories advanced...were almost as numerous as the observers. Even the moon was held to be guilty. But as time has passed...the simple explanation long held by the writer appears to be the most plausible, namely when a long deep gash is cut in the crust of the earth, the shrinkage stresses, present at all times in that crust, are relieved by the inward movement of the rock forming the sides of the opening. When and as soon as these stresses are thus relieved, the movement practically ceases."[10]

All of the duct work and cabling that sent power to Niagara Falls, New York, had to go underground as long as it was inside the Queen Victoria Niagara Falls Park. This meant considerable above-ground disruption during construction. The Cliff House, close to the Canadian end of the Upper Steel Arch Bridge, advertised meals with lager beer. On July 8, 1904, when this photo was taken, patrons could also watch workmen dig up the street to make a conduit manhole, or vault, and prepare to pull cables. (Reproduction by permission of FortisOntario)

The Powerhouse

If excavating the wheelpit was the most challenging construction work, the powerhouse was the most visible. It was the public face of the company. Here electricity is generated and from here it goes out to customers. It is also the major work area, for here the operators control all of the generating and transmission equipment. The floor must be designed to withstand the weight of heavy parts taken out of the pit for service by the two travelling cranes overhead. It was also where visitors came to stand in the visitors' gallery, sign the guest book, but most of all to gaze down on the generator floor where nothing moved and yet everything happened.

When the powerhouse opened in 1905, it was only 296 feet long, big enough to house five generators. It would eventually be lengthened to 600 feet to accommodate 11 generators. Cecil Smith described its outward appearance as "quiet." Architectural historian Mark Fram describes the powerhouse as "an elegant piece of architecture."[11] Certainly it was nothing like the generating station built nearby by the Electrical Development Company, later renamed the Toronto Power Company. That had been designed by noted Toronto architect, E.J. Lennox, who also designed Casa Loma. Built in the Beaux-Arts tradition, it looked more like an oversized bank than a generating station.

Although clad in cut stone, the Rankine Generating Station was thoroughly modern, with a structural steel frame. Good old-fashioned muscle power did a lot of the work, but modern equipment such as the cranes shown here were essential for positioning, raising, and lowering heavy pieces of equipment. In the centre of this photograph, taken March 16, 1904, a self-powered mobile crane runs on temporary rails that could also bring railway cars into the building as it was being constructed. Above that is the cross-beam of one of the two 50-ton-capacity cranes that are still in use today, more than 100 years after this photograph was taken. (Reproduction by permission of FortisOntario)

The unit #1 thrust bearing, shown here on December 30, 1904, carried the weight — about 90 tons — of the rotating parts of the generating equipment. Its design and placement was radically different, and the performance noticeably better, than that of the thrust bearings for the Niagara Falls (New York) Power Company. The thrust bearing supported the weight, or thrust, of the rotating generating system from turbine and shaft to the generator. It was lubricated by high-pressure oil, but sometimes the babbitt bearing surfaces burned out. Replacement was a major task involving heavy work and great skill.

The stone arch under construction in the lower part of this photograph, taken June 14, 1904, was one of 11, one for each massive generator. Heavy timber framework supported individual pieces of stone until the stone arch was finished. A more complete arch is visible in the background. The rails to the left were used for both railway cars and the mobile crane.

(Reproductions by permission of FortisOntario)

When this photograph of the assembly of generator #6 was taken on October 22, 1910, the generating station was only long enough to house 6 units. The two 50-ton capacity overhead cranes are working in tandem to pick up the rotor (the rotating part) and lower it inside the generator housing and stator (the fixed or stationary part). The Canadian Niagara generators were the largest in the world in 1905, but were soon overtaken and today are considered small. However, unlike the radically different generators of the Niagara Falls (New York) Power Company, those of the Canadian Niagara Power Company were essentially modern, with the exception of the long vertical shaft linking them to the turbines. It would have been a good idea to be careful around this job site. The opening cordoned off by the brass railings and the wooden timbers leads to a drop about 12 storeys deep.

The blurred, nearly invisible fly balls on the right side of the governor in this 1908 photograph show that it is operating and controlling the speed of the turbine, and, with it, the generator, which should be kept at exactly 250 rpm (revolutions per minute). The governor, built by Escher-Wyss in Zurich, Switzerland, was based on an adaptation of the fly ball governor invented by James Watt for his steam engines.

(Reproductions by permission of FortisOntario)

The construction site shown in this photo of July 21, 1904 is protected by the cofferdam. Without it, the whole area would be under water. The bridge piers are on the right. On the left are the arches that admit water to the inner forebay and the massive cast-iron inlets that direct water to the penstocks. During normal operation, the top of the arches would be under water or nearly so. (Reproduction by permission of FortisOntario)

Company correspondence suggests that some people, including perhaps the Park Commissioners, favoured a more showy building, rather than the "quiet" look. However, Canadian Niagara got the building it wanted, apparently by not presenting the plans until the building was already under construction. This was not an oversight; it was a careful strategy. On August 3, 1903, A. Monro Grier wrote to Rankine, declaring, "I cannot help but thinking that our plan of campaign with reference to these plans has been justified by the event. Perhaps if we had not forced the hands of the powers that be by submitting to them plans which had already been acted upon to some extent, they might have treated the matter as altogether at large and we might have found ourselves later dwelling in a cross between a mediaeval castle and a modern ice cream parlour, which would not have been 'convenient.' "[12]

Grier was probably thinking of styles popular at the time, such as the "chateau style" that was all the rage for Canadian tourist hotels. The rustic, castle-like style of the powerhouse of the now-demolished Niagara Falls Park and River Railway probably evoked a "medieval castle." The "ice cream parlour" might have referred to something in the style of the white buildings created for exhibitions in Chicago and Buffalo, which did so much to promote electricity.

The Canadian Niagara Power Company powerhouse used local stone with a fine pedigree — Queenston Blue Dolomite from the nearby Queenston Quarry on the Niagara escarpment. It had been used in all of the Welland Canals, Sir Isaac Brock's monument, Powerhouses Number 1 and Number 2 of the Niagara Falls Power Company on the American side, as well as in many civic, religious, and residential buildings.[13] The stone has a bluish tint when freshly quarried, but over time turns to a pleasant gray.

Who was the architect? When the Niagara Falls Power Company designed its key buildings, it had hired one of the most prestigious of all American architectural firms, the New York firm of McKim, Mead and White. This firm designed Powerhouses Numbers 1 and 2 as well as the worker's village Echota, an important part of the company's image and its strategy to attract companies to establish works to use power from the Niagara Falls Power Company. There are similarities between the Niagara Falls Power Company powerhouses and the Canadian Niagara Power Company's powerhouse. But McKim, Mead and White did not design the Canadian powerhouse.

The architect of record was Algernon S. Bell of New York City, who had designed a country house called Skylands in Ringwood, New Jersey, built between 1899 and 1901 for Frances Lynde Stetson.[14] The fact that Stetson had his architect draw up plans for the powerhouse was not unusual. After all, Henry Pellatt, one of the Toronto Power Company owners, hired the same architect for the Toronto Power generating station in Niagara Falls and for his monster house, Casa Loma, in Toronto.

A.S. Bell may have done the main design work, but in a sense, the look of the building is also the result of many engineering decisions. According to Cecil Smith, who never seems to mention Bell in his correspondence, "Speaking generally, the designing was done under the direction of Consulting Engineer [W.A.] Brackenridge and by the designing staff of the Niagara Falls Power Company in charge of A.H. Van Cleve until March, 1904, since which time the writer assumed Mr. Brackenridge's function, but the liberal policy of the company was, that all the main points of design were determined after full discussion by the various engineers mentioned." For example, Smith asked for samples of two kinds of roof tile, red and green, and consulted with others before placing an order for green tile.[15] However the design process worked, the result bore "eloquent witness to the success of the Park Commissioners in ensuring, from the beginning, that industrial development would be compatible with its sylvan setting."[16]

But long before the powerhouse was finished, the company took the opportunity of the cornerstone-laying ceremony in 1901 to try some positive public relations. Alas, as with everything else, nothing went according to plan.

Has Anyone Seen the Duke and Duchess? An Unusual Royal Visit

One of the company treasures is a small oak box, with sterling silver bands and an engraved plaque that reads:

Canadian Niagara Power Company
This box was locked by
His Royal Highness The Duke of Cornwall and York
On October 13th A.D. 1901
In the presence of the Secretary Mr. A. Monro Grier

This is not the inscription that people had been led to expect. On Friday, October 11, 1901, *The* [Welland] *Tribune*, had outlined the royal itinerary under the heading "Will Lay Corner Stone." "It is announced that the Duke of Cornwall and York will lay cornerstone of the new power house at the Falls on Monday morning." But an account of the actual Royal Visit published a few days later mentions their Highnesses viewing the Falls, visiting Queen Victoria Park and the Loretto Convent, and travelling on to Chippawa, and makes no mention of either the powerhouse or the Canadian Niagara Power Company.[17]

A few days later, *The Tribune* partly cleared up the mystery. The reporter, after observing sourly that "Niagara Falls will not see them [the royal couple] again for a long time, and most likely never," explained that "despite all rumors and announcements the party did not come into town [Niagara Falls] on Monday, and a big crowd that waited at the site of the new power house in the park to see His Royal Highness lay the cornerstone was badly sold." The closest the royal party came to Niagara Falls was the railway line where the royal train changed engines before heading off on the Grand Trunk to Hamilton.

In fact, the cornerstone was laid by the decidedly non-royal "Wm. Bartlett, who for the past eight years had been the inspector of masonry for the Niagara Falls Power Company on the American side of the river. He was assisted by James Balston and Harry Uglow, also masons of the Niagara Falls Power Co." Curiously, the ceremony was left in the hands of Douglass, the contractor for the tunnel. Douglass had "asked Bartlett to take charge of the job because he would have none but union men do the work and there is no masons' union on this side of the river." Still, when Bartlett and his assistants came over the bridge to the Canadian side on Monday, they were expecting the Duke. But there was no Duke, only the box he had locked. It took the three stonemasons the better part of the afternoon to set the cornerstone, because there were in fact four large stones to be placed on a concrete base.[18]

What was going on? Were labour troubles brewing? Was there a possible security risk? For one thing, the Royal Visit coincided with the anticipated arrival of the French-Canadians that James Barry had hired to work on the cofferdam. Feelings in town were running high, and the company may have feared some demonstration in relation to the Royal Visit. They couldn't afford another public relations fiasco. And what about not informing Douglass and the stonemasons, not to mention the assembled crowd? Cancelling the event might have been interpreted as anti-labour. Letting it go on as it did cast the company in a strong pro-labour position.

Perhaps there was also the fear of assassination. The Pan-American Exposition held in Buffalo in 1901 had been a marvellous showcase for the wonders of electricity, including long-distance transmission of electricity. The exposition had originally been scheduled for Niagara Falls, New York, in 1899, but was delayed because of the Spanish-American war. In the end, the fair was held in Buffalo. Just a few weeks before the Royal Visit to Niagara Falls, Ontario, President William McKinley had been assassinated at the Exposition. Not surprisingly, after this event, security for dignitaries was tight and plans tended to change without forewarning.

An Uneventful Event?

The story of construction at the Canadian Niagara Power Company site seemed to be a tale of narrowly avoided disasters. Attempted sabotage. An aborted strike. A tunnel that sagged but did not collapse. Wheelpit walls that bent in a bit. And of course, constant delays. But on January 2, 1905, enough was ready to attempt to rehabilitate the company in the eyes of the world with an official opening.

The event was almost as low-key as the laying of the cornerstone. The Chairman and Commissioners of the Queen Victoria Niagara Falls Park Commission attended, but the Premier was absent. No official representatives of the provincial government came. The opening was written up in a few papers, but it wasn't a big story.

After the ceremony and obligatory photographs in the powerhouse, the assembled dignitaries, including the Chairman and Commissioners of the Queen Victoria Niagara Falls Park Commission, trooped up the bluff to the Carmelite Hospice for a ceremonial lunch. It may seem an odd place to hold a celebration, but it was a large public building with a sizable kitchen and dining room, close to the powerhouse. It was also one of Canadian Niagara's first customers and had used electricity for cooking and heating well before the turn of the century. It had also waited patiently during a power outage that had lasted more than four months.

Why so little fanfare? The event was purely symbolic. It was a photo-op, designed to placate the government. No power was generated. No new lights shone, nor did new industries open up that day. The powerhouse was far from complete; the company was still buying and reselling power from the United States and would continue to do so for another six months. At most, what happened was a test of a turbine or a generator.

But on July 27, 1905, Generator No. 1 went into service.[19] Canadian Niagara was finally in the business of generating its own power in its own generating station. The day seems to have passed without commentary or ceremony. No politicians or journalists were present.

A document titled "Canadian Niagara Power Company Kilowatt-hours Generated,"[20] which gives daily production totals, shows how quickly power production increased after that point:

January 1 to July 26, 1905	No power generated
July 27, 1905	6,700 kWh
July 28, 1905	45,000 kWh
July 29, 1905	92,000 kWh
July 30, 1905	97,000 kWh
July 31, 1905	102,000 kWh

The 2nd and 3rd generators came into service on September 11 and September 27, 1905, respectively. Another document, called "Interchange of Power between American and Canadian Plants," contains a monthly summary of the flow of power in two directions. Rankine relied completely on American power until Unit No. 1 came into service on July 27, 1905. By August, Canadian Niagara was exporting more than 5,000 horsepower to the United States. By October, all imports of power from the American side had stopped. Rankine was now playing the role its American owners had created it for: net power exporter to the United States.

Goodbye to William Birch Rankine

The early years had been difficult for everyone. But now the Canadian Niagara Power Company was entering a new phase and it was time to forget how long things had taken to happen. It was also time for different faces to represent the company. Grier was still there, and would remain for many years. But the construction workers went on to other construction sites, the contractors — Barry, Dawson & Riley, Douglass — to other contracts.

Cecil Smith had already left, frustrated with the company's management. Grier and Rankine had piled increasing responsibility on him, without giving him sufficient authority to do his job. He was not even authorized to approve expenditures over $100.[21] He had done heroic work overseeing the construction, but had always been treated as an underling. When he dared to complain to Rankine that his job had been made more difficult because of delays caused by Grier, who insisted on clearing every single decision with Rankine, Rankine wrote Smith a letter that managed to insult both him and Grier. "The misconception which has existed in your mind as to your status may have been due in part at least to the lax manner in which the Company, through its Secretary [Grier], has permitted you to perform certain acts which are not properly within the scope of a Resident Engineer."[22] This, after Grier and Rankine had more or less forced Smith, against his initial reluctance, to take over the job of construction superintendent in 1903.[23] Smith quit in disgust in November 1904, effective December 1904. Ever the gentleman and thorough professional, Smith did not talk about it publicly, nor did he bring it up in his subsequent journal articles about the project. The Company replaced him with A.H. Van Cleve, who was somehow expected to take on Smith's old job, but without relinquishing his work at the Niagara Falls Power Company.

During the early years of operation, the generating station switchboard was on a mezzanine above the main generator floor. Later it was moved to a separate control room on the second floor. This dramatic view of the switchboard taken in 1908 shows the orderly layout that made it easier to grasp quickly what was happening, particularly during an emergency. The fine wood cabinetry and detailing, along with the polished brass rails, bespeak both pride and a strong sense of good design, which one finds throughout the early work of Canadian Niagara. The clocklike device above the switchboard is the synchroscope. (Reproduction by permission of FortisOntario)

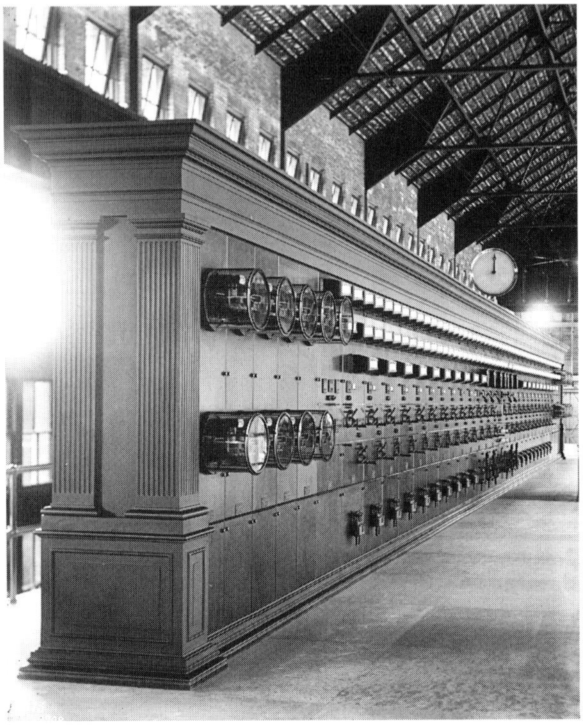

The other departure, a few months after the official opening, was that of Rankine himself. Despite his many other commitments, he had certainly tried to serve the company faithfully. But in his last months, he cut back his work commitments and began to have something of a personal life. A few weeks after the opening ceremony, Rankine married Annette Kittridge Norton, who came from a wealthy Detroit family, at St. Peter's Episcopal Church in Niagara Falls, New York. Even that event was marked by tragedy. Three hours before the wedding, a short circuit in the Niagara Falls Power Company's No. 1 Powerhouse had caused fatal injuries to two workers; a third escaped with serious burns.

After a short honeymoon, Rankine went back to work, planning a longer vacation in the fall. On that holiday, at a lodge at Franconia, New Hampshire, the newlywed husband died of congestive heart failure on September 30, 1905. The outpourings of grief were immediate. It seemed inconceivable that Rankine, who had done so much, should be gone. His death left an enormous gap. Even sadder is the fact that his widow never recovered from his death, and committed suicide in 1921.

On July 11, 1910, the photographer aimed his camera towards the temporary south wall in the working but still incomplete generating station. The building has not been built to its full length. The control panel is on the mezzanine level above the generator floor. The turbine parts, seen to the left and right of the lower part of the picture, will be moved by overhead crane to the south end of the building and lowered into the wheelpit.

Rankine is something of a mystery, and although there are many testaments to his devotion to work, no personal anecdotes appear to have survived. One of the few things that remain is an odd little story, published in the *New York Sun* on June 21, 1903, which takes place at the powerhouse of the Niagara Falls Power Company, in Niagara Falls, New York. Here it is, in its entirety.

This is one of a group of three photographs taken after the control room had moved up to a separate control room on the second floor of the generating station. This layout was even easier to read than the earlier one, which had been well above industry standards. The synchroscope, not visible in this photograph, was slightly to the right of this view.

(Reproductions by permission of FortisOntario)

And the Car Stopped: The Man Who Is Able to Bring Hurrying Motormen to Terms

Buffalo, June 20 — "Finally," said the old Buffalo and Niagara Falls trolley motorman to the novice finishing his course of instruction, "when you see a tall, handsome man in a speckled suit standing in front of a low, gray, stone power house at Niagara Falls — stop."

"Won't it do to just keep my fingers crossed and look over my left shoulder?"

"No. Cut out the dream book and remember what I say — stop."

Inasmuch as the previous instructions had run somewhat counter to this advice, the novice wondered. He had been told that in running from Buffalo to Niagara Falls, it was sometimes wise in order to keep up to the schedule time, to whirr past the frantically waving citizens of Niagara Falls after reaching the boundary line of that city, for experience taught that it was wiser to ignore the impassioned gestures of local people than to arrive at the terminal a minute or so late.

Naturally, the novice wished to know what lay behind the exception.

"Listen to what happened to Bill Smithers," said the old hand impressively. "He was late in getting into La Salle and a teamster delayed him worse. He knew he could never make the Niagara Falls end of it right unless he went through without stopping for locals. So he left with their thoughts a bunch of would-be passengers at Echota and another at the Carborundum works and it looked like making the Falls end right on time.

"So when Bill sees a quiet-looking distinguished man at the power house he stuck to his window with his eyes fixed about two blocks ahead and never a glance to the right to see whether the man would wave his arm off or merely look disgusted. What happened was, that the man never shook his fist or anything but he stepped to one side and raised his hands over his head twice while looking at the power house.

"Bill was just skipping along, but after the man did this it wasn't five car lengths when the motor grumbled and the car skipped uneasily along a few rods and then

The transformer house was outside the Queen Victoria Niagara Falls Park and therefore did not have to meet the aesthetic standards of the Park Commissioners. Nonetheless, Canadian Niagara wanted an attractive building. This 1905 photo of the northeast corner fails to reveal that the south end is not bricked in, simply boarded up. Both the transformer house, or station, and the generating station went into service before they were complete and were enlarged as needed. (Reproduction by permission of FortisOntario)

stopped. The man with the high sign walked leisurely to the car, assisted his friend up and, leaning out the rear window, gave another wave off toward the power house. Bill felt the juice come back and he saw there was no making up the lost time, so he went into town easy, but with never a wonder, because he didn't see the high signs that gray-suited man made.

"When they got to the terminal, Jones came in excited.

" 'Do you know who that was you went by at the power house?'

" 'My friends in Niagara Falls are limited,' said Bill.

"Well, Jones told him that it was no other than W.B. Rankine, and that when Bill was sliding by he signalled to the place where the juice comes from to shut off the current, and the current had been shut off, and seeing that Rankine was W. Caryl Ely's best friend and Ely was president of the trolley line, there might be something doing in blue envelopes next payday.

"So, son, never mind the schedule when you see a distinguished-looking man with a gray moustache standing near a power house. Just stop. If you don't, maybe you'll get the laugh that Bill got. No there wasn't a blue envelope, but the story got round, and Bill had to buy so much he quit."

The old motorman's story was true. W.B. Rankine of the Niagara Falls Power Company had been greatly troubled by the refusal of the yellow cars, as they are called, to stop in front of the power house. One day, having a New York financier in tow, he had watched with no great complacency two yellow cars speed past.

He called out to Electrical Engineer Paul H. Lincoln and told him if the next car didn't stop he would signal him, and upon seeing the signal, Lincoln was to shut off the power. He signalled, and for sixty seconds, every street car in Buffalo, Lockport, Niagara Falls and between these cities stopped at one and the same time — until, in fact, Mr. Rankine got his car.

When the Buffalo trolley men began calling up the Falls to see what the matter was, they got the answer that it was necessary to remember that Niagara Falls was on the map. After the story was told, motormen instinctively seemed to slow down in front of the power house of the Niagara Falls Power Company.[24]

This interior view of the transformer building, taken on September 25, 1905, shows an orderly layout of early equipment. The brackets holding the electric lights once again speak of Canadian Niagara's pride in appearance. (Reproduction by permission of FortisOntario)

Chapter 4

Designing to Get Power to Customers Safely and Reliably

When the Canadian Niagara Power Company generating station started to deliver its own power to customers, it signalled a new era in the history of electricity in the Niagara Frontier. Americans already had access to a good supply of electricity generated on their side of the river. The new supply added to the quantity of power available to them, but more important, as an alternative source, it increased the reliability of the power supply. On the Canadian side, where electricity had been less readily available, the new station represented a large step into the long-awaited world of readily available electricity. Electricity was an emerging mass commodity that families, businesses, and industries wanted. It was seen as the key to the prosperity and industrial growth that seemed to have bypassed the Niagara Falls–Fort Erie corridor. Consumer and political expectations were high.

Transmitting Power to the Niagara Power Company in New York State

Canadian Niagara had been created to generate and deliver power to three distinct customers or destinations — two in the United States and one in Canada. Canadian Niagara's parent company, the Niagara Falls Power Company in Niagara Falls, New York was closest and easiest. Further afield, Canadian Niagara had to deliver power to a related company in Buffalo, the Cataract Power and Conduit Company. Finally, there was the more open-ended task of supplying power directly to commercial, residential, and industrial customers in Canada, initially in Niagara Falls and Fort Erie, but eventually to towns such as Stevensville, Ridgeway, Crystal Beach, and Point Abino. Each destination required different approaches and posed different challenges.

The older American plant (Niagara Falls Power Company) and the new Canadian Niagara Power Company did things quite differently. The former generated two-phase power at 2,200 volts, the voltage at which it was distributed to nearby industries. For longer distances step-up transformers converted it to three-phase at either 11,000 or 22,000 volts. Charles F. Scott, a consulting electrical engineer hired earlier by the Niagara Falls Power Company, had developed the Scott Transformer to convert two-phase power to three-phase power. The phase conversion was important for two reasons. First, three-phase was more efficient for power transmission. Second, it allowed the older American generating plant and the newer Canadian plant to be linked directly so as to operate as one unit. The Canadian Niagara Power Company plant generated three-phase power at 12,000 volts that could be sent directly to the American plant without stepping up the voltage. The output from the Canadian plant could also be stepped up for longer distance transmission.[1]

Inside the Canadian Niagara generating plant, or powerhouse, the 12,000-volt power travelled through cables placed along insulated horizontal grooves or channels that kept the cables properly spaced so that if a short circuit occurred no arc would form between the cables and burn them out.[2]

The Canadian Niagara Power Company generating station was not only in a scenic park, it was very close to Niagara Falls itself. Many tourists would see it. Therefore, the Queen Victoria Niagara Falls Park Commissioners quite rightly insisted that all cables running to and from the station in the park be buried in underground ducts and that the transformer station be outside the park.

The cables left the powerhouse through three manholes (vaults), on the north, west, and south sides, from which points, they ran through underground conduits. Power destined for the Niagara Falls Power Company in Niagara Falls, New York, came out of the north vault, and from there into underground conduits that ran through the park for 6,500 feet to the river, where the cables crossed to the United States on the Upper Steel Arch Bridge. Power destined for elsewhere went to the company's transformer house at the top of the bluff just outside the park.

The conduits and vaults demonstrate careful attention to design, a commitment to going beyond minimum requirements, and an ability to learn from the experience of the parent company. Each cable passed through its own pathway known as a *duct*; all of the ducts taken together are referred to as the *conduit*. Each conduit, whether to the Upper Steel Arch Bridge or the transformer station, carried 32 ducts. At the time, it was common industry practice (including that of the Niagara Falls Power Company) to group all 32 ducts together, because it made construction easier.

Canadian Niagara took a different approach. The Niagara Falls Power Company had found that with all of the cables relatively close together the heat generated in them by the current flow was not readily dissipated. Under some circumstances the heat build up could damage the insulation or even cause a fire. Therefore, in order, as Smith put it, "to provide proper heat radiation," ducts were arranged in four groups of eight — instead of one group of 32 or even two of 16 — in Canadian Niagara's conduits. The Niagara Falls Power Company had put all of their ducts in a single massed group and had suffered overheating problems as a result. Canadian Niagara had learned from the misfortunes of the parent company.

Canadian Niagara's cables travelled underground in

COMPARATIVE TABLE

All Dimensions are given in feet

	American Plant		Canadian Plant
	No. 1	No. 2	
Intake Canal			
Length	1200		271
Width	119-194		282-526
Average depth water	12		14.4
Tunnel			
Length	7481		2165
Height	21		25
Maximum width	18.8		18.8
Wheelpit			
Length	424.47	461	564.3
Width	17.5	17.5	18
Average depth	177,7	177.4	160.6
Penstocks			
Number installed	10	11	7
Diameter	7.5	7.5	10.2
Turbines			
Number installed	10	11	7
Capacity in horse-power each	5500	5500	5-10000 2-12500
Average effective head	140	140	140
Draft tubes	1	2	5-2 2-3
Depth below power house floor	134.5	133.0	122.8
Outside diam. turbine runner	5.33	5.33	5.33
Single or double runner	Single	Single	Double
Discharge	Inward	Inward	Inward
Generators			
Number installed	10	11	7
Capacity in horse-powei each	5500	5500	5-10000 2-12500
Revolutions per minute	250	250	250
Voltage	2200	2200	12000
Phase	Two	Two	Three
Cycles	25	25	25
Number of field poles	12	12	12
Exciters			
Number installed	4	4	6
Capacity in horse-power each	168	200	267
Voltage	220	220	125
Revolutions per minute	550	750	600

The pioneering electric generating station of the Niagara Falls Power Company in Niagara Falls, New York, and the Canadian Niagara Power Company in Niagara Falls, Ontario, were such popular tourist attractions that the companies jointly published a 16-page guidebook titled *Information for Visitors*. This informative table from the 1913 edition provides a statistical snapshot of two generating stations on both sides of the Niagara River.
(Reproduction by permission of John Burtniak)

four-way ducts embedded in concrete. Even if one group of four was directly above another group of four, no group was wider than two ducts, so that every duct was in contact with the ground on at least one face. This configuration allowed for natural cooling. Moreover, transmission by appropriately sized copper cables kept heat gain within an acceptable range.

The company also rethought the approach to underground vaults. Along the route of the underground ductwork, the feeder cables passed through underground vaults. This was the only place where the cables were not in ducts and not encased in concrete. The vaults were underground workrooms. Riggers or cable splicers had to go into them to pull or splice cables. The feed cables were covered in lead, 2 inches in diameter, containing three 19-strand copper conductors, each half an inch in diameter, insulated by paper and gum wrappings.

Where cables had to be spliced, workmen connected individual conductors within the cable, made sure everything was dry, rewrapped the individual splices to reinsulate them, and then covered the splice with lead to make a waterproof joint. It took a great deal of skill, both to ensure that the insides were completely dry and that the lead sheathing joint built up from molten lead was perfectly waterproof. Splicers usually made the joint by pouring molten lead over the splice and then shaping it with their hands while wearing heavy leather gloves.[3]

Splicers needed room to do this work safely. To reduce the chance of accidents, Canadian Niagara's vaults were rectangular and larger than industry practice at the time. The floor was five feet by five and a half feet, and the room was six and a half feet high. Cables placed near the top of the vault were thus safer and easier to pull through ducts or to splice. This layout also allowed the cables to leave the building without any appreciable bends, which might have damaged the lead sheathing and allowed in water. Placing cables well above the floor also kept them as far away as possible from any water that might seep into the vault.

Water is the great enemy of underground circuits — or indeed any electrical circuits — because it is such a good conductor that it can cause short circuits. Even moist air can be a problem for high-voltage conductors. The continuous lead sheathing on the cables helped keep the water out. For further protection, cables were run through ducts and then encased in concrete. Drainage tiles laid alongside the ducts directed water away from the cables.

The layout of underground ducts, the shape and dimensions of the vaults, and many other details say much about Canadian Niagara's design philosophy. There are two common approaches to design. One is based on the assumption that those who use the technology will be able to avoid problems. This approach often leads to catastrophic results when things go wrong, as they invariably do. An alternative approach, known as inherently safe design, is based on the recognition that things will inevitably go wrong at some point, so the designer tries to anticipate what will go wrong and why and then designs to reduce the severity of the results of failure or error.[4]

Although the term "inherently safe design" had not been coined in the early 20th century, many of the design features of the Canadian Niagara Power Company system embodied all that it stands for. When he was writing about design and the choice of equipment, Resident Engineer Smith explained, "in general, every effort was made to design the plant so as to obtain a maximum of reserve in cases of breakdown."

Consider two possible outcomes when one of a number of feed cables catches on fire. If all the cables are in the same area and one starts to burn, it is likely that the others will be damaged and there will be a total power outage. But Canadian Niagara engineers recognized that a fire could break out and therefore tried to come up with an acceptable design that would lead to only a partial loss of power. A curtain wall divided the vaults into two parts or rooms, and soapstone slab partitions were laid horizontally between each group of cables.

Niagara Falls Power Company engineer A.H. Van Cleve succeeded Smith at Canadian Niagara. In 1909 that "The double manhole...has proved very satisfactory. The results of a short circuit are minimized, the cables have a straight lead, with maximum splicing space, and the workmen have an abundance of room. The advantages of this design, as compared with the elliptical and hexagonal manholes frequently used, are apparent." Van Cleve was justifiably proud of the design, but it is not strictly true that the results of a short circuit were minimized. To do that would have taken a vault with 32 fully isolated sections or rooms, one for each cable, rather than two rooms with 16 cables each. Clearly such a solution would have been prohibitively expensive and unworkable. Van Cleve's colleague, electrical engineer H.W. Buck, more accurately described the system: "The manholes are all made double with a concrete fire wall between the two halves, so that in case of a destructive manhole short circuit, not more than one-half the feeders could be involved."

In a generating station, the instruments on the switchboard tell the operator what is happening. But the real test of operator and switchboard comes when things go wrong. Philip P. Barton, General Manager of the Canadian Niagara Power Company, regarded the company's "multiple, auxiliary control, interlocking switchboard" as a "fine example of the best switchboard practice." H.W. Buck explained why to his fellow members of the Canadian Electrical Association:

> This operating board consists of five generator panels, 20 feeder panels, 10 recording wattmeter panels, and three bus bar interconnecting panels. Each panel is distinct and contains no instruments or switches except those belonging to the particular feeder or generator in question. In this regard it is different from most of the switchboard arrangements adopted in modern plants. The so-called "bench-board" system of operation usually installed economizes space, but it has certain disadvantages. The concentration of instruments and relays necessitated in this construction may cause considerable confusion at times of emergency. If instrument needles suddenly begin to slam around on their scales, it is somewhat difficult to make a prompt and correct association in a bench board installation between the instruments and the relay switches involved, which are usually on slabs separate from the instruments. With the panel construction adopted in this plant, possibly more space is occupied by the board itself, but operating simplicity and directness results. Each panel contains within its limits all the instruments and switches involved in any operation which the attendant has to make, and consequently there is minimum probability of confusion and mistake.

A switchboard consciously designed to reduce operator error during times of emergency and possible confusion is another illustration of Canadian Niagara's careful, inherently safe — perhaps the right word is *conservative* — design philosophy.

In the early 20th century, the output of electrical generators grew rapidly. When Canadian Niagara's new generating station went on line in 1905, it temporarily held the record for the world's most powerful generator, 10,000 electrical horsepower. In a 1906 article written for *The Canadian Electrical News*, General Manager Philip P. Barton revealed that getting bigger wasn't a simple matter.

> With the increase in size of generating units, as well as their number, installed in a power station, the problem of positive and rapid control of switching devices became more complicated and at the same time more necessary of solution. It was no longer possible to build switches for manual control and the air break types had to give way to those in which the arc was disrupted in oil. Moreover, with a greater number of generators, a corresponding increase in the number of switching devices and switchboard panels resulted with the consequent danger of errors in switching.

The new oil switches provided reliable technology. But that wasn't enough. Canadian Niagara wanted to reduce the probability of human error by using a design that would make it harder to make a mistake. The all-important type-H automatic oil switches that would cut the power to a feeder line could be activated only if "the synchronizing plug has been inserted in the socket provided. A red and green light are provided...on all panels. The red light burns when the switch cannot be closed, whilst the green light will show when the reverse operation can be carried out." It was an early example of a system designed to give clear information on status and to make a worker pause before taking an important and potentially hazardous step. There was also an emergency board for manually opening the field circuits of all generators to shut them down immediately.

The design of the bench board, instrument layout, and emergency switching, as well as the vaults, the separation of cables, and other design features undoubtedly drew on experience — both good and bad — at the parent Niagara Falls Power Company. At the same time, the newer Canadian equipment had to work with the older American equipment. At a time when many generating stations were designed to stand alone, the Canadian Niagara plant was part of a system, what might even be described as an emerging grid.

On the Canadian side, the 10,000-horsepower generators were twice as big as anything the American parent company had. However, the selection of these large generators was not based merely on the belief that bigger is always better. The larger-capacity equipment allowed the company to generate power with fewer generators and build a smaller building for a given amount of power. Even so, the company chose the 10,000-horsepower generators only after determining that the electrical systems of both companies could handle the resulting amount.

Transmitting power from Canadian Niagara to the Niagara Falls Power Company in Niagara Falls, New York, was a relatively easy task. But a larger market waited in Buffalo. Sending power to Buffalo would be more challenging — and far more lucrative in the short term.

Transmitting Power to Fort Erie and Buffalo

From the earliest days of planning, the Canadian Niagara generating station had been envisioned as operating in parallel with the Niagara Falls, New York, station to supply power to Buffalo. The Niagara Falls Power Company had been sending power to Buffalo since 1896. And with Canadian Niagara sending power across the Upper Steel Arch Bridge to the parent company since mid-1905, the solution might have seemed obvious: send more power across the bridge and then follow the existing path to Buffalo. But this approach was never seriously considered.

First, it might have been difficult, if not impossible, to get permission to send more power across the bridge. Second, the company wanted to enter the Fort Erie–area market for electricity, so it had to get power there somehow. Third, the Canadian route from the Canadian Niagara station to Buffalo was shorter than the American route, 16 miles vs. 22 miles.[5] And finally, even though Canadian power could be sent to Fort Erie through the United States on existing lines and then brought back to Canada with a new line spanning the river between Fort Erie and Buffalo, such an arrangement would have added to system vulnerability, just the opposite of what the companies were trying to do. Fort Erie would add to overall system reliability by providing an alternative pathway to Buffalo.

Another Niagara Falls Power Company subsidiary, the Cataract Power and Conduit Company, had built a new Buffalo power station, Terminal House B, which, along with the older Terminal House A, received 22,000-volt power (then regarded as high-voltage) from Niagara Falls, New York, for distribution in Buffalo. The two terminal houses were linked by underground cables. However, with all the power coming into Buffalo from one route, if the power to A went out, so did the power to B. And if B supplied power to Fort Erie, then Fort Erie would be without power too. But with power from Canadian Niagara crossing at Fort Erie and going directly to Terminal House B, the Buffalo market would have more reliable service. So would Fort Erie, because it could draw power from Buffalo if the line from Niagara Falls, Ontario went out.

An official pamphlet for the Canadian Niagara and the Niagara Falls Power Company generating plants told visitors that with an alternative route to Buffalo, "the transmission lines on both sides of the river can be interconnected at the Buffalo end, making almost impossible any serious interruption to the Buffalo service."

The link between Canadian Niagara's generating station and Buffalo depended on four elements: the right of way; Canadian Niagara's transformer station; pole lines to Fort Erie; and the river crossing to Buffalo.

This map from *Information for Visitors* shows that the Canadian route to Buffalo is shorter than the American route. The combination of two major transmission routes to Buffalo meant greater security of power supply. However, there were plenty of power outages caused by storms that showed absolutely no respect for international borders. (Reproduction by permission of John Burtniak)

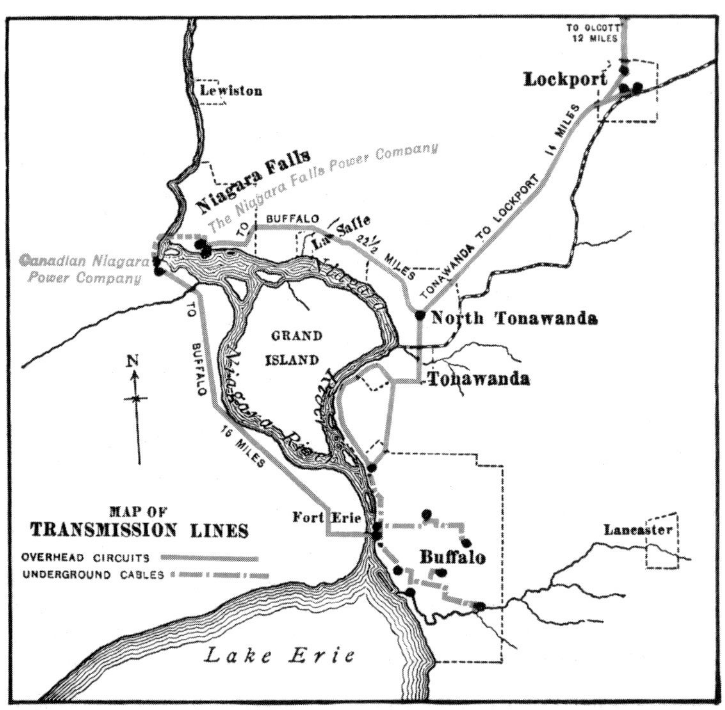

Step One: The Right of Way to Fort Erie

Whether for local distribution or long-distance transmission, securing a right of way is essential for any electrical utility. During its early years, when it had hoped to enter the Toronto market, Canadian Niagara had negotiated either a partial or a complete right of way to the city, largely through farm lands. Securing a right of way to Buffalo was much easier. The Michigan Central Railroad line had played an important role in delivering construction materials and equipment to Canadian Niagara. The line ran along the bluff, past the transformer station, and on to Bridgeburg, near Fort Erie. It then crossed into the United States over the International Railway Bridge (designed by the same Casimir Gzowski who was Chairman of the Queen Victoria Niagara Falls Park Commission). Canadian Niagara negotiated a 30-foot-wide right of way along the route of the Michigan Central to Bridgeburg, from which it is a short distance to Fort Erie and the river crossing.

Step Two: The Canadian Niagara Power Company Transformer Station

Because of the Park Commissioners' prohibition on any aboveground cables inside the park, power from the Canadian Niagara generating station ran underground, either to the Upper Steel Arch Bridge or to the transformer station, which stood on a large tract of land that the company owned on the bluff just outside the park. From there, the company could run overhead lines to various destinations.

Although the transformer station had a practical job to do, and although it was outside the Park, the designers took care with its appearance. The steel-framed building was made of pressed brick, with stone sills and Ludowici roof tile. In addition to its many windows, the central portion of the roof was raised along its length so that windows (sometimes called clerestory windows) could be inserted on either side to provide natural light throughout the station. There were offices near the front of the building and an entry large enough to accommodate fully loaded freight cars.

A 15-ton Whiting electric travelling crane stood in the 38-foot-wide main or central room, which was flanked on either side by rooms 20 feet wide. The 12,000-volt underground cables entered the transformer station through the east room, which housed the bus bars and switches. From there, power could take two possible paths. That destined for local use went to air-cooled transformers that stepped it down from 12,000 to 2,200 volts to be transmitted overhead. In 1909, this local power represented only a small load. Most of the power was destined for the 22,000-volt Fort Erie–Buffalo line, which went through passages underneath the concrete floor to water-cooled, oil-type, step-up transformers in the central or main room. Simply by changing the connections, these transformers could also produce higher voltages, should that be a future requirement.[6] From there, the high-voltage power went through overhead arches in the curtain wall to the high-tension switches. The wires passed out of the building through specially designed windows.

The generating station and the transformer house both came on line before construction was finished. The transformer house was initially equipped to handle the output of six generators[7] and conventional practice called for one transformer per generator. However, Canadian Niagara connected two 1,250-kilowatt transformers to each generator. Each transformer handled half the output of a generator at the powerhouse. This design gave the plant more flexibility by making it easier to isolate circuits and bus bar sections. Also, the company, and virtually everyone else, could see that transmission and distribution voltages would continue to rise. Canadian Niagara designed for future change even though it required a higher initial investment. As H.W. Buck explained in *The Canadian Electrical News* in 1906:

> The arrangement of windings on the 1250 k.w. transformers are somewhat unusual.... The transformers are constructed with five similar coils, two of which constitute one winding and three the other winding, either of which can be used interchangeably as primary or secondary. By the various combinations of coils...any of the following ratios can be obtained with the full copper efficiency of each coil.
>
> 12,000 volts delta to 24,000 volts delta
> 12,000 volts delta to 36,000 volts Y
> 12,000 volts delta to 41,5000 volts Y
> 12,000 volts delta to 62,500 volts Y

Voltage and current are inversely proportional; that is, as voltage rises, current or amperage falls. The higher the voltage, the greater the need for insulation. Similarly, the higher the current, the larger the amount of conducting metal — be it copper or aluminum — needed in the wires, and the greater the costs. In the Canadian Niagara transformer station, all the high-tension wiring was insulated for the highest voltage, 62,500 volts, and the conductors had sufficient current carrying capacity for operations at full output at the lowest transmission voltage of 24,000 volts and the highest accompanying current.

Opting for a three-phase system, rather than the older and, at the time, more common two-phase system, decreased costs, because the required switches and wiring were simpler, and also because the three-phase system required about one-quarter less in transmission copper than two-phase. Three-phase was a much more efficient carrier.

Another part of the transformer house was the water system. Although it no longer stands, a 12-storey tall water tower, or standpipe, once stood on the bluff beside the transformer station, silent testimony to the fact that water did more than just generate electricity. The Rankine site had its own waterworks system in two distinct parts: an operating component that worked every day, and a backup system to provide cooling water to the transformers if the other system broke down.

The company was limited in how much electricity it could generate, but not in how much water it could draw from the river. Therefore, the generating station had water-powered auxiliary equipment that saved the company money. After all, why run equipment with electricity that you could otherwise sell, when you got water for free? In fact, for many years, Canadian Niagara supplied industrial water to Carborundum, Queen Victoria Park, and company-owned factory sites between the transformer station and Chippawa.

However, cooling transformers was the most important single use of the water. While transformers are very efficient, they still generate heat, and with the higher voltage needed for long-distance transmission, they had to be cooled by circulating water outside the tank containing the transformer wires, which as we have seen, were bathed in insulating oil to prevent short circuits.

To assure a ready supply of water, inlet pipes led to two 3,000,000-gallon-per-24-hour, single-lift centrifugal pumps, each driven by a 174-horsepower turbine. One of the pumps was connected to an electric motor, in case the turbines were not working. The inlet pipes were also connected to a penstock, so that if the smaller inlet pipes became choked with ice, the massive penstock would supply water. A 24-inch main carried the water outside the powerhouse, where 12-inch pipes led to the transformer station and the standpipe at the top of the escarpment. Monumental as it was, the standpipe was there as a reserve. If for some reason Rankine could not deliver water, there was enough water in the standpipe to cool the transformers for 48 hours.

Inside the generating station, the pumped water flowed through cooling coils in the generators and went to hose outlets at the various wheelpit decks and the powerhouse floor, where it could be used to extinguish fires, even though, as A.H. Van Cleve put it, "such preparations would seem to be in the nature of making assurance doubly sure, as the power-house and wheelpit decks are absolutely fire-proof, practically the only wood being that in the southerly temporary end of the power house, upon which a sprinkler system is installed."

Oil was another key ingredient in safety and good design. The low-voltage transformers were air-cooled, the high-voltage ones water-cooled and oil-insulated. Care had to be taken that no water entered the insulating oil or that power surges or heat build-up didn't char or burn the oil, which would then lose its insulating qualities. When this happened, the oil had to be replaced and often some of the wiring as well. To prevent loss of oil, the central room of the transformer station contained a lead-lined oil drainage system. Moreover, in the generating station, there was an elaborate system for reusing oil. Thrust bearings on the generator shafts were lubricated with oil under high pressure, but other equipment such as the exciter plant, which generated 125 volts DC current, not just for exciting the turbines but also for operating cranes, switches, various gate and pump motors, and lights, had oil bearings supplied with oil from the oil chamber.

There was a complicated system of motor and Pelton-driven (water-powered) Quimby screw oil pumps, oil tanks, filter tanks, and cooling tanks, arranged in two circuits, one for low-pressure and one for high-pressure oil. In the high-pressure system, a special pump located at each thrust bearing raised the oil pressure at these bearings, after which the oil flowed back to the oil chamber to be cooled and clarified.

Reclaiming and reusing oil was more than an economy; it represented a genuine commitment to clean river water. Even today, people working in the generating station proudly point to design details that ensure that if any oil and water mix, as they do in some areas, the water-oil mix is captured and the oil removed before the water returns to the raceway and begins its journey back to the river.

Step Three: Getting Power to Fort Erie

In May 1906, a construction crew started work on the 22,000-volt transmission line from Niagara Falls to Fort Erie. As soon as they left Niagara Falls, the men were out of tourist country. The Niagara Parkway cabins that would serve later generations of motorists had not yet been built, and the crew slept in tents. So did the engineer, the camp cook, and the cook's assistant. It was the kind of self-contained construction crew one associates with northern Ontario.

Poles had to be firmly anchored and this usually meant deep foundations for both poles and guy wires. Digging was difficult, time-consuming, and expensive, because much of the soil along the right of way was hard clay. The system devised by Canadian Niagara to minimize digging called for five concrete foundation piers, four of which were for guy rods. Each foundation pier was 5 foot 5 inches deep, 12 inches square at the top, and 18 inches square at the bottom. The bottom part of the guy rod had a steel plate. Each assembly was embedded inside one of the four poured concrete piers. A fifth pier that was not tapered stood in the centre to carry the pole. Each of the four 1-inch diameter guy rods carried both longitudinal and transverse stresses. Each had a turnbuckle to adjust tension on the guy rods and to position the pole vertically.

The economical design of the poles and how they were secured reminds one of the rigging of a sailing ship, a place where there is no room for breakage or excess material. Because 20-foot lengths were easier to transport than 40-foot lengths, each pole consisted of two 20-foot lengths of 4-inch steel pipe, which were assembled on site. The two pieces were put end-to-end with a four-armed casting between them. This casting was known as a spider; each arm of the spider projected about one foot out from the pole. These carried another type of guy rod called a sway rod. Each of the four sway rods went from the top of the pole, past a spider arm groove and down to the bottom of the pole. When tightened, they stiffened the jointed pipe. This unusual pole could use such economical, small-diameter pipe — only 4 inches — because it had to carry downward or compression forces only; the anchored adjustable guy rods resisted transverse or horizontal stresses. There were two U-shaped metal cross arms, one above the other at the top of the pole, with holes drilled in them to carry the mounting pins for the insulators that ultimately carried the conductors or wires.

Vandalism, which was a constant problem with telegraph wires, was equally a problem with electric wires. Taking a lesson from the Niagara Falls Power Company, Canadian Niagara chose an unusual insulator made of a compound called electrose. H.W. Buck noted, "This material is a very good insulator, is very strong mechanically, and is entirely free from cracks and other defects which are common in glass and porcelain. Similar insulators have been used on the Buffalo transmission of the Niagara Falls Power Company for the past three years, and they are the only insulators on those lines that have caused no trouble. It is impossible to shatter electrose insulators by stone throwing, and they will frequently turn a rifle bullet without being damaged seriously."

The entire pole assembly was sturdy. Experiments with the guy rod and concrete pier anchoring system used by Canadian Niagara revealed that the guy rod would break before the footing or pier lifted up, and it would take quite some doing

to snap a one-inch-thick guy rod. According to Van Cleve, "The pole adopted is economical and durable; it is quickly erected and easily adjusted. It has no thin sections to be destroyed by rust, it requires no field riveting and but little bolting. The parts could readily be galvanized, but it has not been thought necessary. Its limiting feature is the necessity for good anchorage."

The towers were built to carry the lines or conductors, which in turn had to be chosen carefully. The conductors were made of 37 strands of aluminum, a lightweight conductor. Spacing the poles was an important matter, because aluminum, like any other metal, stretches under its own weight, depending on the temperature. At the coldest time of the year, wires contract, so care had to be taken that the wires did not contract so much that they snapped. In the hot summers, wires stretch, and it was important that they not sag dangerously low and cause a short circuit or an accident. Ice and snow loads also entered into calculations regarding sag and breaking point for the conductors. Writing in 1909, after the line was already in service, Van Cleve felt that the 300-foot average spacing between poles could have been safely increased to 400 feet. Perhaps it could have been, but in the early years of the electric industry, the best approach was to stick to conservative engineering principles.

The first lines between Niagara Falls and Fort Erie were strung on the ingeniously designed spider poles, which were intended to combine ease of transportation, ease of erection, and minimal footprint. In this late 1907 photograph, the cross-arms on the single pole are mounted so that the lowest one carries four wires and the upper two. This made sense from the point of view of reduced swaying in a windstorm, but it was harder for linemen working on the poles, who had to get past the more congested area of four wires or cables to get to the less congested two. This was later reversed, with two wires below and four above. In some instances, two spider poles were placed side by side and joined together. This made it easier to work on the lines. (Reproduction by permission of FortisOntario)

Most of Canadian Niagara's right of way was 30 feet wide. The base of the pole required a space 14 feet by 14 feet when it was complete, but more space during construction. There were some exceptions. Four reinforced concrete poles were used in the Village of Chippawa, where there was not enough space for the standard poles; and steel was used at railroad crossings.

The transmission lines to Fort Erie and on to Buffalo were designed to carry 50,000 horsepower, or one-half of the planned output of the Canadian Niagara generating station. However, since the station was being built in stages and the transformer house was being equipped in stages, the line could also be built in stages.

Generator No. 5 went into service May 17, 1906. The line to Buffalo had to be ready to carry 25,000 horsepower when it entered service. The two pole lines were fundamentally identical; each had two circuits of 12,500 horsepower for a total of 25,000 per pole line and 50,000 for the system overall. Building the second pole line a few years after the first saved the company money before it started to earn revenues from the Buffalo market. There are also some interesting differences between the two lines, which shed more light on Canadian Niagara's approach to design.

When the crews were erecting towers and stringing lines between the generating station at Niagara Falls and the town of Fort Erie, the tourist industry outside of the immediate vicinity of the Falls was in its infancy and the crews slept in tents. But tenting is not always the same as roughing it. Here, in a photograph taken July 15, 1908, the tent has a wooden floor, suitcases, proper bedding, a chest, and campaign-style portable desk. This was probably the Chief Engineer's tent. (Reproduction by permission of FortisOntario)

In many companies, those who do physical work complain that engineers, managers, or both, won't make necessary changes that would simplify their lives. This appears not to have been the case at Canadian Niagara. On the first line, the pins holding the insulators were malleable iron; on the second they would be steel. But the cross-arms tell us something more important. In Van Cleve's words, "The cross-arms are ship channels; the longer arm is below on the first line, and will be above on the second line. The latter arrangement is not as satisfactory structurally, but offers certain advantages from the operating standpoint." In other words, a design decision was taken that made workers' lives easier.

Consider the implications of repositioning the cross-arms. The longer cross-arm carries four lines and the shorter, two lines. The longer cross-arm and associated pins, insulators, and cables weigh more than the shorter cross-arm. Theoretically, the heavier part should be closer to the point at which the pole touches the ground. That is the voice of physics and engineering speaking. But what about the men who were called out in the middle of storms to work on the cables? They had to climb past four cables to get to the top two cables, even though it would be safer and easier to climb past two cables to get to four. Someone at Canadian Niagara clearly listened to those men. That was the difference between the first and second lines to Buffalo.

Once the cable lines arrived in Fort Erie, and before the power proceeded to Buffalo, the company had to provide service to the town itself. In Fort Erie, the 22,000-volt line was connected to a local substation. Starting in August 1907, the company started to erect a transformer station to supply street lighting to Fort Erie at the southwest corner of Queen and Goderich Streets. The building was 18 feet by 24 feet, made of steel and brick, and had a corrugated iron roof.[8] Here power was stepped down to a lower voltage for distribution to local homes, businesses, and institutions, and eventually street lighting.

AMERICAN ENTRANCE TO PROPOSED BRIDGE.

When electric power generated on the Niagara River arrived in Fort Erie and other communities, it didn't happen the way Alonzo Mather had once hoped. Mather, a Chicago industrialist and Fort Erie benefactor, wanted to link Fort Erie and Buffalo with the bridge shown here. The bridge would have had eight horizontal waterwheels suspended underneath to capture the power of the swift current to drive generators. In what might have been an excess of optimism, Mather claimed the ice would be no match for sturdy engineering and an ice boom. He could never get permission to build the bridge. (Reproduction by permission of the Buffalo and Erie County Public Library, Buffalo, New York.)

Step Four: Taking Power Across the River to Buffalo

The 22,000-volt lines that had travelled from Niagara Falls to Fort Erie now had to cross the river. There was no way to do that that unobtrusively.

An underwater crossing was out of the question: too expensive to build and maintain. The wires would have to go above the river, which at that point was deep, fast, and 1,650 feet wide. It was also a busy navigation channel. Thanks largely to the Erie Canal, Tonawanda was the largest lumber market in the U.S. at the turn of the century. The industries of Buffalo, the eighth largest and one of the richest cities in the United States, seemed to have an insatiable appetite for almost anything that could be shipped along the canal. To keep the ships moving, both Canada and the United States insisted on 130-foot minimum clearance above the river — that meant 130 feet in all temperature, weather, and wind conditions.

There were also unstated psychological constraints. People needed to feel confident it was safe to let high-voltage cables pass over a navigable river as well as Black Rock Harbor, the Erie Canal, and the New York Central and Hudson River Railroad tracks. On the Canadian side there was another concern. The cables had travelled from Niagara Falls at about 40 feet above the ground, but closer to the river the lines had to pass right over Fort Erie, and residents would be concerned about safety.

SIDE ELEVATION OF ONE SPAN OF PROPOSED BRIDGE.

One of the horizontal water wheels on Alonzo Mather's proposed Fort Erie–Buffalo bridge. (Reproduction by permission of the Buffalo and Erie County Public Library, Buffalo, New York.)

In some parts of Fort Erie, a bluff runs roughly parallel to the Niagara River. As Bertie Street travels away from the river, it rises up this bluff on what is known as Bertie Hill, close to Goderich and Queen. The engineering design team decided to put the conductors on three towers: one on Bertie Hill west of the settled area of Fort Erie, one at the river's edge, and a third near the terminal house in Buffalo. The span crossing over Fort Erie would be 1,667 feet from the first tower to the second, and that over the river 2,192 feet from the second tower to the third, with, of course, the 130-foot clearance. As Van Cleve notes, somewhat ominously, "The combination of spans of this magnitude, with the large amount of current to be carried, had never been attempted."

First, the engineers had to choose a conductor, because its weight would be an important design constraint on the towers. There were only three commercially available options at the time: steel, copper, and aluminum. As the engineers saw it, steel would have been the easiest to erect, but it was not a good enough conductor for their needs. Copper was too heavy. Aluminum won out. It was not perfect, but it had the best combination of conductivity, light weight, and tensile strength.

The cables across the river carried the same power as those across the land from Niagara Falls, so the cross-sectional area of the cables would be the same. But because the river crossing cables would be subjected to far higher stresses, the manufacturer suggested more strands — 61 instead of 37 — which would be wrapped differently, in a concentric configuration. One has to remember that both Canadian Niagara and the cable manufacturers were heading into uncharted territory. The decision about how to wrap or lay the strands would come back to haunt them.

Next, the engineers had to determine the height of the three towers. When a cable or string is run between two supporting points, it falls in a shape known as a catenary curve. The bottom of this curve had to be at least 130 feet above the river. But in hot weather, the cable would stretch and sag closer to the water. This meant the tower had to be high enough to allow for hot weather, even though in cold weather, when the cable contracted, it would be well above the 130-foot minimum. Well above, of course, unless each cable became so encrusted in ice that it doubled in weight and sagged closer to the water's surface. The engineers also had to build in a margin of error, because materials may not behave exactly as predicted. And the higher the tower, the greater the expense. But, the engineers asked themselves, what if, while allowing for a margin of error, they could compensate for the stretching and sagging, contracting and tightening? They answered the question by designing an ingenious counterweight system to keep the bottom of the catenary curve at a constant height.

Once this system had been worked out, the team could decide on the height of the towers. The Bertie Hill tower would be 61 feet high, and the Canadian high tower 202 feet, 9 inches. On the Buffalo side, where the land was higher, the tower needed to be only 150 feet, 9 inches. This meant that the tops of the two higher towers were at about the same level.

Getting the conductors up and properly installed and connected was a complicated operation. The usual practice for getting a cable across a river is to string a lighter-weight cable across at first and use that to draw a heavier one across. For one of the suspension bridges across the Niagara River, the builders had offered a reward to the boy who could fly a kite with a string attached across the river to the other side, where it could be attached to a heavier string to haul across even heavier rope, then cable. For the line to Buffalo, the operation started with a manila rope that was taken across by a tug. The rope then pulled a steel cable across. Finally, the steel cable was used to haul the aluminum cables across. Communications between the two sides of the river were important. When signal flags failed, the company installed telephones to link the three towers. Timing and coordination were crucial.

After all that work, the aluminum conductors proved to be troublesome. There were good reasons for choosing them and there was no reasonable alternative. But the aluminum was so soft that it had to be protected from scratches and it could not be hauled under stress. Moreover, the strands had a tendency to unwind. The manufacturer had recommended against the conventional way of positioning the various strands in what was called a "rope lay," in which strands wrapped around each other, because they feared one strand might cut into adjoining strands under stress, but had not realized that the strands might separate if they were not twisted.

Another problem was that aluminum conductors vibrated in still or nearly still air. Telegraph or telephone wires sometimes had the same problem, but not to the same extent. Where the conductors were attached to the high tower on what is known as a saddle, the stress caused by the vibrations broke many of the strands. The engineers tried unsuccessfully to dampen the vibrations. Finally, they replaced the aluminum conductor with galvanized chain for about 13 feet on each side of the saddle; the current was carried around the chain by an aluminum jumper cable. Even then, the connector used to join the conductors to the drop line for the counterweights damaged the aluminum conductor over time and had to be replaced.

In the end, when all these adjustments had been made, the line performed remarkably well. In 1909 Van Cleve reported that the conductors were "operating successfully, and are daily carrying to Buffalo large amounts of power for the operation of industries and public service utilities. During a recent sleet storm, which practically doubled the weight of the cables, the apparatus operated exactly as designed and the conductors were not injured, although, if they had not been counterbalanced, injury to them would have been certain."

The cables worked well for decades before they were replaced. But shortly after they were installed, engineers had developed copper-plated steel-wire conductors and other types of conductors that would have worked better if they had been available at the time.

But then, aluminum was the best available choice. It even had some symbolic significance. The Pittsburgh Reduction Company, later Alcoa (Aluminum Company of America), had been the Niagara Falls Power Company's first major industrial account; this company had proved to others that it was worth moving to Niagara Falls, New York, for the power, service, and industrial land offered by the Niagara Falls Power Company. And aluminum was a new material for a brave new world based on science and technology, just as new as electricity. They were both objects of faith, both seen as keys to a better world, both seen as something that would benefit ever-increasing numbers of people.

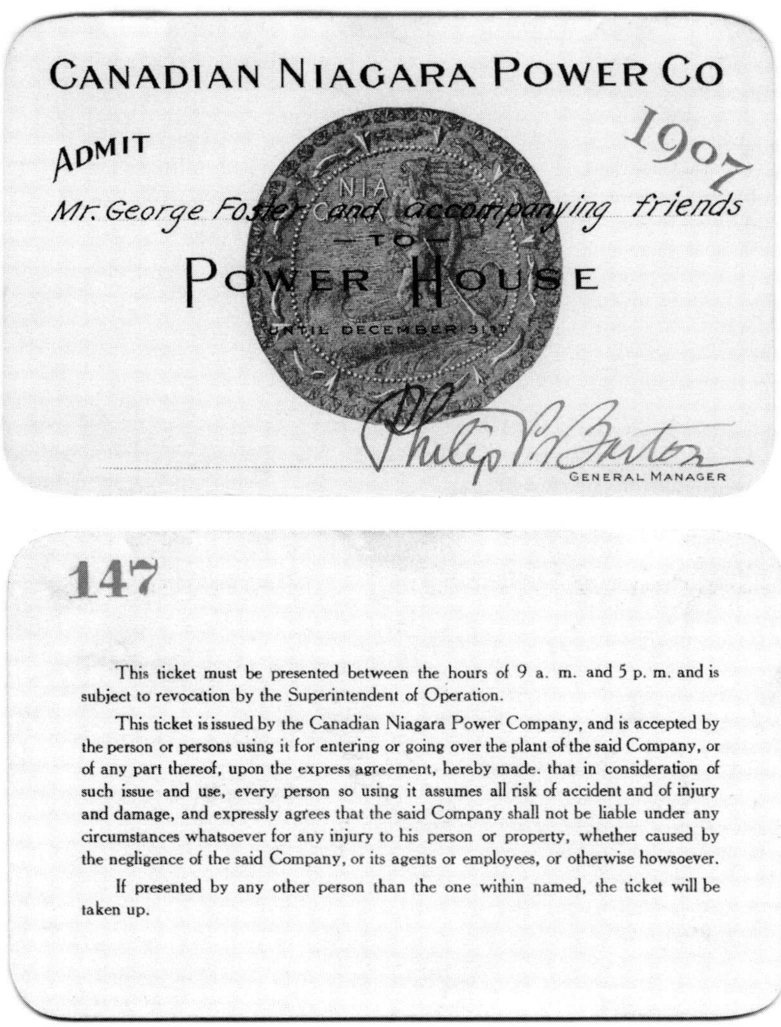

CANADIAN NIAGARA POWER CO

ADMIT 1907

Mr. George Foster and accompanying friends

— TO —

POWER HOUSE

UNTIL DECEMBER 31ST

Philip P. Barton

GENERAL MANAGER

147

This ticket must be presented between the hours of 9 a. m. and 5 p. m. and is subject to revocation by the Superintendent of Operation.

This ticket is issued by the Canadian Niagara Power Company, and is accepted by the person or persons using it for entering or going over the plant of the said Company, or of any part thereof, upon the express agreement, hereby made. that in consideration of such issue and use, every person so using it assumes all risk of accident and of injury and damage, and expressly agrees that the said Company shall not be liable under any circumstances whatsoever for any injury to his person or property, whether caused by the negligence of the said Company, or its agents or employees, or otherwise howsoever.

If presented by any other person than the one within named, the ticket will be taken up.

Many tourists paid to enter the generating stations of the Canadian Niagara and the Niagara Falls power companies to look at the equipment. This money was given to local hospitals. Some visitors were lucky enough to have a pass. This 1907 Canadian Niagara Power Company pass bears an image of the company seal. It was signed by Philip P. Barton, General Manager for both companies. (Reproduction by permission of private collector)

The town of Crystal Beach and the amusement park, Buffalo's Coney Island, waited patiently for electric power from Canadian Niagara, which arrived in 1914. The amusement park had its own steam-powered electric generating plant to supply power for lights as well as electric motors for everything from rides to an air compressor. The card with the writing on it was mailed from Buffalo, July 8, 1905, at 10:50 p.m. The horse-drawn carriage took people to and from the docks. The second postcard is undated, but the photo was taken a few years later. Long dresses, covered arms, hats and parasols were about more than modesty. Suntans were most unfashionable, since they implied that one worked outside.
(Reproductions by permission of John Burtniak)

Extending Power to Local Communities

Chapter 5

Canadian Niagara's primary purpose in bringing electricity to Fort Erie in late 1906 was to send it on to Buffalo. There was power for the town too, and residents hoped it would bring the prosperity that usually seemed to go elsewhere.

The story of the arrival of electricity in Fort Erie is actually one of arrival in three separate communities — Fort Erie, Bridgeburg, and Amigari — which amalgamated in 1931 to form the Town of Fort Erie.

It might seem astonishing today, but before the Erie Canal opened in 1825, Fort Erie "was a larger commercial centre than Buffalo…a distribution point for manufactured goods and imported articles…from Europe to be sold to settlements in New York State, the Ohio Valley and western Upper Canada."[1] But the opening of the Erie and Welland Canals drew trade and prosperity away from Fort Erie, just as it did from Niagara Falls and Niagara-on-the-Lake.

Fort Erie tried to fight back. A ship canal was proposed in the 1830s, but never built. Things looked up in 1854, when the Buffalo, Brantford and Goderich railway arrived. But transportation connections were not enough. The people of Fort Erie understood that good energy sources were part of the recipe for commercial prosperity. After the incorporation of Fort Erie as a village in 1857, a group of local businessmen reasoned that they could attract farmers from the surrounding rural area if they built a grist mill. And once in town, the farmers and their wives would go to Fort Erie's taverns, stores, and businesses. So a number of businesssmen put up the money and built a three-storey, wind-powered grist mill on the hill on Goderich Street. They hired a miller, Joseph Wells from Hull, England.

Shortly after he was invited to Fort Erie, Wells bought out the businessmen and the mill on "Wells Hill" became a local landmark, visible from far out in Lake Erie. But eventually the march of technology caught up with Wells. Newer mills offered their customers finer "patented flour." His business declined, so Wells moved the mill from the top of the hill and turned it into a steam-powered feed mill. When his wife died in 1890, he tore the mill down and sold the land for building lots. He moved to St. David's, operated two watermills, and eventually returned to Hull.[2]

A mill was nice, but a railway bridge to replace the ferries across the Niagara River would be even nicer. When it came time to build a bridge, the location chosen was a mile north of the village. The International Railway Bridge, designed by Sir Casimir Gzowski, opened in 1873 in Bridgeburg. By the 1890s, Bridgeburg had surpassed Fort Erie in terms of population and importance as a commercial centre.

The third part of Fort Erie, Amigari, was bounded by Gilmore and Concession Roads and the rail line. Local MP Christopher Bunting wrote to the Postmaster General asking that it be given a post office and called Amigan, after his birthplace in Ireland. Unfortunately, somebody's bad handwriting made it come out as Amigari.[3]

A Chicago Industrialist's Vision of Fort Erie

By the late 19th century, Fort Erie had acquired a champion — Alonzo Mather, a Chicago industrialist. He had made a fortune out of a new type of railroad car with a mechanized feeding system for the Midwestern cattle destined for the slaughterhouses of Chicago, the meat-processing capital of the world. Mather had a passionate and generous interest in Fort Erie. In 1893 he proposed to link Fort Erie and Buffalo with a combination bridge and electric generating station. Mather envisioned an ornate, multi-laned bridge for "steam trains and electric trolleys, carriages, wagons, and pedestrian traffic" that was wider than anything that has ever crossed the Niagara River to this day.

He wanted his bridge to be a "memorial to 19th-century inventors." As he put it: "Why should all the monuments in the land be erected to soldier and statesman, whose laurels would be lost had not the inventor given them rapid means of destruction, publication and communication?"[4] The decoration on the new bridge would honour 19th-century inventors who had made the modern world, men such as Sir William Perkins (aniline dyes), Dr. Richard Gatling (rapid-fire guns), and Sir Humphry Davy (the safety lamp).

The most unusual feature of Mather's vision of the future was suspended from the bottom of the bridge: a string of generators stretched across the river. Surviving drawings (pages 86–87) show an ornate bridge with suspended horizontal water wheels to power electric generators. He figured that the rapid current that allowed sailboats to pass upriver only in a good stiff wind could generate enough electricity to make the bridge self-financing. Moreover, the bridge would not divert water away from the river, so the water could be used downstream for other power projects.

Mather's bridge was never built. Conspiracy theorists see that as the work of New York State power interests who didn't want the competition. There might be another explanation: it probably wouldn't have worked. It had an even more powerful and more unbeatable enemy than any New York power consortium — ice. The ice load that once ripped a bridge out below the falls, that more than once filled the inside of a powerhouse, and that has formed many spectacular ice bridges below the falls, enters the river at Fort Erie. Even though Mather's bridge had an ice boom, the proposal was technically questionable. Mather optimistically assured everyone who would listen that ice would not be a problem. But a few years after his proposal, the sinking of the *Titanic* made people a little less arrogant about encounters with ice.

Meanwhile, however, Mather had bought up land on both sides of the river for his bridge as well as for the docks, warehouses, and elevators that would be needed for the international harbour he was going to build. He wanted to make Fort Erie everything that Buffalo had become. But he couldn't do it without electricity, a bridge, and a proper harbour. Although his bridge was never built, in 1927 the Peace Bridge opened, thanks in part to his vision and donated land. The Mather Arch and Park grew out of his generosity, as did the modern lighting system installed on the bridge and approaches in the 1940s, after he left $125,000 for that purpose in his will.[5]

Street Lighting for Fort Erie

Before electricity, Bertie Township used natural gas for lighting and power. Its first commercially successful drilled gas well dates back to August 1889. By 1890, both Fort Erie and Bridgeburg had piped natural gas,[6] and two pipelines were laid on the bottom of the Niagara River. As early as February 1891, Buffalo streets were lit by natural gas from Bertie Township.

Even before the Canadian Niagara power lines reached Fort Erie, the town had started its quest for electric street lighting by announcing on July 9, 1906, that tenders for an electric light system would be accepted until August 1. Canadian Niagara submitted the winning bid. Then residents on Queen and Princess streets complained that they weren't getting any of the 25 arc lights. Town Council said they couldn't afford any more. It turned out they couldn't even afford the 25 they had planned. Canadian Niagara's offer lapsed.

The Council called for new bids, and in April 1907 agreed that the town would supply the arc lights, Canadian Niagara the power, and H.D. Robertson of Toronto would supply and install the electric equipment, including poles. B.E. Norrish, the Robertson foreman, wanted to start work on June 27 or 28, but there were more problems. Norrish was worried about a shortage of workmen and the town was hit by an outbreak of NIMBYism: Not In My Back Yard. People wanted the lights near their residences, but not the poles. The *Bridgeburg Review* even wondered where one could get the "pole-less electric lighting for the street" that people seemed to want.

By July 15 everything seemed to be on track. A gang of ten men was busy working for H.D. Robertson and a few weeks later Canadian Niagara erected its transformer station at the northwest corner of Queen and Goderich Streets. By October the building was nearly completed, but the arc lamps had not arrived. So many cities and towns were installing electric lighting that demand far exceeded supply, unless one could call in a few favours. Canadian Niagara's parent company, the Niagara Falls Power Company, used its influence, and the lights arrived on October 22.[7]

On a cold night early in November 1907, it seemed as though everyone in the area was bundled up in warm clothes, waiting for the streetlights to go on at 5:30 p.m. The village choir had started things off 15 minutes earlier with a few selections, including "There's a Bright Day Coming." The choristers kept on singing, and singing, and singing; 5:30 p.m. came and went. There didn't seem to be a bright evening coming. Rumours and speculation ranged from "the current froze" to "the engineer's asleep at the switch." Finally, around 6:30 p.m., the lights went on — figuratively, not literally. There would be no lights. The next day, the townspeople learned that William M. Baldwin, the eighty-something postmaster, had obtained an injunction to keep the lights off.

For the previous twenty-five years, Baldwin had not worried about the legal implications of the fact that his verandah extended about three feet outside his property line on Niagara Street. But an electric light pole two inches *inside* his property line was an entirely different matter. He claimed that "he might be liable, should the wire happen to drop from the pole and fall on someone just as he or she was walking by his property. The court had agreed."[8]

It took a week to straighten out the legal niceties. The big night arrived on Thursday, November 14, and this time no one was disappointed. Newspaper reporters waxed eloquent about the brilliant light from the "electric fluid" that banished "shadows…from the village streets" which were "as brilliantly illuminated as in the full glare of the noon-day sun." The future seemed equally bright.

Of course there were some problems. The components of the lights burned out before replacements arrived. Eventually new longer-life carbons — 140 hours of service instead of 12 — arrived, and Fort Erie settled into the joys of dependable street lighting.

Arc lamps, however, required considerable maintenance and threw off a harsh light with intense pools of light and equally intense shadows. Incandescent bulbs, developed simultaneously in England and the United States by Joseph Swan and Thomas Edison, respectively, provided more convenient and more pleasing light. In June 1920, Fort Erie ordered new 100-watt incandescent lamps; 125 new lamps replaced 28 arc lights.

Fort Erie became the bridgehead for expansion into other parts of the Niagara Peninsula. A typescript headed "Some Facts About Canadian Niagara Power Company, Limited," found in the Fort Erie offices of the company, gives a brief chronology of the expansion of power in the area.

1906 — Started operations — Supplied street lights to Fort Erie only [not quite accurate]
1910 — First supplied power in Bridgeburg for commercial use
1911 — First supplied power in Fort Erie for commercial use
1912 — First supplied power in Erie Beach for commercial use
1913 — First supplied power in Ridgeway for commercial use
1914 — First supplied power in Crystal Beach for commercial use

Bridgeburg

Bridgeburg voted overwhelmingly on January 3, 1910, to give Canadian Niagara "the privilege of entering the village" to provide electric service. While Fort Erie had had electric streetlights first, Bridgeburg got the more modern 32 candlepower lamps, 50 of them, at a cost to the village of $20 each a year, or $1,000 altogether. (Given the rivalry between Fort Erie and Bridgeburg, Fort Erie voted in 1913 to install 25 new incandescent lights that were brighter than Bridgeburg's — 200 instead of 32 candlepower — in the northwest section of the town as well as Garrison and Lake Shore roads, also at a cost of $20 each a year. In 1914, the company reduced the annual price to $15 per light, and reduced residential power costs too.)

By September 1913, the promotional pamphlet *Bridgeburg: The Gateway to Canada's Abundant Prosperity* trumpeted that "its natural advantages in the way of water shipping facilities, its unsurpassed railway advantages, the cheapness of Niagara power, complete waterworks and sewerage plants, electric street lighting and the fact that it is within half an hour's ride of the heart of the city of Buffalo, makes it one of the most desirable of towns, both for manufactories and for purely residential purposes." Six railroads offered more than 50 scheduled daily train stops, in addition to the "dummy" passenger rail car, which ran across the International Bridge every 40 minutes from six in the morning to midnight.[9]

For all practical purposes, Bridgeburg was a suburb of Buffalo and most people wanted to keep it that way. Electricity would help do the job. So would the substantial tariff barriers between Canada and the United States. The pamphlet extolling Bridgeburg's virtues promoted it as "an ideal location for firms having a United States plant in Buffalo, as both plants can be operated with the same executive force" and went on to list 22 American companies already located in Bridgeburg. There were only 12 non-American manufacturers in town.

Many of the 22 American companies listed in the pamphlet are now gone. The Mentholatum Company survived into the beginning of the 21st century (it is now gone), but the Oxypathor Co., listed as a maker of medical appliances, disappeared long ago. According to an exhibit at the Buffalo and Erie County Historical Society, the oxypathor was one of Buffalo's "local flops." It was "a device to channel magnetic energy through the body to cure lumbago."[10]

Despite the promotional efforts suggested by the pamphlet, the arrival of electricity does not appear to have ignited any great industrial boom in Bridgeburg. Rather, electricity and the tariff barriers helped the town to keep the industry it had and avoid decline as an employment centre.

Electricity Fuels the Ambitions of a Small Town Newspaper

In 1895 a newspaper known as *The Canadian Free Press* started up in Bridgeburg. The name of the paper was soon changed to the *Bridgeburg Review* and it published a sub-edition, *The Fort Erie Times*. The printers eventually replaced their old press, which had operated with a foot treadle, and their hand press, with a drum-cylinder press run by a gasoline engine. But the engine was unreliable and kept breaking down.

Nevertheless, the business continued to grow and in 1904 moved into the newly built brick Review Building on Jarvis Street. In 1909, electricity — the easy-to-manage, reliable power source — gave the company the confidence to acquire a new press "which was among the largest of its type built and of which there was only one other of the same size in Canada, at that time."[11]

With a reliable source of power, the Review Company could now print daily programmes for the Fort Erie racetrack. The racetrack needed the programmes only a few hours after it gave the printer the list of horses, jockeys, and other information. There was no room for delay. One day, when a nearby poletop transformer failed, production dropped from several thousand per hour to a few hundred, as printing plant staff had to turn the heavy press by hand.

In 1919 the Review Company bought its own linotype machine, and five years later bought another. A single linotype operator could set as much type as five experienced typesetters working by hand. Both machines generally worked full time during the day; sometimes the company even put on a night operator to keep up with the work. In 1932, *The Fort Erie Times*, *The Bridgeburg Review*, and *The Bertie Township Herald* merged to form *The Times-Review*. Electricity was one of the foundations upon which *The Times-Review* and the Review Company built their success.

Ridgeway and Bertie Township

Other parts of the area took longer to adopt electric lighting and power. Ridgeway had to wait until 1913-14. The townspeople had been waiting since 1901, the year of the Pan-American Exposition in Buffalo. For six months that year, Buffalo was the world's showplace for electricity. Searchlights visible as far away as Toronto scraped the night sky. Even when the searchlights weren't on, there was a glow in the sky, what we now call light pollution. But back then, it was enough to give the people of Ridgeway a good case of what we might call electricity envy. Volunteers at the Old Bertie Township Hall in Ridgeway, which is now the Fort Erie Historical Museum, remember their parents and grandparents saying that at night groups of people would go to the top of Cherry Hill — the highest spot in the area — just to look at the glow in the sky, the electric glow, coming from Buffalo.[12] Electricity must have seemed so near and yet so far.

Electricity was the 20th century must-have technology, as railways had been in the previous century; without it, your town would sink into oblivion.[13] In 1906, the Bertie Township Council sent delegates to a Toronto meeting to encourage the Ontario government to generate power at Niagara Falls and distribute it to Ontario towns. Meetings such as these led to the creation of Ontario Hydro, but Ontario Hydro didn't bring power to Ridgeway. Canadian Niagara did.

Electricity meant poles and wires, and that means a right of way. The options were to buy it, rent it, or trade for it. Bertie Township seemed to like trading. R.M. Disher observes that "by 1914 a By-Law was passed for Electric Lighting in Ridgeway. The Canadian Niagara Power Co. was to furnish free electricity for lighting the Town Hall for the privilege of putting their wires on the Township road allowances."

The arrival of electricity changed the appearance of business establishments. In 1903, Ed Buck's barber shop on Ridge Road in Ridgeway was lit by natural gas. He probably also used natural gas to heat water for shaving his customers. (Reproduction by permission of Fort Erie Historical Museum, Ridgeway)

This undated photograph of N. Boulding Choice Meats on Ridge Road north of the Bank of Commerce clearly shows that electricity has arrived. Electric lights hang from the pressed metal ceiling. Changes made to counters and displays indicate that the store was largely retrofitted for electricity. Parts of the wooden and glass display case are blocked by the more modern refrigerated meat counter. (Reproduction by permission of Fort Erie Historical Museum, Ridgeway)

It was a mutually satisfactory arrangement; the Township had done something similar earlier, when it put in natural gas pipes. In 1891 Council had pipes "put in for natural gas as fuel for heating the Hall and Lock-Up [jail]. Cost $30.00." In return for being allowed to run natural gas lines, "The Bertie Gas Co. (and then the Provincial Gas Co.) provided free gas for the Town Hall and also all school houses in the Township."[14]

Crystal Beach

Canadian Niagara brought power to Crystal Beach in 1914. The amusement park was an early customer. In most cases, the history of late-19th-century and early-20th-century amusement parks is intimately tied in with the history of electric street railways. Streetcar lines usually paid a flat rate for electricity. This meant they had unused power on the weekends and evenings when demand fell. To increase the demand, and make the most of the electricity they had paid for, they built amusement parks at the ends of trolley lines. Visitors paid to ride the trolley and paid to get into the park. But Crystal Beach did not fit this pattern. It had been built primarily to serve the citizens of Buffalo, who escaped the city by boat. For many years, Crystal Beach was referred to as Buffalo's Coney Island. Some visitors from the surrounding area came by horse and buggy or by bicycle, but there never was an electric street railway to Crystal Beach, only an odd sort of monorail, the Peg-Leg, officially the Ontario Southern Railway, which operated for just three years, 1896 to 1898, between Ridgeway and Crystal Beach.

According to one source, the first electric arc lights were installed in the park by J.E. Rebstock in 1897. A steam engine installed in a local planing mill drove the generator.[15] The generator in the park was so unusual that it was an attraction in itself, just as the Canadian Niagara generating station was in Niagara Falls. Visitors could look through a glass window to watch the large leather drive belt turning the generator. Unfortunately, one day the belt flew off, crashed through the window, and killed a child who had stood mesmerized by the wonders of modern electrical technology.[16]

Electricity provided both lighting and power for the expanding Crystal Beach park. But the most dramatic impact of electricity had to await the arrival of Canadian Niagara, when any business or home could get electricity at the flick of a switch, not by running one's own electric generating plant. The many photos showing the town and amusement park in a blaze of light represent power transmitted from Niagara Falls by Canadian Niagara.

Electricity and Commerce in Niagara Falls

In Niagara Falls the impact of electricity was dramatic. Of course the town was larger and got electricity earlier than the communities near Fort Erie. Niagara Falls, New York, had secured electricity even earlier than its Canadian namesake, and with it had received the lion's share of industry (and, as time would reveal all too clearly, the lion's share of the pollution). The Pittsburgh Reduction Company, an aluminum refiner and manufacturer, signed on as the Niagara Falls Power Company's first big customer. There was even more joy, except in Pittsburgh, when the company, later known as Alcoa, relocated its entire operation to Niagara Falls purely because of the cheap power. Other manufacturers soon followed.

On the Canadian side, there were hopes that cheap power would attract a similar aluminum industry. Those hopes were dashed, to some extent by delays in developing power on the Canadian side, but mainly because of a larger power development in Shawinigan, Quebec, where shipping costs for the aluminum ore, bauxite, were lower. It took more than cheap, reliable electricity to attract industry. The Niagara Falls Power Company had amassed a large quantity of land on which companies could locate. On the Canadian side, A. Monro Grier and William Rankine also tried to use land and electricity to attract industry, although much more of the land near the river was taken up by park.

Niagara Falls, New York, became an industrial centre based on the new industries made possible by electricity as a mass commodity. These new industries may be grouped primarily into three categories: electrometallurgical, most notably aluminum; electrochemical and abrasives; health food and cereals. In Niagara Falls, Ontario, the aluminum industry did not materialize, but the other two industrial groups were very important.

Carborundum

Carborundum, the artificial abrasives manufacturer, arrived in Niagara Falls, Ontario, actually Chippawa, even before Canadian Niagara was generating power. It is still a customer.

During the 1890s, some experimenters had wanted to find out how electricity could be used in the chemical industry. Dr. Edward Goodrich Acheson was an inventor and industrialist who had helped Thomas A. Edison perfect the carbon arc filament in Edison's original incandescent lamp. In 1891 he was working in his own shop in Monongahela City, Pennsylvania, trying to use electricity to bring about chemical reactions to create an artificial abrasive to replace naturally occurring substances such as emery, sandstone, and corundum. Many manufacturing processes relied on grinding action for shaping, cutting, and polishing. The potential market was huge, but artificial abrasives were still in their infancy.

Using a crude electric arc furnace, Acheson created a new substance from carbon and silicon, carbide of silicon, or silicone carbide, later known commercially as Carborundum. It was an entirely new substance, not an imitation of something else, whose "sparkling, bluish, diamond-like crystals…cut or scratched the surface of…glass as readily as would a diamond."[17] Further experiments on a larger scale led to commercial production of this new abrasive, which at only $880 a pound could polish gemstones every bit as well as diamond powder at $1,500 a pound. It was such a superior abrasive, which cut so quickly and cleanly, that for specialized uses such as grinding valves, which had to be very smooth and very accurately shaped, it could compete with natural emery, which sold for five cents a pound.

To increase production in Monongahela City, Acheson formed his own 135-horsepower electric generating company. The process took enormous amounts of electric power. So it is no surprise that by 1894 he was in Niagara Falls, New York, with larger furnaces and a long-term power contract with the Niagara Falls Power Company for 1,000 electric horsepower. Because of high tariffs for goods imported into Canada, in order to sell profitably into the Canadian market, Acheson formed Canadian Carborundum Co., Ltd. to manufacture in Niagara Falls, Ontario, and later, Shawinigan Falls, Quebec. The latter eventually became the largest plant of its type in the world.

The long-term business relationship between Canadian Carborundum and Canadian Niagara Power Company is an important milestone in the generation and transmission of reliable electric power. During Canadian Niagara's early years, many customers complained about power fluctuations or power outages. No one liked this situation, but it was more serious for some customers than for others. Carborundum production requires immense quantities of power continuously over an extended period of time. One company publication, *Carborundum Brand Products for the Stone Working Trades*, described the production process:

> Each furnace has a capacity of 2,000 electrical horsepower. The duration of a furnace run is thirty-six hours, therefore there is a total consumption of 72,000 hp.-hours for each furnace run.… The heat generated in the high resistance core of the furnace, by the consumption of such an enormous amount of electrical energy is tremendous. In this terrific heat of about 4,000 F., the crystal masses are formed. After the current is turned off, the furnace is allowed to cool, the walls of the furnace are taken down, the outer crusts removed and the crystal masses exposed and removed.[18]

If during that thirty-six-hour run, the power supply was interrupted for more than a very brief period, an entire batch could be ruined. Continuous power was even more important for making some types of grinding wheels that had to be heated in a kiln for about fourteen days, after which the kilns were broken open and the products removed.

Reliable electric power from Canadian Niagara enabled Canadian Carborundum to manufacture a wide range of products, which in turn supported a range of other industries. The anonymous author of *The Romance of Carborundum*, an early Canadian Carborundum publication, notes that "To describe the work of the manufactured abrasive in industry would be to enumerate almost every product of the useful arts. In the metal industries it ranges from the grinding of Victrola needles to locomotive parts; from fountain pens to plows and harvesters; in the stone industries it has revolutionized the working of marble, granite and sandstone; it is indispensable in the finishing of fine leather and shoes; for the paper industry it is made into huge pulp stones for the grinding of wood pulp; it surfaces and bevels glass; it shapes pearl buttons and precious stones...polishing jade...[and] hulling rice." In the automobile industry, effective grinding contributed to the interchangeability of parts, which kept prices down. "Practically every part of the automobile must be ground at some stage of its manufacture. Take away from the automobile industry manufactured abrasives and the other products which the chemist has made available to it by the electric furnace such as aluminum, alloy steel, and high speed steel, and the labor cost of building a car would become prohibitive. The industry would cease to exist on its present lines."[19]

Just as the electrically powered Review Company saw itself as being in the business of building business for others, Canadian Carborundum helped build business for others. But another type of industry supported by electricity, shredded wheat, illustrates another aspect of electrical history, one where myth and image are every bit as important as electric power.

Shredded Wheat: Cooked by Electricity, So It's Good For You

Few of us today are likely to regard corn flakes as a health food, even less so modern sugar-saturated cereals. And yet, corn flakes, along with many other dry, prepared cereals, came out of an age intensely concerned with diet and health food. The names of two of the great pioneers of the era, Kellogg and Post, are memorialized on millions of boxes of cereal every year. Another name, Henry Perky, never appears on cereal boxes, yet he too was one of the pioneers. In terms of Niagara Falls history, Perky was the greatest of the health food cereal pioneers. His name isn't familiar, but his products are: Shredded Wheat and Triscuit.

Perky's is an unlikely success story. He was rather unsuccessful teacher, editor, exhibition organizer, inventor, and promoter.[20] And he was sickly; his delicate constitution gave him life-long indigestion. But that delicate constitution led to his eventual triumph. In 1890 he saw a man in a Nebraska hotel eating boiled kernels of wheat. Perky, always on the lookout for something he could safely eat, tried it, liked it, and soon his diet consisted largely of boiled whole wheat. Like all converts, Perky wanted to share his new miracle diet with others.

Along with a partner, William H. Ford, he took the boiled stringy mass and cut it into narrow ribbons or shreds. This was shredded wheat. Oddly enough, Perky and Ford thought people would want to make their own health food and so in about 1895, instead of selling the food itself, he started selling machines made by his Cereal Machine Co.[21] Not surprisingly, the idea didn't catch on.

Then Perky discovered that toasting improved the flavour of the food, and probably the appearance and texture as well. He opened his first shredded wheat bakery in Worcester, Massachusetts. Suddenly, his shredded wheat biscuits caught on. Perky felt that he could sell more if people only knew more about food and nutrition. So he got into the education business and installed a cooking school in his factory. To spread the gospel, graduates fanned out across the country praising the virtues of health food and above all else shredded wheat made by the Shredded Wheat Company. Perky's business was growing so rapidly that New York financiers noticed him. The Shredded Wheat Co. was a success. Perky was a noted health food lecturer, but he still needed a gimmick to boost his invention into the big time. He found it in Niagara Falls. A New York lawyer by the name of William B. Rankine helped too.

The Niagara Falls Power Company was providing low-cost electricity to Niagara Falls and Buffalo. The Pan-American Exposition, dedicated to extolling the great future of electricity, was due to open in Buffalo in spring 1901. The area would be filled with tourists avid for new experiences. On December 14, 1900, the *Niagara Falls [New York] Gazette* nearly filled its front page with a 120-point type announcement of a "Crowning triumph for Niagara Falls." It explained in more modest type that "It sounds like a dream, reads like a fairy tale, seems too good to be true — but it is positively true."[22]

Ground was broken that afternoon for what would become a show factory that wasn't even called a factory. Shredded Wheat would be made in Niagara Falls in a palace of light, a Conservatory of Natural Science, with more than 800 windows and more than 30,000 panes of glass; a lecture theatre seating 1,000, along with banquet halls and dining halls; dinners that were free for women employees and a dime each for men; and even 13 bathtubs, 13 marble showers, and 104 sinks for employees.[23]

Perky was now the charismatic president of the Natural Food Co., head of a recently incorporated business backed by Niagara Falls, Buffalo, and New York capital, with $10 million in stock. And the vice-president of the new company? None other than William B. Rankine.

The company extolled the purity and health benefits of its shredded wheat, particularly the fact that it was baked using electricity. To reinforce the notion that this was no ordinary product and no ordinary factory, it was built in a desirable residential area (several beautiful mansions were torn down to make way) with a view of the park and river. Cool water drawn from the river helped air-condition the building in summer. Workers had a model work environment with what may have been the first-ever official coffee break built into their schedules.[24]

The factory, which first started shipping shredded wheat in August 1901, became an important part of the community. It supplied more than employment. It drew about 100,000 visitors a year to tour the plant, some of whom lingered long enough to be enlightened by lectures. But the employees and residents of Niagara Falls and area were the real beneficiaries of the 1,000-seat lecture hall. It was the site of numerous public lectures put on by a wide range of organizations and individuals. In addition to contributing to public enlightenment, the facility helped to attract conventions.

The next business success was the shredded wheat product known as Triscuit. The *Niagara Falls [New York] Gazette* presented this new product as a tribute to electric baking, evidence that "electric baking has been proved a success." [25] Somehow the fact that it was baked by electricity seemed to add as much nutritional value and promise as the fact that it was made from whole wheat.

Shredded Wheat in Niagara Falls, Ontario

Shredded Wheat was more than a food; it was a symbol of all that was modern and desirable. In 1904 the manufacture of the great symbol started in Ontario, when the Canadian Shredded Wheat Company purchased a former chain foundry on Centre Street in Niagara Falls and converted it into a shredded wheat plant. The plant was managed from the start by Edward Wallace until 1926, when he left to bring the gospel of shredded wheat to Great Britain.[26] Canadian Niagara supplied electric power to the Canadian Shredded Wheat Company, and no doubt William Rankine played an important part in bringing another customer for electricity to Niagara Falls, Ontario.

The branch plant was not as large and luxurious as the home plant on the other side of the river. After all, the first Niagara Falls plant had been intended to dazzle the world. This one was there simply to make the product, but to the same exacting standards, "using the same machinery and the same process of manufacture that are used in the home plant." The Canadian plant had the same emphasis on "perfect sanitary conditions" and manufacturing under "conditions of cleanliness which guarantee the wholesomeness and purity of the product."[27] The Canadian plant allowed the American owners to avoid Canadian import tariffs and gave Ontario farmers a new market for their wheat. In 1906, the year of Perky's death, shredded wheat consumption in Canada increased by 51 percent. More than that, a local book on Niagara Falls industries described shredded wheat as "the favorite breakfast food [of] discriminating Canadians."[28] Statements such as these represented an important change in mindset: good, healthy food was equated with industrial food. Electricity was equated with modern industry.

In 1928 Nabisco (The National Biscuit Co.) purchased all the Shredded Wheat plants. In the same year, the company moved into a new plant on Lewis Avenue. In 1993, Kraft acquired the company, but throughout all the changes, Canadian Niagara continued to supply the power.

A Legal Tussle

With all the activity of getting the generators on line, building transformer stations and transmission lines, and supplying power to industry, it may sound as if Canadian Niagara's legal battles had ended and all was quiet on the regulatory front. This was not the case. From 1906 right into the 1920s, the question of how to calculate fees for generating electric power remained unsettled. By that time, a new and aggressive player had entered the picture — the Hydro-Electric Power Commission of Ontario, later Ontario Hydro.

When Canadian Niagara lost its monopoly on generating electricity from water running through the Queen Victoria Niagara Falls Park, two other companies had received water rights. Electricity was a commodity that had to be priced and so was the water that generated it. Each company paid a basic annual rent of $15,000, for any amount of power generated up to a threshold level set at 10,000 horsepower for Canadian Niagara and Ontario Power and 20,000 horsepower for the larger-capacity Electrical Development Company. This basic amount was less than Canadian Niagara had paid in its early years. Rents were calculated and paid twice a year. Everything seemed straightforward. As each company raised production above the threshold level, the lease called for additional payment termed "excess rental." This is where the trouble started. Not everyone agreed on the meaning of the wording that dated from 1899.

The Park Commissioners turned to an expert, Dr. John Galbraith, Principal at the University of Toronto School of Practical Science (Engineering), for help. Galbraith spelled out four possible methods for charging for water usage.

1. *Increasing rental – Peak power* bases the rental upon the maximum peak load developed, and remains at that level until a higher peak load is reached.
2. *Increasing rental – Average power* takes the maximum average daily load instead of the peak load, and increases the rental as this amount increases.
3. *Fluctuating rental – Peak power* increases or decreases the rental, based on daily peak load.
4. *Fluctuating rental – Average power* increases or decreases the rental according to the daily average load.[29]

Sir Aemilius Irving, Solicitor for the Park, felt that the agreement called for Option 1, the "Increasing rental — Peak power" system, which just happened to be the most lucrative and the most predictable for the park.

At the time, only Canadian Niagara had passed the minimum rental threshold, but the agreements with each of the three companies were based on the same pricing principle. The government passed an Order-in-Council in 1907, authorizing the Park Commissioners to use the "Increasing rental — Peak power" system. Canadian Niagara and the Electrical Development Company were billed accordingly, but they claimed that the agreement they had signed called for Galbraith's "Fluctuating rental — Average power," which just happened to result in the lowest possible rent. They appealed to the Lieutenant-Governor-in-Council for a hearing.

On February 15, 1908, the parties gathered at the Clifton Hotel. The Canadian Niagara Power Company, the Ontario Power Company, and the Electrical Development Company all brought their own lawyers. Wallace Nesbitt, K.C., and A. Monro Grier, K.C., represented Canadian Niagara.[30] The Commissioners reproduced the entire transcript of the hearing in their annual report for 1908,[31] It makes fascinating reading.

The two sides were arguing over commodities that couldn't be stored up for later use. The park had water flowing by, and if the power companies didn't use that day's allotment of water, it was gone, never to return. They couldn't take twice as much out the next day. The companies produced electricity that they too couldn't store. If no one wanted the electricity they were capable of producing, they didn't generate it. Each party viewed the situation differently because each was selling a different commodity with no shelf life.

The real issue seemed to be what the power companies should pay the Park Commissioners *for*. The park naturally wanted excess rental based on the power the company *could* generate with the equipment it had and the water it had the right to draw from the river. That is, what the company had the *potential* to sell. On the other hand, the power companies wanted to pay excess rental based on the power they were *actually* able to sell. This meant that if the power company, or companies, had a bad day, week, or month, the park would have one too.

Sir Aemilius Irving argued for Galbraith's idea that "The rental is to increase within the terms of the agreement on the day on which the power generated becomes greater than that generated on any previous day, and is to be based on this increased power until the day on which a greater power shall be generated." Recording instruments would measure each successive "high water mark" of power generation. Under this scenario, excess rental could increase but never decrease, because excess rental was based on the assumption that your potential power output is equal to the highest you have ever achieved. There was even a debate about the meaning and measurement of horsepower, and its relation to kilowatt-hours.

The Clifton Hotel meeting solved nothing. The matter slowly worked its way through the courts, first in Ontario, including the Ontario Court of Appeal, and then, on final appeal, at the Privy Council in London, England. In the interval, the power companies paid according to the average, rather than the peak power system. In 1912, the Privy Council ruled in favour of the Park Commissioners. The companies were billed for the accrued differences between the two systems of calculating rents. It was a significant amount of money.

Canadian Niagara Power Company	$ 52,435.37
Ontario Power Company	$ 50,274.33
Electrical Development Company	$ 92,402.76
Total	$195,112.46

The Park expected to receive this as revenue for 1913. Each company would also be expected to pay excess rental annually, according to the interpretation of the leases given by the Privy Council.

Canadian Niagara was the first to pay up, in late December. The other two companies refused to pay, claiming the Privy Council decision did not apply to them. It was an inexcusable example of bargaining in bad faith. All three companies had participated in the process, but now two of them refused to abide by the outcome. The Park Commissioners were in a difficult position.

The power companies had more financial and legal clout than the Commissioners, who just wanted to get on with running their park. Instead, they had to consider a compromise with the two companies which, if accepted, would have meant refunding some money to Canadian Niagara, because all three companies were supposed to be treated equally. By this time, the First World War had started. The two recalcitrant companies, the Ontario Power Company and the Electrical Development Company, posed as patriots, demanding to be allowed to generate more power to help with the war effort, but without paying more for the privilege.

Then, in 1916, it was discovered that although all three companies were using the same instruments to measure power output, they were using three different systems of measurement, which would make comparisons among them very difficult.

To add to the problem, on August 1, 1917, the Hydro-Electric Power Commission of Ontario acquired the Ontario Power Company. This event could have opened a door to the resolution of the problem of the overdue rental, but it didn't.

Ontario Power and Ontario Hydro had cooperated, one might say conspired, to withhold payment from the Park Commissioners. They cleverly set up the deal so that settlement of past rental amounts was the responsibility of the former owners of the Ontario Power Company. The whole mess was not settled until the courts forced a resolution in the 1920s, by which time the Hydro-Electric Power Commission of Ontario owned both recalcitrant companies.

It was an odd situation. The Queen Victoria Niagara Falls Park had been created by the Ontario Government in a way that made it the weakest player in tough negotiations. Meanwhile, the representatives of Ontario Hydro, who liked to pose as public benefactors and talk about the "people's power," treated the representatives of the people's park shabbily. But odd things often happen at Niagara Falls.

Extending the reach of electric service also meant extending Canadian Niagara's need to respond to emergencies. When a disastrous fire struck downtown Ridgeway on September 27, 1948, Canadian Niagara linemen rushed to the scene to cut off electrical service so that falling lines and live wires in burning buildings posed no threat to firefighters. Later, Canadian Niagara helped rebuild the burned-out electrical system. (Reproduction by permission of Fort Erie Historical Museum, Ridgeway)

Compared to the seemingly informal arrangement at the Rankine powerhouse, security was much tighter at the Adams Generating Station in Niagara Falls, New York. (Reproduction by permission of FortisOntario)

The First World War and the 1920s

CHAPTER 6

The First World War

There were many reactions to the outbreak of war in Europe in 1914. Many people rushed to sign up. It was the patriotic thing to do. Some even feared the war would be over too soon if they didn't hurry up and get there. Few, if any, could foresee the years of horror ahead, or the high-level political and military bumbling. Those at home did their best to cope and contribute to the war effort.

Canadian Niagara's Victoria Cross Hero

Three days after the war broke out, Graham Thomson Lyall left his job at Canadian Niagara to join the 19th Lincoln Regiment. He was soon posted to the Welland Canal Field Force that had been ordered by the Governor General to guard the canal, international bridges, and ferries, along with Niagara's hydroelectric generating stations.

Lyall had studied mechanical engineering in England before emigrating to Canada in 1912, where he lived in Welland and then Chippawa. He joined the 81st Battalion, C.E.F. (Canadian Expeditionary Force), which was headquartered in Toronto at the Exhibition Grounds, but recruited heavily from the Niagara area. Lyall sailed for England on May 16, 1916.

Canadian battalions and regiments were often broken up when they arrived overseas, and Lyall was sent to join the 4th Canadian Mounted Rifles, which had been virtually wiped out when all but 76 out of 702 men had been lost at Sanctuary Wood in Belgium. Lyall attracted favourable attention during the Battle of the Somme and in 1917 at the Battle of Arras, which included the capture of Vimy Ridge. He was clearly officer material and soon found himself in Officers' Training School at Bexhill-on-Sea in England. He graduated as Lieutenant Lyall in the 102nd Battalion, C.E.F., part of the 11th Canadian Infantry Brigade.

He saw action at Ypres and then Amiens. But it was when he was sent to breach the Hindenburg Line on the Canal du Nord that he won the Victoria Cross. At one point, with 13 prisoners already in tow, his much-diminished platoon

Industries turning out war materials needed every available kilowatt of electric power. This made generating stations an ideal target for enemy saboteurs. Armed soldiers were posted to guard Rankine against enemy agents. (Reproduction by permission of FortisOntario)

was held up by machine-gun fire. When he and a few others reached a strong point, Lyall rushed ahead alone, killed the enemy officer in charge, and captured 45 prisoners and five machine guns. Shortly thereafter, he captured another 47 prisoners. In consolidating his position, he managed to protect the remainder of his company. That was September 27, 1918. A few days later, October 1 and 2, he captured a total of "three officers, 18 other ranks, 26 machine-guns, and one field gun, exclusive of heavy casualties inflicted."[1]

In March 1919 Lyall travelled to Buckingham Palace, where King George V presented him with the Victoria Cross. He never returned to Canadian Niagara, and we don't know if he stayed in touch with any of his former workmates. He settled in Scotland, married Elizabeth (Elsie) Moffat Frew, and worked in the construction industry as Managing Director of Aerocrete (Scotland) Limited.[2]

Guarding Against Sabotage

When war broke out in 1914, the Canadian Niagara Power plant suddenly became even more important than it had been in peacetime. Mechanized wars depend on a long supply chain that stretches from farms, forests, and mines to powerhouses, offices, and factories and all the way to the battlefront. Damage to any part of the supply chain reduces the ability to wage war. Enemy agents and saboteurs looking for a place where a small amount of well-placed effort could do major damage to the supply chain feeding the Allied War Effort needed to look no further than the Niagara Frontier. First, with its hordes of tourists, agents would find it easy to blend in. Second, it offered a tempting range of targets: power plants and electricity-dependent industry on both sides of the river as well as international road and railway bridges and the Welland Canal, one of the most important links between European battlefields and the natural and industrial resources of the North American Midwest.

Government and industry reacted quickly to the possible threat from enemy agents. The newly formed Welland Canal Police Force moved into position at bridges and powerhouses and along the canal. Strong lighting and barbed wire were also used to protect key sites. One Fort Erie soldier, who wrote from "Somewhere in Belgium" on September 8, 1916, picked up on barbed wire as a common link between home and battlefront.

"So they have barb [sic] wire all around the power houses now have they? I get enough of the darn stuff. Am with the scouts now and have to go out in no man's land where it is pretty thick in some places. It's pretty hard on the

With the outbreak of the First World War, former tourist sites were now classified as strategically important parts of the war effort. Powerhouses, the Welland Canal, and key bridges were cordoned off by barbed wire barricades. Parts of the Queen Victoria Niagara Falls Park were closed and park revenues fell, but war required that everyone make sacrifices. (Reproduction by permission of FortisOntario)

hands and clothes. Was right up to the German wire a few nights ago. Could hear them talking quite plainly."[3]

The Queen Victoria Niagara Falls Park Commission also swung into action to protect the region's power supply — after all, the three major Canadian power plants in the Niagara Region were all on park property. The summer encampment for the park's military guard was "immediately north of the Jolly Cut upon the old cricket crease." The Park Commission also turned the Administration Building into living quarters for the guards during the winter of 1915.[4] In addition to the security provided by the Welland Canal Police Force, companies also hired their own guards. In November 1918, eight of Canadian Niagara's 127 employees were classified as guards.

Signs such as this appeared in homes and stores to let people know that one was a supporter of the war effort. (Reproduction by permission of the Fort Erie Historical Museum, Ridgeway, Ontario)

The War Effort Needs More Electricity and Money

Patriotism was the word of the day, but not everyone agreed on what it meant. With Ontario Hydro still in its youthful, aggressively expansionist mode, Adam Beck tried to use Canadian manufacturers' urgent need for electricity as an excuse to ban all power exports to the United States and take virtual control of Canadian Niagara. Beck and the Ontario government, which pressured Ottawa to block export permits, seemed to forget that American manufacturers were supplying essential goods to the British and that such a move would harm the war effort. British diplomats reminded Ottawa of this fact and Beck's proposal died.

Even before the First World War, Canadian Niagara had faced a serious market threat when the American government passed the *Burton Act* of 1906 to cap the amount of power the company could export to the United States. The next year, Ottawa followed up with the *Exportation of Power and Fluids and Importation of Gas Act*, which compelled Canadian power exporters to obtain a licence, and gave the government the option of taxing exports up to $10 per horsepower per year. Fortunately, the government never exercised this option, which would have crippled the company.

When war broke out, the government's main reaction was to ensure that as much power as possible was available to Canadian industries. "An order of the Federal Power Controller directed that all the power companies should develop electrical energy to the full capacity of their works."[5] All three power companies — Canadian Niagara, the Ontario Power Company, and the Electrical Development Company — increased production beyond their legally allowed limits. Although the Parks Commission would have liked the thorny issue of excess rentals ironed out before allowing excess production, they allowed it in patriotic good faith.[6] They even went

Throughout the war period Victory Loan parades allowed companies and their employees to express their support for the war effort and compete with each other in making imaginative floats and displays. It is believed that Ontario Hydro employees created this exuberant float, which paraded on Armistice Day, November 11, 1918.
(Reproduction by permission of John Burtniak)

so far as to allow a temporary aboveground power line within the park, as well as other non-scenic elements, and quickly approved Canadian Niagara's request to install more generators and turbines.

When the United States entered the war, the demand for power soared. "The President of the United States, by orders signed by the Secretary of War under date of December 28, 1917, requisitioned the total quantity and output of electrical power produced or capable of being produced by the two generating companies at Niagara Falls, New York, as well as the power transmitted into the United States from Canada."[7] By that time, Canadian Niagara had increased its power production by a wide margin.

At the outbreak of war, Canadian Niagara could produce 74,000 horsepower from seven generators, the most recent of which had come into service July 28, 1913. Under the Federal Power Controller's orders, the company ran the existing equipment flat out. As well, three new units were installed and brought on line during the war: No. 8 (10,000 horsepower) on September 7, 1916; No. 9 (10,000 horsepower) on October 22, 1916; and No. 10 (11,000 horsepower) on January 13, 1917.

In order to provide Canadian manufacturers with the power they needed and reduce the amount sent to the United States, Canadian Niagara put in a new underground transmission line between the powerhouse and Stamford Township. And as the war progressed, the company steadily increased the power it sold to Ontario Hydro. As of July 26, 1916, only 12,000 horsepower, or 16.6 percent of output, went to Ontario Hydro. But that allotment rose steadily to 50 percent or 44,000 horsepower by December 17 of that year. In 1917, when Canadian Niagara brought a tenth generator on line, the figure went up to 50,000 horsepower. But even in the first few years after the war, when Ontario Hydro and the province of Ontario faced a serious electrical power shortage, Canadian Niagara continued to increase the power it sold to Ontario Hydro, rising to 58 percent of production, or 59,000 horsepower, as of January 1, 1920, and then to 67 percent or 68,000 horsepower, on November 30, 1920.[8]

In addition to installing more generator units and running flat out, Canadian Niagara also took two major steps to improve the reliability of supply of electric power from the generating station. To ensure that power surges, particularly from lightning strikes, did not damage equipment, the Parks Commission gave the company permission to build "chambers for reactance coils adjoining the power house on the westerly side." But ice posed the biggest threat to reliable output. Each time the company installed another generator, the company needed to draw more water from the river. But its additional demands lowered the water level and made the ice problem more serious. The other two Canadian power plants on the upper river — the Ontario Power Company and the Electric Development Company — also took more water and increased ice problems for Canadian Niagara. The story of Canadian Niagara's efforts to deal with ice and fluctuating water levels during and after the war is told in Chapter 9.

Meanwhile, other Niagara Frontier industries contributed to the war effort by using electric power to produce war goods. Carborundum, Canadian Niagara's largest industrial power customer, made grinding wheels, needed in the production of everything from rifles and ammunition to ships, tanks and army boots. Almost every kilowatt of electrical power that Canadian Niagara sent to a company like Carborundum went indirectly to the battlefront.[9]

Wars cost money, and the Government of Canada borrowed money from citizens and corporations through interest-paying Victory Loans or Bonds. In 1917, Canadian Niagara invested $250,000 in the Victory Loan campaign. The company also helped individual employees contribute through payroll deductions. The company maintained a blackboard that named everyone who contributed to the Victory Loans. Support for the war ran high, but a little competitiveness between different shifts and job classifications probably helped Canadian Niagara achieve 100% employee participation rate for the Second Victory Loan campaign. The company also made donations to various war-related charities and causes, such as the Sailors' Week Campaign Fund of the Navy League of Canada.

Countless off-the-job hours also flowed into supporting the war effort. Ann Augusta Miller of Fort Erie, who was born in 1848 and lived to October 25, 1932, started keeping a diary in 1870. During 1917 and 1918 she makes nine references to war bonds and many more to knitting socks to send to the men overseas. On November 11, the day the war ended, she put quotation marks around the word "Peace" in her diary, and noted, "Sent 4 pairs socks."[10]

The First World War officially ended on the eleventh hour of the eleventh day of the eleventh month in 1918. After the celebrations, thoughts turned to what to do after the war. For some, peacetime meant learning to live without those who would never return. For the more fortunate, peace meant being reunited with family and friends. It also meant returning to work. The Park Commission wanted to provide work for the large numbers of veterans who would return from overseas. The Commission also hoped the return of peace would let it ease back on the efforts devoted to increasing power production in the park.[11]

The Canadian government called upon citizens to help finance an expensive war by purchasing Victory Loan bonds. Bond holders would eventually be repaid with interest. Companies competed with each other to exceed the targets set for them and to have as high a participation rate as possible. Posting the names of subscribers certainly encouraged individuals. This blackboard in the Rankine station showed that of the 158 employees listed, 31 worked for the American parent company, the Niagara Falls Power Company. (Reproduction by permission of FortisOntario)

Finding a Use for All That Power

Contrary to many people's expectations, the return of peace did not bring the expected fall in industrial demand for electricity for light and power. Businesspeople and the public in general had grown increasingly accustomed to electric power for industry, office machinery, streetlighting, and other uses. Many had invested heavily in electrical equipment for war industries and now wanted to use that machinery for peacetime production.

At the same time, electric utilities, including Canadian Niagara, had invested heavily in generation, transmission, and distribution networks. (The new Adam Beck generating station, the largest hydroelectric plant in the world at that time, was nearing completion in the Niagara Gorge; it would be officially opened in December 1921.) The expanding system could provide more power than customers needed, and it could do so at little or no additional cost. Utilities needed to sell more electricity to recover their investment.

How could one encourage greater electricity consumption? At the 1919 summertime convention of the Commercial Light and Power Sales Committee of the Canadian Electrical Association (CEA),[12] held at Alexandra Bay, New York, and presided over by A. Monro Grier of Canadian Niagara, industry representatives considered the question carefully. The use of electricity by heavy industry was stable. The biggest new market lay with domestic consumers. By 1920, about a third of Canadian homes were wired, and electricity was available to the residents of nearly all urban centres with a population of more than 5,000. Electric light was popular, and one small electric appliance — the iron — was already well established. But irons did not use very much electric energy. Growth in sales demanded bigger appliances that used more energy.

Electricity is now so common in Canada that we forget there was a time when it was so new that people not only had to be convinced of its value, but also had to learn how to use it in the form of electrical appliances. The participants at the 1919 meeting knew that sales of cars had skyrocketed when Henry Ford had produced the first affordable automobile. But in 1919, the electrical industry was still waiting "for what might be termed the 'Ford of the Electric Range Business,' and if a range could be sold…at about $60.00, it would add a great stimulus to its use. This range could be sold to people of very moderate means, such as mechanics, factory workers, etc." When dealing with the rich, people the CEA called "discriminating buyers of more extensive means," "the higher price ranges are easily sold and are actually in great demand."[13] But to succeed dramatically, the industry had to enter new markets — the much larger group of people who were not wealthy, but could take advantage of electrical appliances.[14]

A CEA member from Trois-Rivières, Quebec, reported on ways of getting around consumer resistance to buying high-priced electric stoves and ovens. Allowing customers to pay for the stove over a longer period of time certainly helped. But their merchandising department had come up with an even cleverer psychological ploy. They "had been very successful in using a two-plate cooker, which is a rather cheap proposition, with a little portable oven to be used on top, as an entering wedge. When once these are installed and used, the lady of the house will not go back to any other method of cooking, and she will, some way, persuade her

husband to put in a larger range when she has demonstrated to him that she cannot do the necessary cooking on the small two-plate burner."[15]

The trick was clearly to convert women to the benefits of electric cooking in the first place and let the momentum of rising expectations do the rest. "It is electric *cooking* rather than the electric *stove* itself that we are selling these days." Advantages included the fact that electric stoves were much easier to keep clean, and that it was easier to control the heat with an electromechanical thermostat. Electric ovens were also better insulated, so that the heat remained inside the oven, rather than heating the surrounding air. All these benefits were carefully considered at the CEA meeting.

But what about the cost? Anyone switching to electric cooking was switching from wood, coal, or gas, which could be either artificial or natural. Coal prices were rising and this had been a factor in promoting early hydroelectric power generation in Ontario. In many areas, wood was becoming increasingly expensive. But gas was another issue. When A. Monro Grier asked his colleagues about how the "actual cost of running compares, for instance, with gas stoves" he couldn't get a clear answer. The chair of the committee pointed out that "it is very difficult to get the actual comparison because the conditions are so different in different towns." That was a fact that the industry had to live with. Other responses made it obvious that to some extent, perceived costs were more important than actual costs. Another pointed out that even in an area where gas was cheaper, "in practically every case the electric range was put in and the gas range was taken out."

The words "taken from an aeroplane" were there to let postcard buyers know this was something absolutely modern. It also shows that much of what is now a very scenic part of the Queen Victoria Niagara Falls Park was a construction zone in 1920. The long black line behind the Canadian Niagara powerhouse is a temporary aboveground enclosed water pipe or penstock, sending water to the Ontario Power Company generating station below the Falls. (Reproduction by permission of John Burtniak)

Clearly it was not a straightforward economic issue. Nor was it solely a technological decision — technology alone rarely, if ever, drives consumer decisions. One member suggested that the electric industry could learn from the gas industry, which had set up programs aimed at "educating their customers how to use gas stoves." He went on to add that he understood "from agents who sell electric stoves that the great fault with all the people who use them is that they have been accustomed to using gas stoves, and they start to use electric stoves in the same way."[16] Consumer education would be needed. But more than that, consumers would need to experience for themselves the differences between the old stoves and the electric models.

Those two days in Alexandra Bay seemed to influence Grier's thinking. Certainly Canadian Niagara's eventual electric stove campaign used many of the ideas discussed at that meeting.

Selling Electric Cooking

Although the hoped-for "Ford" model did not appear for some years (Westinghouse did not begin mass production of stoves until 1926), the quality of electric stoves did improve in the early 1920s, with developments in the technology of switches, thermostats, and heating coils, and better enamelling of the interior of the oven.

By 1922, Canadian Niagara was ready to enter the domestic appliance business. In late November, its customers received something special with their power bill. It was an invitation "to attend a special demonstration of electric cooking during the week beginning Monday, Nov. 27th," and a chance to "spend a pleasant hour learning about clean heat." Since home electric ranges were still relatively rare, the invitation stressed, "You must see an electric range actually cooking to appreciate its value." The demonstrations would be "conducted by courteous women who will cheerfully explain how electric cooking conserves food, lightens labor and economizes both time and money." There would be no fast-talking hucksters. The idea was to explain and convince, not to intimidate. The message was convenience: just "'twitch the switch' — instantly electricity begins to serve you! It lights your home, and keeps it clean. It cooks your food, washes and irons your clothes, and does many other odd jobs about the house."[17]

Since few people had cars, the company held separate demonstrations in Fort Erie, Ridgeway, and Bridgeburg. We have no reports of what happened at those demonstrations, but it is quite likely that women vastly outnumbered men and then had to convince husbands or fathers of the urgent need to "enjoy the comfort and economy of clean, electric cooking." As an added inducement, Canadian Niagara offered to pay the cost of installing and wiring the Westinghouse Electric Range for customers who placed their order before December 15, 1922. And to make electricity even more attractive, the company lowered the cost of current by 25 percent.

But there were many homes with no electric wiring at all. There would be no electric ranges or other convenient appliances for them until their homes were wired. As part of the same campaign, Canadian Niagara offered installation and wiring for a Westinghouse Electric Range on a time payment plan: "We will pay your contractor for wiring your house, and bill you for the amount in twelve monthly payments." Those who wanted to enter the electrical world more slowly could still get the time payment plan for home wiring without buying a range.

With so much that has changed since the 1920s, it is easy to miss the significance of the time payment plan offering. It was revolutionary. General Motors had been the great pioneer in time payment plans and that, along with a wider selection of colours for cars, had allowed General Motors to jump ahead of Ford in sales.

The campaign played out by mail and in newspaper ads. There was also an even more personal offer. The bottom of one of the mailed flyers carried the question, "Did you forget to mail that blue card?" This was a postcard, for which the customer had to supply a 2-cent stamp, requesting more details about the electric range and house wiring offers. Since getting basic electric service or an electric stove were big steps, Canadian Niagara added an extra suggestion to help people feel more comfortable about it. The bottom of the card read, "If you are acquainted with one of our employees and prefer to talk with him, put his name below." The employee might be a neighbour, someone who went to your church or service club, or played in your local baseball or hockey league, or it might be the meter reader you had come to know and trust if you already had basic electric service. It was the kind of personal service that a company could offer in a small market.

This invitation from the *Fort Erie Times*, October 19, 1923, marked the beginning of a new era.
(Reproduction courtesy of the Fort Erie Historical Museum, Ridgeway, Ontario)

Nearly a year later, during the week of October 22, 1923, Canadian Niagara welcomed everyone to a "House Warming!" to celebrate "the Opening of our New Office and Showrooms, Goderich and Queen Sts., Fort Erie." The centrepiece was cooking and appliances, along with the time payment plan.

> Come and see Electric Cooking!
> Taste the products of an Electric Range — or bring your
> Roast, bread, or pie and have it roasted in a —
> Westinghouse Electric Range
> The Clean, Cool, Carefree Way to Perfect Results!

A joint Canadian Niagara and Westinghouse ad placed in the *Fort Erie Times* to catch the 1923 Christmas trade showed four gift suggestions: an electric iron; a two-slice, one-side-at-a-time electric toaster; an elaborate coffee maker; and one-element hot plate.[18] These were gift or "starter" appliances. They were intended to introduce someone to the world of household electricity or to extend the role of electricity in a home that already had electric lighting. However, the electric range was the most desirable household load builder. "Cook by Wire," a November 1922 Canadian Niagara ad in the *Fort Erie Times*, added a further incentive — what we would now call keeping up with the Joneses: "Take advantage of our recently announced, large discount by installing an electric range. There are already on our circuits fifty-seven ranges operated by satisfied customers."[19] In other words, the buyer would be joining a select group.

Demonstrations, classes, and special offers eventually led to the widespread use of electric stoves during the 1930s. Some drawbacks remained — electric stoves took longer to heat up than gas stoves, there was an additional fee for hooking them up, they were still more expensive to buy and, in areas where electricity rates were high, more expensive to use. But they had one special feature — they were perceived as modern. As one historian of technology has noted, "The single most prominent appeal in all the range advertising was to modernity, a subjective quality that sold ranges by connoting fashion and status rather than denoting actual material characteristics."[20]

Learning to Live Better Electrically

By 1929 the *Niagara Power Service* article "Canadian Villages Reach High Kilowatt-Hour Average" proudly proclaimed that in the town of Bridgeburg one in four domestic customers had an electric range and in Fort Erie one in five households with electricity had an electric range. In the eyes of an electrical industry publication, this was an admirable achievement, particularly for villages normally overshadowed by Buffalo in nearly everything. "The slogan 'The Town of Electric Homes' may well be applied to these sister towns located just across the river from Buffalo on the Canadian shore."

The same article also praised their power consumption. Fort Erie customers used an average of 1,464 kilowatt-hours of power per year and Bridgeburg customers 1,922. This "exceedingly high average of kilowatt-hour use" was regarded as "exceptional, especially in towns of the size of these Canadian villages." [21]

Was this a case of electrical envy and local pride? Both towns had for decades looked enviously across the Niagara River to Buffalo. Once smaller than Fort Erie, Buffalo had grown to become what Fort Erie had once thought it would be — a major city of continental significance. Buffalo was one of the greatest and richest American cities and it had electrified in the 1890s.

In 1901, thousands of people had flocked to Buffalo to see the Pan-American Exposition, that great showplace of electricity. John Stickles of Ridgeway recalled his early years by dividing them into two periods: before and after he visited the Pan-American Exposition with his parents.[22] They may also have been among those from Ridgeway who gathered at Cherry Hill in the evenings during 1901, just to look at the glow in the sky coming from the electric lights of the Exposition. In 1901 Buffalo was bathed in electric light and Fort Erie and Bridgeburg were still waiting for Canadian Niagara to begin generating its own power. So when Canadian Niagara finally gave them electricity, the people of Bridgeburg and Fort Erie were quick to make the best of it. Electricity transformed local businesses and, with the aid of laws and import duties that encouraged foreign firms to manufacture in Canada, both towns successfully courted a number of American companies. The two towns prospered and the townspeople saw themselves as electrical leaders. They wanted to live up to that image.

The high level of electricity use is even more unusual in light of the importance of natural gas in the area served by Canadian Niagara. For example, Mrs. Shirley Beam, who was born in Stevensville in 1926, recalled that for her family the Depression wasn't much of a worry, since "we had free rights to a gas well on another farm that heated our house." Such shared gas wells and free gas were common. Moreover, as she recounted, "We didn't have electricity so we didn't have big electric bills. It was gas, gas lights."[23]

Mr. Sumner Beam, who was born in 1898 near Black Creek, spent his early childhood years in a home heated with wood and coal. Then his family moved to another farm and the gas well drillers came by and boarded in their house. When the drillers struck gas half a mile away, "Why, of course we had gas then in the house. Gas for heating and everything. There was no limit on the amount of gas you could use, and of course a lot of it was wasted...there was no end to it." They "never had electricity, no electricity at all." Farmers who let the gas company drill on their land got fifty dollars a year, plus unlimited free gas if the company struck gas.

For people like this, electricity was for work, not homes. When Mr. Beam worked at the shipyards on the Niagara River near Black Creek, he got his first look at electric welding. It was something new and he recalled people had the usual doubts about new technology. They looked at electric welding and said, "'Oh, that wouldn't stand up at all, it would break loose.'" There was also the fear of new technology, particularly electrical. Mr. Beam remembered that a generator was used to power the welding equipment, but he didn't get very close, because "they had a lot of us scared to go near it because there was so much electric there."[24]

Ruth Carver, who was born in 1903, remembered growing up with coal oil lamps. Even in the 1920s, she and her husband still didn't have electricity. Just before her fifth and sixth children (twins) were born, she got a hand-powered washing machine. "That took a lot of strength. Sometimes I thought I would rather do it by hand, but you'd get along faster, and then I had the hand wringer, too, so I could wring them out. I didn't have to wring them by hand." When the Depression started, two more children came along just sixteen months apart and she got a new washing machine. It had a gasoline engine underneath and a "foot crank" to start it. It was a dangerous setup. As many early car owners knew, if the engine backfired, that is to say fired and turned the engine backwards before the crank was withdrawn, it was a recipe for disaster — a broken wrist or a broken leg. "When we got the electric in I sure was happy over that! No more having to crank that [washing machine] over." Electricity made an enormous difference. "Sure was a happy day when it came! I think I appreciated the electric more than anything." The house had some electric lights, but she "didn't insist on a lot of outlets. I had one in the kitchen and for the washing machine and that was it."[25]

Electricity entered people's lives in different ways. The diary of Mrs. Ann Augusta Miller, which has been preserved in the Fort Erie Historical Museum, provides a glimpse of the transitions that people went through as the 20th century progressed. Mrs. Miller didn't go in for lengthy writing or wasting paper; almost all of her daily diary entries were one line long and short on punctuation.[26] However, the brief entries offer a revealing picture of life on a farm along the Niagara River near Black Creek. In June 1906, she mentions that a crew stringing power lines from Canadian Niagara to Fort Erie was camped in the neighbouring farmer's field. But the crew didn't connect her house to the line. At that point, she and her husband were using natural gas from a shared well on the next farm. Electricity entered their lives in 1919, but it wasn't in the house. Not long after they bought a car in October 1919, Mrs. Miller began to mention chores related to car batteries. Early car batteries were unreliable and fragile. When it was cold, people took the batteries out of the car. Sometimes repairmen had to come to charge them, fix them, or replace them. Or one could take the train to Bridgeburg, get the battery charged there, come home on the train, and put the battery back in the car.

In November 1921, the Millers had a furnace put in, but it appears to have been coal. In June 1923 they bought a coal oil stove. It seemed to work well, but it meant repeated trips to town to buy coal oil.

In 1924, they bought a radio, a "big radio." That meant more batteries to take care of. Finally, in 1925, men started setting up poles and stringing wires, inspectors started coming around, and on June 24, while she was at a Falls picnic, electric fixtures went in. The next day, she wrote "lights on" in her diary. From that point on, things happened quickly. Within a week, the Millers were in Bridgeburg looking at Canadian Niagara's ranges and they had one by July 10. Before the month was out, she could proudly write, "Used electric iron for first time." And the next week even the shed and cellar had electric lights. Of course the car batteries still required extra work and trips to town, and the radio demanded its regular diet of A and C batteries, but hydroelectric power was more reliable, although the following year the power went off on a "windy and cold" January 28 and came back on January 30.

When Mrs. Miller wrote that she "used electric iron for first time," she didn't say she owned one. She probably used it at Canadian Niagara's showroom in Bridgeburg, where a month earlier she had looked at ranges. Once you had electric lights and an electric range, what better Christmas presents for a homemaker than small electric appliances? The diary entry for December 24, 1926, was: "Chas [Mr. Miller] to Bridge[burg] got elec. iron and toaster." A Christmas gift, probably. The diary entry for the first day of 1927 read "Chas took elec. check to mail box." On September 26, 1928, she wrote "got electric heater" but we do not know if it was an electric water heater or a space heater. A diary entry suggests that by June 1930, she had an electric vacuum cleaner and then in 1931 there is a mysterious entry — "put electric machine in cellar" — probably a washing machine.

Clearly Mrs. Miller liked convenient electricity instead of clumsy batteries that required trips to town. But at the time of her death in 1932, she still had to put batteries in her phone and she was apparently still using the battery-powered radio.

People adopted electric technology at varying paces and with varying priorities. After all, Adam Beck, who certainly did a lot to promote electricity, had a cigar factory in London, Ontario, but he never got around to fitting it up with electric power to replace the steam engines.

The Peace Bridge

The Peace Bridge is one of the great symbols of international cooperation, trade, and travel between Canada and the United States. It opened officially on August 7, 1927, with an assemblage of heavy-hitters from Canada (Prime Minister William Lyon Mackenzie King), Britain (H.R.H. Edward, Prince of Wales, and his brother, Prince George), and the United States (Vice President Charles G. Dawes). On that day, 20,000 people flocked to the bridge site and an estimated 50 million listened to the radio broadcast of the opening. By that time, a lot of people had already been over the bridge, which had opened to traffic on March 13, 1927. Chief Engineer Edward Lupfer made the first official automobile crossing.

The Peace Bridge provided Canadian Niagara with new customers, because it accelerated what some call the "American colonization of the Lake Erie shore" by cottagers. These Americans, some of whose cottages were really palatial homes — particularly those in Point Abino, an early gated community — were an important part of Canadian Niagara's customer base.

The Peace Bridge also brought at least one new employee. Louis Rouse had worked as a construction labourer on the bridge. In the late 1920s and early 1930s, a number of first-generation Canadian Niagara linemen were retiring, and Rouse was one of the new hires to fill the gap. Linemen were climbers in those days, and Louis's son Norman still has his father's spurs and belt. Louis loved the job and the men he worked with. He stayed at Canadian Niagara for 42 years, retiring as line supervisor.[27]

Louis liked to take snapshots. One of his photographs, with the Peace Bridge entrance in the background, captured a transformer that must have been eight or ten feet tall. It was on a float trailer with wide solid wheels that was chained to the back of the truck. The truck was trying to pull it up a slight rise. That didn't work, so another truck was chained to the first truck. Still nothing. Reverse gear usually has the lowest ratio and so when a third truck was added, it was positioned backwards. Finally the transformer was on the go again. Louis caught it all on film. The background shows the Bertie Street tower and the high tower for sending power to Buffalo, as well as part of the huge Imperial Oil sign advertising Premier Gas that was aimed at American tourists arriving across the bridge.

Jack Irish worked in the service station under the huge sign from 1935 until after the Second World War, when he and his brother Duncan opened the hardware store on Queen Street where Canadian Niagara employees cashed their weekly pay

For decades, people had been dreaming of a bridge linking Buffalo and Fort Erie. It finally happened in 1927, with the opening of the Peace Bridge. It meant different things to different people. It was a new customer for Canadian Niagara, and quite a challenge for the crew that had to bring in a large transformer on a float trailer with extra-wide wheels so it would not sink into the road. When a slight rise in the road was too much for one truck, another was fastened to the first. Still not enough. A third was added, this one in reverse to get even lower gearing. That did the job. (Reproduction by permission of Norm Rouse)

cheques. Imperial Oil had built the station as a showplace to introduce Americans to a brand of gas available only in Canada. Imperial Oil expected customers to be given first-class treatment. Irish recalls, "They frowned on your doing any work that would prevent you from being right out there to take care of the customer when he came in." The road to the bridge was paved, but the station lot was gravel in the early years. Jack would start each day by raking the gravel. Then he would go out and sweep up anything on the road in front of the station.

The station had to live up to the image created by the huge red, white, and blue sign made by E.L. Ruddy in Hamilton. It was more than just a good-looking sign that used lots of Canadian Niagara electricity. It was modular, a design that allowed for easy maintenance and repair. The entire sign was made up of numbered sections, each of which was replaceable. Jack and his older brother Duncan, who ran the station, had cards that showed a picture of the sign broken down into numbered sections. When a section of neon tubing burned out, they circled the numbered sections that needed replacing on the card and sent it off to E.L. Ruddy. A trucking company such as Direct Winters or Consolidated would bring in the part. Then Jack or Duncan would phone Chester Bearss, an electrical contractor in Ridgeway, to come out and put in the new tubes. It all seemed extremely modern.

The station is an interesting example of how electricity worked its way into local businesses. Aside from the impossible-to-miss neon sign, electricity started off with a rather modest role: lights and an air compressor. Electric pumps and equipment came later.

To ensure that American visitors knew what kind of gas to buy after they crossed the Peace Bridge from the United States to Canada, Imperial Oil, makers of Esso Gasoline, put up a huge lighted billboard that used a lot of electricity from Canadian Niagara. Jack Irish, who worked there with his brother Duncan, thinks they would have sold more gas if they had used the smaller, American gallon instead of the Imperial gallon. Some Americans thought the gas cost too much per gallon. This sign replaced the one shown in the previous photograph. (Reproduction by permission of Jack Irish)

The first pumps Jack worked were the old-fashioned Clear Vision pumps that had a big glass cylinder with numbers marked on the side to show the level of gas. The pump had a long handle. "After the transaction was finished, you took the handle and pumped it back up to full again." Eventually, the hand pump would be replaced with an electric pump to raise the gas. Still, the only way to measure the gas was by looking at the marks on the glass. In the 1940s, odometer style electric gear-driven indicators were installed on pumps to measure the amount of gas pumped out, and coloured balls behind glass turned as the gas flowed by. Most people asked for a fill-up, but sometimes the larger Canadian gallon caused some confusion among Americans who couldn't understand why Canadian gas was so expensive. These were the ones who would say, "I just need enough to get over the bridge."

The Peace Bridge service station didn't do major repairs, mostly tires and oil changes. Still, it had an outdoor hoist raised by electrically operated hydraulic pumps. It didn't always work well. Jack remembers, "It was designed for a weight of a car, but we would occasionally put trucks on it. You'd have a truck up there and you'd be greasing and the damned thing would drop about 10 inches...bang! while you were under there greasing...safety was a little casual then." Still, this was perhaps better than the arrangement at Canadian Niagara, which consisted of a pit with a ladder. The car or truck would straddle the pit with the mechanic underneath. If anything went wrong, it could be hard to get out quickly. No one at Canadian Niagara was ever seriously injured in the pit, but injuries and deaths did occur in similar pits elsewhere.

The station also had an electric grease gun. At first, Jack thought it was a great idea. "You just take an extension cord and go out and do the job." A small electric pump in the handle forced the grease out the tube and nozzle. "There was only one trouble with it. Not every time, but sometimes, when you'd give the thing a squeeze to get the grease out, you'd get a shock." Jack sent it back to the manufacturer in Hamilton several times, but the problem was never fixed. Somehow the electric grease gun ended up with "an old mechanic in Font Hill. ... He found a bare wire in it right away." Whenever the wire touched the body of the gun, whoever was using it knew.

But that was about the extent of electricity making the life of a mechanic easier. There were no impact wrenches or power machines for taking tires off rims. That was just straight muscle work.

Meanwhile, the Peace Bridge revived dreams of prosperity for the Fort Erie area. "In two years, real estate values tripled as farmers sold their land in 40-foot lots to promoters and developers who laid out subdivisions everywhere within 15 miles of the bridge. Golf courses, private clubs, and commercial ventures took shape as investment capital was raised.

"Among the proposed projects was the International Beach Club, 2 miles from Point Abino, encompassing 75 acres of grounds, and offering...a gymnasium, private beach, fishing and horseback riding. Another proposed project was the Garrison-Goderich Building in Fort Erie, with shops on the ground floor, office on the second, a public meeting hall for 800, and the Canadian Arcadia 'where nightly those in quest of an outing can dance to good music.' The Peace Bridge Brewing Company was proposed for the old shipyard site, a $750,000 plant to employ 100 men. Proposed for the entrance to the Peace Bridge was Haddon Hall, an 8-storey fireproof building with 125 rooms and a dining room seating 400."[28]

These grand projects had one significant thing in common: not one was ever built. The Depression of the 1930s saw to that. During the Depression, 6,000 building lots reverted to Bertie Township for unpaid taxes. Farmers farmed again and the speculators left town.

But the Peace Bridge continued to bring in tourists, some for the day, some for the season. The tourists and the businesses serving them helped fuel the demand for more electricity from Canadian Niagara.

The Peace Bridge dramatically increased the number of American cottagers and tourists along Lake Erie from Fort Erie to Point Abino. Land prices soared and brochures multiplied until the Depression brought everyone back down to earth. Some areas, such as Crescent Park, did develop, but only after a few decades had passed. Crystal Beach was one of the main destinations for cottagers or vacationers. Canadian Niagara supplied the electricity to the cottagers, as it did to the Crystal Beach Amusement Park.
(Reproduction by permission of the Fort Erie Historical Museum, Ridgeway, Ontario)

When the Queen Street Service Centre opened, it was the pride of Fort Erie. With its corner turret, orange clay tile roof, and expansive show windows, it was as distinctive outside as inside. More than a building, it demonstrated Canadian Niagara's commitment to Fort Erie and the beginning of a company focus shared between Fort Erie and Niagara Falls. The opening of the Peace Bridge in 1927 and the Service Centre in 1928 brought visions of unparalleled prosperity to Fort Erie. On the far right the taxi stand and garage for car repairs and "X-tra Mileage Motor Oils" link electricity with the other great technology of the era: mass-production automobiles. The decorated Christmas tree in this photo, taken December 28, 1928, was brought in for the occasion and marked the birth of a Canadian Niagara tradition – linemen finding and bringing in a large tree for the Service Centre. (Reproduction by permission of Sandra and Frank Flake)

The New Service Centre and the 1930s

CHAPTER 7

A Local Landmark

"On April 2nd [1928] one of the most beautiful electric service buildings on the continent will open its doors to residents of the Canadian Niagara Frontier," proclaimed a *Fort Erie Times* reporter.[1] There was good reason for the excitement. The new building on the corner of Queen and Goderich was both beautiful and useful. Along with the recently-opened Peace Bridge and Canadian Niagara's high tower to carry power cables across the river to Buffalo, the service building soon became one of Fort Erie's most recognizable and well-loved landmarks. "Its graceful tower fittingly serves as a symbol of our future," the reporter wrote.[2]

Fort Erie had patiently waited for and then enthusiastically embraced electric power. Canadian Niagara had given nearby Niagara Falls the Rankine Generating Station, but until 1928 Fort Erie had lacked a distinctive architectural landmark related to electricity. Now the town had its reward. This building, which no longer stands, is fondly described by employees and retirees as "Spanish-style," largely because of its red clay roof tiles. It appears to have been influenced by the architecture of the Pan-American Exposition of 1901 in Buffalo.[3]

The reporter who saw Canadian Niagara's Service Centre as a "symbol of our future" noted that the building existed "to help residents of the Canadian Niagara border obtain the utmost use of cheap electricity in the home. For instance, a complete model kitchen has been installed, including every helpful aid from electric dishwasher to cake mixer and ironer. A domestic science expert will be in charge ready to show you their wonderful convenience — how they can relieve you of household drudgery."

"In the main foyer with its graceful domed ceiling are displayed charming bridge and table lamps just as you might arrange them in your own home. There too you will find all the newest electric servants designed to do your sweeping and cleaning, washing and ironing, cooking and refrigerating."

The showroom and displays were separate from the kitchen where the cooking classes took place, which was on a raised section of floor, two steps up from the general seating area, so that the audience could clearly see the demonstrations. It had two stoves, a sink and a refrigerator. The demonstrations also included appliances such as

At night, particularly at Christmastime, the centre shone and stood out like a beacon. From a distance, it looked like a lighthouse, only instead of pointing to shore, it pointed to a bright electrical future. Downriver at Niagara Falls, people claimed "Electricity is King and Niagara is its Throne." Now it had a throne in Fort Erie. (Reproduction by permission of FortisOntario)

electric mixers. Regrettably, the earliest known photo of the demonstration kitchen was taken in 1953, a full quarter century after it opened.

Doris Dunn — Aunt Dorrie to some — worked at the Queen Street Service Centre from its earliest days. Her sister Vera was already working for the company when she started. Another sister, Ann, a trained dietitian, was also a company employee. Both Dorrie and Ann toiled at compiling recipes that used electric appliances, many of which were considered exotic for the day. Dorrie seems to have been best known for the cooking schools and the meals she prepared for the board of directors' meetings. Ann carried out research on the nutritional value of different foods. This information became part of the cooking classes.

Audrey Troup was another trained dietitian who worked at the service centre for several years in the late 1930s and early 1940s; she left in 1943 to have a baby. She had studied at MacDonald College in Guelph. Her job involved carrying out surveys to find out customers' needs and preferences, providing advice and information over the telephone, and creating recipes. She remembers that cooking classes were offered in the fall and winter, and that 25 to 30 women attended each one.

Audrey also provided advice on other domestic uses of electricity, such as the placement of electric lamps. On occasion, she even did home calls to help customers who were not sure how to use electric stoves. She and the other dietitians shared a company car — a station wagon — for this purpose. During the war years, she also provided advice on canning, which was considered an important part of the war effort.

Shirley Athoe's mother was one of the many Fort Erie–area women who attended Canadian Niagara cooking classes and enjoyed them. "She'd come home and try the things she had learned." Children liked them too, because mothers brought home samples. They also brought home recipe cards with specific instructions for electric cooking.

The recipe cards were a direct link between Canadian Niagara's cooking school and the families and homes that the company hoped would come to use electric stoves and a growing list of electric appliances. Dozens of these cards have survived to leave a clear picture of how electricity appeared to be the only way to be truly modern in the kitchen.[4] Many of the recipes were for quick and easy fancy dishes or desserts, which carried the message that electric cooking was partly about convenience and partly about demonstrating one's modernity while entertaining (and impressing) one's friends.

The word *electric* appears constantly in the recipes. Ingredients had to be frozen "according to directions of electric refrigerator," or beaten "with electric beater." One didn't just bring something to the boiling point, one did so "using HIGH heat of surface unit on electric range." One didn't just "simmer rhubarb with water until tender" but did so "using LOW heat of surface unit on electric range." And when it was time to put something in the oven, it was to "bake in electric oven" at a predetermined temperature.

Electricity even offered new cooking operations. One recipe that called for broiling carefully explained what the term meant. This was a new technique in the 1920s and 1930s. Cooks couldn't broil using ovens heated with wood, coal, or gas, because these ovens didn't have an element that radiated heat directly downwards onto the exposed food. But with upper and lower electric heating elements in ovens, cooks could prepare food using one or both of them. Broiling was one of the new ways of cooking that came with modern electric ranges.

Inside the new Service Centre, the marvels just kept on coming. The random marble floor and tile mosaics were the talk of the town as was the almost endless array of electric goods for sale, ranging from radios and refrigerators to lamps and hair curlers. And every year, the interior was decorated for Christmas. (Reproduction by permission of FortisOntario)

Other features in early ranges have not survived in later models. For example, some recipes called for the use of a "well cooker." Shirley Athoe's mother had one: "My mother's stove top had a big hole in it that this big pot fit into. The burner was down more than a foot, maybe 14 inches and the big pot just fit in it." Compared to a large pot sitting on top of a stove, this arrangement threw less heat into the kitchen and was less likely to spill or burn fingers, particularly children's fingers.[5]

The recipes also reflect a certain scientific precision that was characteristic of the time, such as the instructions to "cook sugar and water…until syrup spins long threads (234 degrees F)." Here we have the traditional cooking test of appearance, blended with the scientific test of measurement.

Selling Appliances

On April 27, 1928, only a few weeks after Canadian Niagara opened its new Service Centre, the *Fort Erie Times* put out one of its biggest electrical ads to date.[6]

This was a generic ad adapted to Canadian Niagara by prominently listing the company name, phone number, and address near the bottom of the ad. The bottom border read "Niagara Power Service" which was part of the "Buffalo, Niagara & Eastern System," the name of the parent company at that time (it had bought out the Niagara Falls Power Company).[7] This ad is an important clue to how Canadian

"Newspaper ads helped Canadian Niagara gain more customers and encouraged them to buy more electric appliances. This ad appeared in the *Fort Erie Times* on April 27, 1928 only a few weeks after the new Service Center opened. At the time, many electric utilities sold appliances directly to their customers but usually only carried one or two brands. Canadian Niagara carred a wide range of brands and models. (Reproduction by permission of Fort Erie Historical Museum, Ridgeway, Ontario)

Niagara helped promote the use of household electric appliances. The style of the ad, along with the code number CN-22 in the lower left corner, indicates that it is a standard ad created by the larger company for use by the subsidiary company Canadian Niagara. This meant that a relatively small company had access to the entire publicity network of the larger company. This access gave Canadian Niagara attention-grabbing advertising copy at very low cost.

Canadian Niagara also had access to advertising from appliance manufacturers. Earlier ads had concentrated on Westinghouse appliances, but this ad featured the most popular ranges "made by Westinghouse, Gurney, Hotpoint and Moffat." A small company was able to offer goods from four major manufacturers, and the ad stressed that it was "in no way a clearance of obsolete styles — they are the very latest, most up-to-date models. They come equipped with high-speed heating units in oven and on surface plates. They carry a guarantee of one year's free service." Canadian Niagara and its customers benefited by being part of a much larger electrical organization. The cooking schools also benefited from the relationship with its American parent in that some of the recipes came from the larger and wealthier company across the border.

The stoves certainly were top quality. In 1936, when Canadian Niagara electrical janitor Leonard Heximer married Margaret Fretz from Ridgeway, Leonard's brother bought the couple an electric stove from the Queen Street service centre. It was — and still is — an attractive four-element Moffat, with a stove-top-height oven. Nearly 70 years later, it is still used every day in the Heximer household, although three of the four elements on top have been replaced.

Appliances were also sold from a smaller store in Crystal Beach, which was opened in 1928, shortly after the Fort Erie Service Centre. This store, which was at first open only during the summer months, did particularly well selling refrigerators to summer residents.[8] It was also a place where customers could pay their bills.[9]

The display windows in the Queen Street service centre were an important advertising and public relations tool for the company. Few surviving photos show how they were decorated, but we do know that for many years professional window decorators came from the parent company to install Christmas displays that attracted great attention. Audrey Troup remembers one Canadian Niagara employee, Paul Bloomfield, who did some of the Christmas displays. "He could design anything and make it look beautiful." The windows also displayed electric appliances and products throughout the year. Lamps were an important item, and surviving photos from the parent company show window displays of lighting with information on the importance of both proper wattage and placement of lamps. Canadian Niagara likely had similar promotional material either in its window displays or as part of the cooking school programs.

Harvey McAteer began working at the Fort Erie office in 1948 or 1949, when the company was starting to get out of selling electrical goods, but he remembers "a showcase about eight to ten feet high and about fifteen feet long, angled front, sliding glass doors. There must have been 45 lampshades in there, on shelves, different sizes and different shapes and different colours for all the lamps. But that was kept closed, because they'd get dirty very easily." He also remembers that appliances came in "by the boxcar-load…there weren't many trucks in those days, either. It was all train." He and his wife Shirley were careful shoppers and knew that if you bought something from Canadian Niagara, "you got a good price and it was on your bill and you could spread it out."

Canadian Niagara not only sold the appliances but offered servicing too — "four guys doing maintenance work on fridges, stoves, fans, lights, shavers, everything — anything you wanted, they had it." If you bought it at Canadian Niagara, Canadian Niagara maintained it. Eventually the company got out of selling and servicing electrical goods, for the same reason they had entered it in the first place — a concern for good community relations. As local businesses gradually began to sell and service electric appliances, Canadian Niagara did not want to be seen as unfair competition. In Fort Erie, Chris Wren, a former company employee, opened Wren Electric

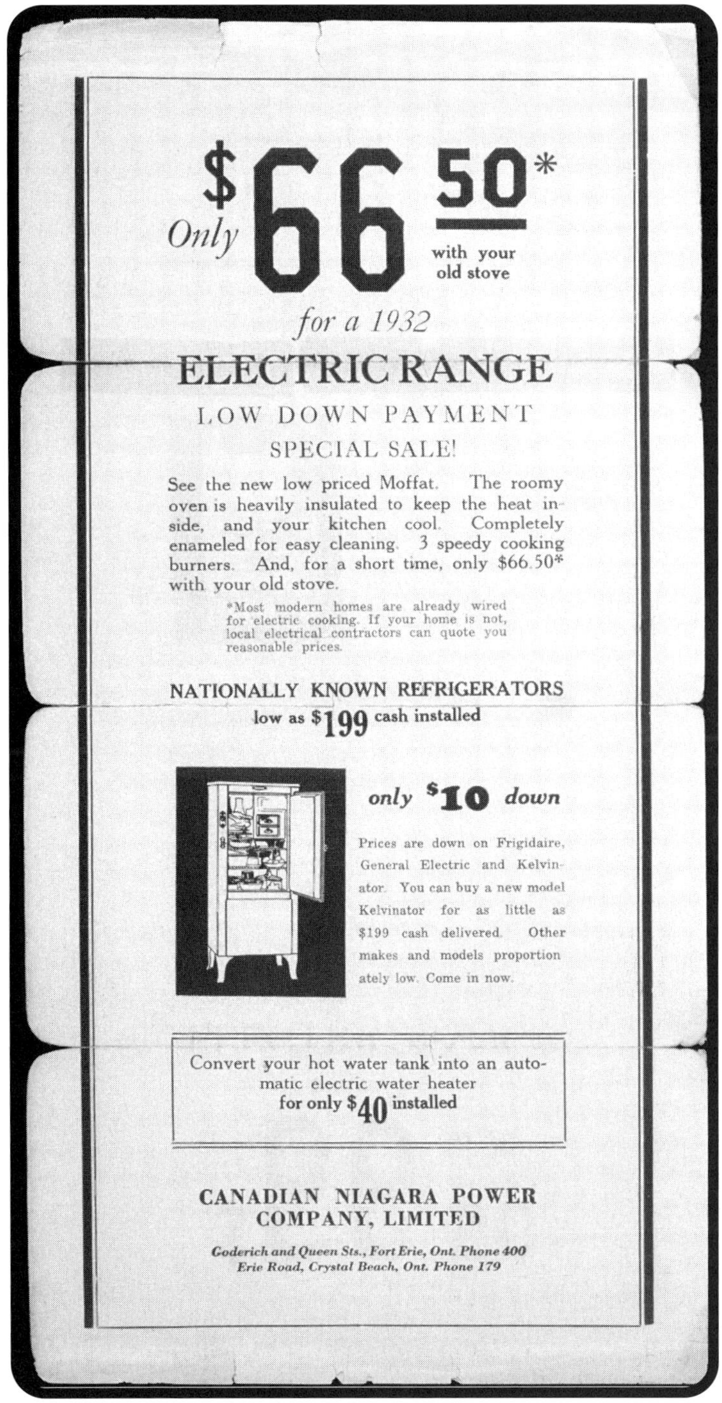

In 1932, electricity bills from Canadian Niagara were sent out as postcards so this clever piece of advertising could not have been included with bills. It may have been hand-delivered to customers. In addition to displaying good graphic design, it offered customers a $10-down, time payment plan, a trade in for an old stove, and a choice of name-brand appliances.
(Reproduction by permission of John Burtniak)

As the number of customers increased, so did the number of poles and wires. Canadian Niagara's Reo Speedwagon helped business in several ways. This truck carried the crew and equipment, towed the trailer with new poles, and carried an advertising sign that reads "25$ for your old stove. Cooking demonstration this week." The crew wore the work pants, work shirts and caps or felt hats of the day. Hard hats and protective clothing would come many years later.
(Reproduction by permission of Beverley and Ken Barnhart)

and leased window display space in the company's service centre. Windows once used to promote Canadian Niagara's appliance business now advertised a locally owned sales and service business.

Meanwhile, another part of the building had been put to a new and interesting use. In the company minutes for November 12, 1936, is a brief note: "Resolved then adopted that the Company lease to John E. Lawson [vice-president at the time] in consideration of an annual rent of $100.00 living quarters comprising a living apartment on the second floor of the service building at the corner of Goderich St., for a term of one year, to commence November 1, 1936, and thereafter from year to year, until cancelled by either party."[10] Lawson had no intention of living in the building. This was a long-standing legal fiction: if the upstairs was classified as a dwelling, it would be legal to serve liquor there without a special permit, which made for more lively board meetings and corporate entertainment. Lively upstairs meetings were a cherished tradition at the Queen Street service centre.

The End of a Landmark

When Shirley Athoe joined the company in 1957, it no longer sold appliances, but the big open floor space was still there. When a new manager came in a few years later, he found the existing work space for the office workers was too crowded. He also wanted a separate office. Shirley recalls that a "big part out on the floor that had originally had appliances and stuff on it, was not being used for anything by this time, because they no longer sold them. So he decided that they'd make that into the office for all the girls." The construction work cut off the old demonstration kitchen, which was no longer used, and the addition of a new wall turned it into a staff coffee room.

The Board of Directors met in the Service Centre for the last time on July 24, 1987. It was a sentimental moment and they reviewed the history of the building that traced its roots back to when Canadian Niagara was "acquiring property and rights of way for the Bertie Hill and Queen Street towers" and "a number of lots were also purchased in Bridgeburg [Fort Erie]. One of these lots, at the southwest corner of Queen and Goderich, became a home for the local CNP manager, and another, at the northeast corner, was to be the site of the first local office." After the sale of the building for $550,000, the company rented it until November, when the last few employees moved into the new Service Centre on Bertie Street.

The closing of Queen Street ended a chapter in Canadian Niagara's history. Shirley Athoe felt disoriented. Decades after her first day of work there, she recalls it as one would an old friend or a loved one. "I just admired the building so much, it was just a unique place to work. And everybody that walked in the door would go, 'Ahhh.' To me, it should have been a museum or something." She remembers childhood trips to Toronto to look at the wonderfully decorated Christmas windows at Eaton's and Simpson's; they were enchanting but temporary, and couldn't measure up to the "good materials...the tiles, the marble, everything" in the Service Centre.

Shirley was one of the last to work in the old centre, which stayed open for several days after most employees had been transferred to the new centre on Bertie Street. Shirley was there on the final day, a Friday. "I cried. I think we all did. We were very sad. I spent the last day in the building...we had to keep it open until five o'clock. I didn't even have a chair to sit on...people came in at the last minute to pay their bill. I stayed there for that purpose. And there was nothing else going on, the building was totally empty." At the end of the day "we went out and the doors were locked and the keys were turned over." She didn't know how upset she had been until Monday morning, when the books didn't balance and she realized, "When we left I forgot to take the money that was in a drawer.... I left the money right there.... We had to make special arrangements to go back and get the money."

New owners converted the Service Centre into the Power House Border Café. A few years later the building burned down. Shirley Athoe was one of the many who grieved at the loss. On the day it happened, just as she was leaving for work, her son called to say there was a fire down on Queen Street. She "drove right by the tower at the top of the hill and there it was burning. I just sat there and cried. I was overwhelmed.... I was never late for work, unless I had a good reason. And it didn't even occur to me that — hey, you're going to be late for work.... I went into work late and a lot of them hadn't even heard about it. And they said, 'What's the matter?' I said, 'The power company burned down...on Queen Street.' They couldn't believe it. That was so sad...you felt like you were a part of history, just working there."

For a second time the life had been sucked out of the Queen Street Service Centre. There would be no third time. The wreckers descended on the site and tore apart the charred carcass with the copper window frames. A landmark was gone for good. But in the 1930s, no one could foresee the end of Queen Street. The company faced other challenges.

Takeover Threats and Public Relations

Although the country was in the grip of the Depression, the need for power continued, and Canadian Niagara's records do not suggest a company that was in financial danger. The company routinely announced dividend payments for its shareholders throughout the 1930s, and even provided money for unemployment relief "in and for the various municipalities, towns and villages in which this corporation conducts its operations,"[11] thereby reaching out to its own customers, many of whom were in difficulties. Only at the Depression's lowest point, 1933, did the level of power generation drop significantly.[12] This suggests a sudden, but short-lived, drop in demand, probably from the company's industrial customers. Since Canadian Niagara consistently charged low rates for electric power, domestic use generally remained stable throughout the Depression — too many customers had become accustomed to using electric lights and cooking with electric stoves.

Any threats to Canadian Niagara came from another source: the trend towards public power. One Niagara Falls resident, Alonzo Robertson, a contractor, suggested that Canadian Niagara should be a municipal utility. At a meeting of the Rotary Club on March 20, 1934, he proposed establishing "a trust fund for the purchase of the assets of the Canadian Niagara Power Company in Canada." He blamed high Ontario Hydro power rates and low Ontario Hydro tax income for the level of municipal debt and suggested that with his proposal, the municipality would soon free itself from debt.[13] No one seems to have taken up his idea, which didn't make sense at a time when money was so tight. Canadian Niagara already paid full municipal taxes, unlike Ontario Hydro or municipally owned utilities, and it ensured that its customers paid the lowest rates for electricity in the region.

More serious takeover proposals came from the Hydro-Electric Power Commission of Ontario. From time to time, a Hydro spokesman would publicly mention the possibility of buying Canadian Niagara. The Commission's chairman, T. Stewart Lyon, apparently did so at an annual banquet of Hydro employees at the Brock Hotel in Niagara Falls in April 1936.[14] Almost from its creation, Ontario Hydro had been buying power from Canadian Niagara, as it did from many other private or municipal generating companies. Over time, Ontario Hydro bought up many of these smaller producers, and may have assumed it was only a matter of time before it swallowed Canadian Niagara.

It would have been an unpopular move. At the time, Ontario Hydro was not particularly well regarded, either locally or elsewhere in Ontario. Locally, it had made enemies by refusing to pay adequate taxes to Stamford Township, where it had large land holdings. Farther afield, it had tarnished its reputation by "secret" agreements with Quebec power companies to buy power it did not need, resulting in higher-than-necessary power prices for consumers.[15] Ontario Hydro was often in the news, and the stories were seldom complimentary. Some journalists seemed to make a career out of writing about its scandals and improprieties. Everyone appreciated the electricity it provided, but it was an easy target for criticism.

By contrast, Canadian Niagara had a good reputation and excellent customer relations. An article that appeared in the *Niagara Times-Review* puts it very clearly:

The new weir which opened in 1937 raised the water level at the inlets into the forebay. Now Canadian Niagara had a weir, as did the Toronto Power Company. Nonetheless water level and ice problems persisted. (Reproduction by permission of the author)

That remarkable development, the Ontario Hydro Electric System, is held up as a shining example of public enterprise…[and] those living in areas served by the 'Hydro' are often unaware that Ontario has other distributors of electric power.

This was impressed on us during a recent conversation with a new resident of Fort Erie who moved here from Toronto a few months ago. He expressed his amazement at the extent of what he described as the 'friendly feeling' of Fort Erie citizens toward the Canadian Niagara Power Company. As an ardent advocate of public ownership, he just could not get over it. Here was a privately owned company, enjoying a local monopoly in the sale of a virtually essential service and, instead of being hated as a 'commercial octopus,' the people of Fort Erie actually liked the Power Company and all its works! Our friend demanded that we explain 'how come.'

We did what we could; but, after all, whether a corporation or an individual is highly regarded usually depends not on one thing. It is more likely to be the result of a long series of good deeds extending over a period of many years. In our judgment, there is no single factor which can be cited as the cause of the Power Company's good public relations. Rather, it is the result of long years of [service provided] courteously and efficiently — which is generally conceded to be a good course for any business to pursue. But the Canadian Niagara Power Company has actually pursued it.[16]

Good public relations are indeed good business, and Canadian Niagara benefited from them. For example, in 1935, Fort Erie held a vote to approve a franchise allowing Canadian Niagara Power Company to use the streets of Bridgeburg, which was now part of the Town of Fort Erie, for the next 50 years; this was a renewal of a previous 25-year franchise. Voters overwhelmingly supported the company, endorsing the franchise by 513 votes to 115. As the newspaper noted, "This action of local voters confirms the general feeling of satisfaction felt towards the service and rates offered by Canadian Niagara Power Company."[17]

Of course, the company didn't mind reminding Fort Erie residents how well off they were. In an advertisement that appeared the date of the vote, the company pointed out that the average power rate for domestic consumers was $1.75 per kilowatt hour, compared to $1.77 for the rest of Ontario and $2.18 for the rest of Canada. It also pointed out that many Fort Erie residents were employed by the company, that the municipality benefited handsomely from the taxes paid by Canadian Niagara, and that the company provided free power for lighting streets and public buildings. Finally, the company even offered a chauffeur service for voters on voting day: "Telephone 400 and a car will be sent promptly to take you to and from the polling place."[18]

So why did 115 people vote against the bylaw? An internal memo written by J.E. Lawson as a post mortem on the vote suggested "many did not understand the By-law, having the opinion that it gave the Company the right to increase their rates." He added that in one area, "the biggest part of the population is on Relief and are therefore ready to vote against anything," and that in another, "the railroad workers were against the By-law, and it is hard to understand why they should be."[19]

Railroad workers were particularly hard hit during the Depression, and this may have accounted for their contrariness. Walter "Sid" Morris started with Canadian Niagara as a meter reader in Crystal Beach in 1939. Two weeks after he started, his father, who worked for the railroad, told him that he could get a good job as a fireman on a steam locomotive. Sid replied, "Dad, you know what it was like during the Depression. The railroad was pretty rough. But people will always need power. I think I'll stay at Canadian Niagara." He stayed for 43 years.

New Construction and a Nearby Disaster

Before 1939, Sid Morris had worked for Northern Construction on a project for Canadian Niagara, so he knew the company. Canadian Niagara had contracted with Northern Construction to build a 900-foot weir beside the intake for the powerhouse. "Its purpose is to restore the level of water in the river at this point to that which existed in 1905 when the station was built. A protracted period of abnormally low water levels in the Great Lakes, together with diversions of water at upstream points by power plants built after the Rankine Station was completed made the weir necessary. But the weir will serve a double purpose because it will also redistribute the flow of water over the Canadian side of the Horseshoe Falls and preserve the scenic beauty of that famous cataract by supplying a greater quantity of water near the Table Rock shore line."[20]

The construction was as challenging as the original construction of the forebay. In December 1936, workers built a pier consisting of 51 separate steel frame structures, extending out into the river to support the tracks for the construction crane. The pier incorporated stop logs that could be lifted as needed to allow ice to pass through. Behind the pier, a cofferdam was installed by lowering steel frames into the river and surrounding them with steel sheets. The cells created in this way were pumped dry, and then the river bed was cleaned down to the solid rock. Workers drilled into the rock to insert steel dowels to hold the weir, which was made of concrete. The whole thing was completed by December 1937.

A month later, January 25 and 26, 1938, the river was in the grip of an ice jam, which packed the river below the falls, destroying houses and docks on the riverbanks and filling the powerhouse of the Ontario Power Company at the foot of the falls with ice. Canadian Niagara, above the falls, was much less affected by the ice and merely lowered its output for a few days. By January 29, it had returned to full capacity.

Meanwhile, however, the pressure of the ice in the river was threatening the Upper Steel Arch (also known as the Falls View or Honeymoon) Bridge. On January 26, workmen tried to remove the ice pressing against the bridge abutments but had to abandon the effort. Bridge traffic was halted, and at 4:20 pm on January 27, the ice pushed the bridge off its abutments. It buckled into a giant W shape, with a sound that a local reporter described as a "mighty belch," "Those two sections at the ends slowly sloped downward in a shower of powdery snow, dirt, tumbling rock. Then the two sections inside these began to rise as though in a frantic effort to escape that mangling death below, attempting to soar skyward. It was a last, futile gesture. Slowly — it seemed minutes though it was but split seconds — the giant folded into the masses of ice."[21]

The bridge, which was uninsured, had once carried the trains of the International Railway Company, as well as several power lines. The IRC president quickly announced that a new bridge would be built on the same spot, even before engineers had figured out a way to remove the debris. Meanwhile, the collapse and debris were grand public spectacles for residents and tourists alike. Special trains even brought tourists from Buffalo to see the sight.[22]

The Changing of the Guard

The Depression years were difficult for people on the Niagara Frontier, but the records do not suggest that anyone was laid off from the company, and several new people were hired. For example, Leonard Heximer, whose father George had been a station operator at the Rankine powerhouse, started with Canadian Niagara in 1934 as an electrical janitor.

Nevertheless, for Canadian Niagara, the 1930s were a period of transition. In 1933, A. (Alexander) Monro Grier retired as president and became chairman of the board. He had joined the company in 1901 as secretary and solicitor, had become vice-president in 1905, and had been president since 1919. Grier, like his predecessor Wallace Nesbitt, was a lawyer; his successor Alex D. Robb was an engineer who had worked his way up from the position of mechanical inspector. Grier was a link back to the days of Rankine and to a tradition of litigation and financial battles; Robb was a link forward to a company that was more focused on technology.

Grier came from a background of money and culture. His father had been a physician, one brother was a portrait painter, and the other ran a school of landscape

The Niagara Frontier is famous for its winter storms. It might be snow or it might be ice, hail, and freezing rain, but every decade or so, it seemed as if there was a storm worse than anything that had come before. The snow storm of March 1936 left a heavy wet snow load three feet deep that crushed the roof of the Peace Bridge Arena in Fort Erie. Lineman Louis Rouse worked during that storm but also found time to take some snapshots.

The lineman was used to climbing poles, but climbing a snowbank this high must have been a new experience.

and marine painting in St. Ives, Cornwall. Grier belonged to the Arts and Letters Club and the Empire Club in Toronto (although he lived for a time in Niagara Falls, he seemed to have lived in Toronto while he was president), and was a noted public speaker. During the First World War he addressed audiences on patriotic topics; later and through his retirement, he spoke on more informal subjects, such as Gilbert and Sullivan. Grier's voluminous correspondence, which is preserved in the National Archives of Canada, is often amusing to read, with a blend of sarcasm and dry wit.

Grier remained active in his retirement — one year later, when he was 74, he married for a second time (his first wife had died in 1913) and he stayed on as chairman of the board until 1947, when he was 86. He lived to be 94, dying at his Toronto home in 1954.

Alex D. (Alexander Duffield) Robb was to be president for only three years — he died suddenly of complications from appendicitis in 1936 at the age of 58. He had been born and brought up in Pennsylvania, and attended Lehigh University. He worked his way up through the company from the position of mechanical inspector to those of assistant superintendent, superintendent, vice-president, and president over the course of 30 years. He had even married a fellow employee — a Miss F.C. Durdan, who had worked for Canadian Niagara for 15 years herself.[23] According to his obituary,

Louis Rouse knew the danger that came from the snow load on the lines. This photograph gives an idea of just how much snow a thin conductor can carry. Eventually, the weight would become so great that the line would fall. The biggest danger was from ice, which builds a heavier load than snow and often brings down branches or entire trees, which tear down power lines as they fall.

(Reproductions by permission of Norm Rouse)

> His work was his passion. His companies [he was also president of the Niagara Electric Service Corporation, the Niagara Junction Railway, and the Niagara, Lockport and Ontario Power Company, and vice-president of the Niagara Falls Power Company, the Buffalo General Electric Company, and the Tonawanda Power Company] were more than mere businesses to him, they were living institutions with the possibilities of growth and usefulness that inspired his enthusiasm and drove him to the full extent of his capacity in their service.[24]

His successor was Norman R. Gibson, who remained in office from 1936 to 1946. He too was an engineer, who had worked for Canadian Niagara since 1918 and had been chief engineer before becoming president.

However, the changing of the guard was taking place at all levels of the company. The minutes are filled with announcements of the retirements of long-serving employees, some of them well past the usual retirement age of 65. Here are a few typical examples:

John Schoenacker, 72, served for 21 years, retired in 1932
George R. Heximer, control room operator, 60, served almost 28 years, retired in 1934
Fred Lock, shift foreman, 65, served for more than 38 years, retired in 1936
Charles J. Smith, shift foreman, 70, served for more than 37 years, retired in 1936
Cornelius Callaghan, thrust deck attendant, 80, served for more than 31 years, retired in 1936
Thomas C. Pearson, governor attendant, 65, served for more than 30 years, retired in 1936
Herman Robida, pipefitter, 79, served for 42 years, retired in 1938
David M. Weaver, mechanic, 65, served for 36 years, retired in 1939

The list goes on. Many of these men had been at Rankine since it opened. Given the spate of retirements, the company approved a new retirement plan in 1938, rather than buying individual annuities for retiring employees. The new plan introduced a standard formula for calculating pensions. The company had already brought in an employee insurance plan in 1933.[25]

In 1937, the Niagara Power Pioneers was formed to honour long-serving (more than 25 years) and retired employees of Canadian Niagara and its parent American company (by this time the Buffalo, Niagara and Eastern Power Corporation had been acquired by Niagara Hudson, which in the 1950s became part of Niagara Mohawk). A program from a dinner held in 1943 shows a list of more than 100 names of working employees who had served between 25 and 45 years at the company, and almost 50 retired employees who had served more than 25 years each.

But it wasn't just the employees who were retiring. The equipment was beginning to feel its years, and in January 1939 the company dismantled and scrapped an excitation plant in Rankine that had been installed in 1904. People were getting tired too, tired of the seemingly never-ending Depression. But none wanted what followed, even though it helped pull the economy out of the economic slump of the 1930s.

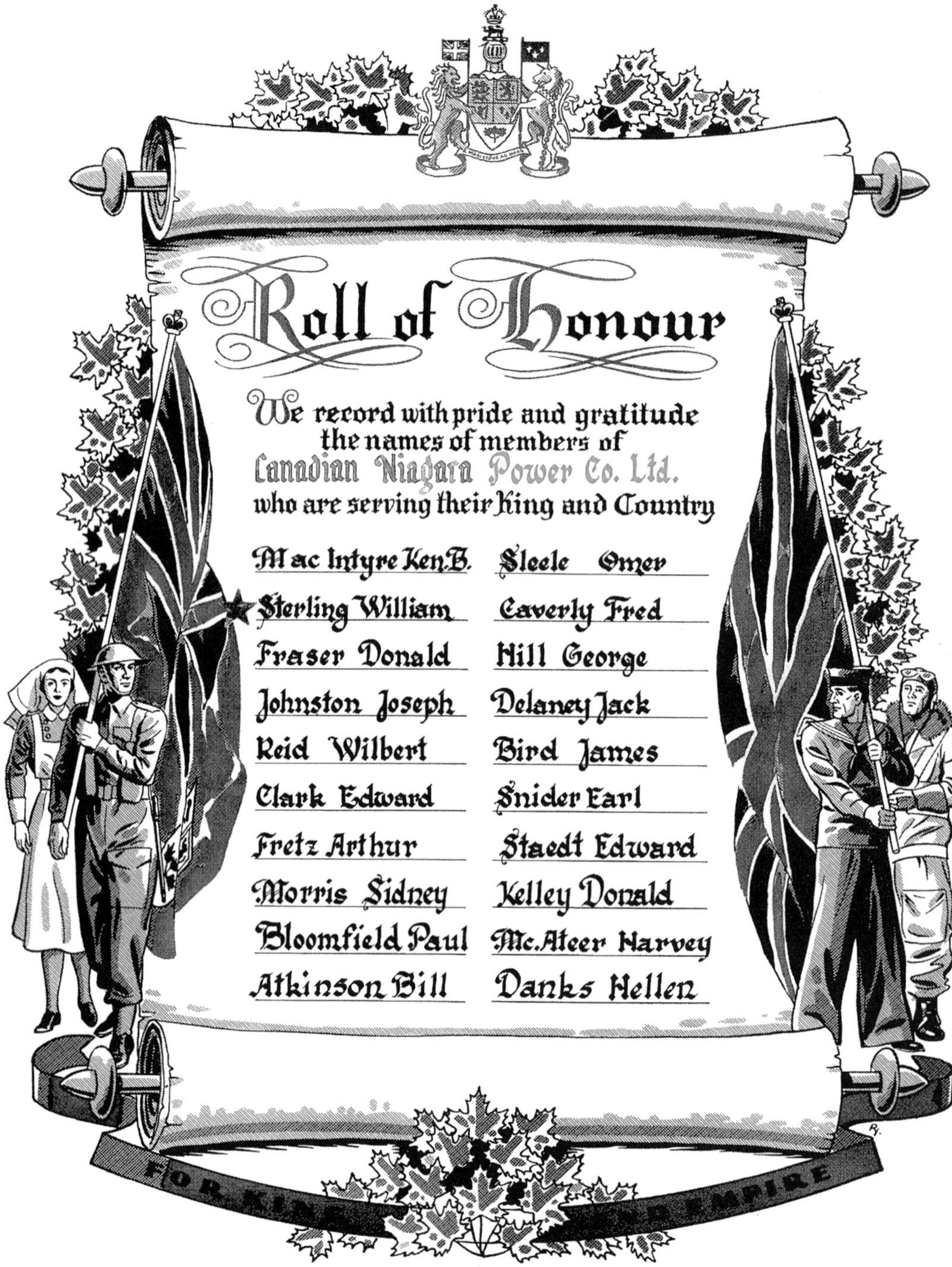

The Roll of Honour lists the Canadian Niagara Power Company employees who served in the forces during the Second World War. The star beside the name of William Sterling, Scotty to his friends, indicates that he was killed in action. The Roll of Honour does not mention Arthur Tilton but his death was also deeply mourned. He was the husband of telephone operator Josie Tilton. (Reproduction by permission of FortisOntario)

The War Years, 1939–45

Chapter 8

The outbreak of the Second World War put the Queen Victoria Niagara Falls Park Commission in a difficult position. Unlike many parks, it had no government revenues or endowment funds. It depended heavily on tourist dollars, but these had fallen during the Depression and had been further reduced after January 27, 1938, when ice tore out the Upper Steel Arch Bridge (also known as Falls View Bridge, Honeymoon Bridge, or Upper Bridge) and made it harder for American tourists to visit the Canadian side of the Falls. Now, since the park housed major wartime strategic assets — all the generating stations on the Canadian side of the Niagara River — parts of the park were considered prime targets for enemy sabotage. Tourism relied on open access; high security meant restricted access. But in the minds of the Park Commissioners there was no question: tourism, and reduced revenue, had to give way to the war effort, just as it had in the First World War.[1]

First came the decision "to barricade the main roadway in front of the Park Restaurant and reaching to Table Rock House, which is the largest revenue producer of the Commission's Concessions." Table Rock House stayed closed throughout the winter of 1939-40. But this was the time of the Phony War, when very little seemed to be happening overseas. The Commissioners began to feel that they had overreacted. So in spring 1940, at a cost of about $10,000, the barricades were removed to allow traffic to reach Table Rock House.

The easing of restrictions was short-lived, as the Phony War turned into war that was all too real. As the Commissioners reported, "when Germany invaded Holland and Belgium and carried on total war against France and Britain, the old barricades were re-erected…it being felt essential that no chances should be taken which might result in the Power Houses being sabotaged." The Commission would try to continue "its programme of preservation, restoration, commemoration, beautification and attraction," but its highest priority would be the "protection of the vital Hydro-Electric plants which are located within its domains."[2]

Life on the Niagara Frontier had changed. And Canadian Niagara was one of the key players.

Don Kelly, who joined the company in 1941, remembers the civilian guards, who were later replaced by armed military guards at all of the power stations as well as the Welland Canal, the Chippawa–Queenston power canal and the bridges. The Niagara Parks Commission reported, "During the summer months, a battalion of active service men, in training in the Niagara training area, was at all times maintained within the grounds of Queen Victoria Park. During the winter months, the tented city was abandoned for more comfortable quarters, but thorough patrols were maintained at all vital points" in the park.[3]

Was sabotage likely? Successful measures against sabotage (or, today, terrorism) sow the seeds of doubt. Were the guards really needed? Were the hydro plants really in danger?

Russell "Sandy" Sanderson was one of the many who could hardly wait to get overseas and fight the enemy. While guarding German prisoners in Holland in April 1945, he met one who spoke English, had been to Niagara Falls, and was familiar with the generating stations, including Canadian Niagara, as well as the power canals and the Welland Canal. Sandy was astounded by the depth of his knowledge. The prisoner had been chief steward on a German passenger ship that sailed regularly to New York City. As soon as it docked, he would take the train to Niagara Falls to do his other job: photographing industrial sites on both sides of the river for German intelligence. He had not volunteered for the job; he had been told to do it. But as he explained to Sandy, things hadn't worked according to plan. Intelligence wasn't always organized properly in Germany, and his information had never been used.

Working Flat Out

Even if they've had a late night and too many beers, soldiers are expected to snap to attention bright and early the next morning. Similarly, after a decade of economic depression and high unemployment, the onset of war forced industry to snap to attention. Canadian Niagara was called on to help meet the ever-increasing electric power needs of both Canadian and American industry. (Although the United States didn't enter the war until 1942, after the bombing of Pearl Harbor, American industry supplied essential war materials right from the start of the war.) But just as power demands rose, intense patriotic feelings called Canadian Niagara employees into the services. The combination put a heavy burden on the company.

Every industry felt the impact of the war. In Fort Erie, Fleet Aircraft was the most visible wartime producer, and a Canadian Niagara customer. The company, which had been founded in 1930, specialized in the design and manufacture of airplanes for military training and civilian flying. It made fuselages and wings for the Lancaster bomber and employed nearly four thousand workers, most of them women.[4] Early in the war, the company also made aircraft such as the Fleet Finch, which was used in training but later phased out and replaced with the Fairchild Cornell.[5]

Canadian Niagara truck No. 41 stands on the cobblestones in the storehouse yard above Rankine station in Niagara Falls. The company houses visible on the right have since been demolished. Beyond the houses was a guard house put up during the Second World War. On the left is Willard Benson, a much-admired welder who was the first man at Canadian Niagara who could do arc-welded repairs on turbine runners in place overhead. Next to him is Lester Myers, the purchasing agent whose job was made more difficult by wartime regulations and restrictions. Next is Henry Schangula, a storehouse janitor at Niagara Falls, and on the right is Sid Morris, home on leave from the services. (Reproduction by permission of Sid Morris)

The 1943 photograph of Freddy Gibbs using a compressed air grinder or wire brush to clean a generator down for repairs represents what everybody had to do during the war: keep on working. (Reproduction by permission of FortisOntario)

The need to push further and faster than you ordinarily would was part of the wartime experience. Reports from the Rankine Generating Station give this sense of effort. July 1942: "We are running the plant wide open now and squeezing out the last ounce of power."[6] March 1943: "We are now feeling the shortage of manpower at the Plant, but, never the less, we are running full blast turning out kilowatts-hours to defeat the Axis." July 1943: "The plant is still hitting on all eleven [generators], 24 hours a day and seven days a week."

The generating plant had not been designed to have all 11 generators working at the same time. The 11th was a spare to fill in for whichever generator was down for repairs or maintenance. Using all 11 generators made it harder to plan for routine maintenance. In September 1942, the men in the powerhouse had to overhaul No. 8 unit. Then No. 6 showed its age and demanded attention. "On October 27th [1942], we had a breakdown in the armature winding of No. 6 Generator. This was an old original winding, so it was decided to install a new one.... The job was completed by November 9th, and the unit was put on load on November 10th just two weeks after we shut it down." The next month it was No. 9 Generator, which blew several coils. "Then it was discovered there was a ground on the field poles. One of these was removed with much difficulty, repaired and put back on. After the machine was started, another ground developed in the field, but this was repaired without taking the rotor out."

Until relatively recently, Canadian Niagara had a tradition that whatever needed doing, employees could do it. The original "spider" poles carrying power from Niagara Falls to Fort Erie had been modified to make them tall enough to carry a third set of lines. However, in the winter of 1940–41, the men were using a rental crane to erect new towers to replace the old. Here they are putting a box across to join the parallel towers, which they will then build higher, before stringing the cables. (Reproduction by permission of Sid Morris)

The new year, 1943, brought "another break-down on No. 8 Generator. One coil burned off in the iron and the flame spread to several adjacent coils, burning the insulation quite badly. Tom Sills and his gang spliced in one new half coil and had a nasty job taping the other burned coils without removing them from the slots."[7]

Another chronic problem was winter ice. When the ice in the forebay got too thick, it reduced, or even stopped the flow of water through one or more of the penstocks and the turbines they fed. Sometimes so little water came through that the turbine and the generator it drove stopped. Now, a generator is the reverse of an electric motor. In an electric motor, electric power is fed in and the motor turns. In a generator, the turbine provides the rotary motion and the generator produces electric power. In peacetime, when ice stopped the turbine from turning, electric power from the American side would be sent into the system to make the generator behave like an electric motor. It would then move the turbine, allowing for ice to be cleared away. But wartime regulations forbade Rankine Station from importing power from the United States.

Retired operator Don Kelly remembers what he and others called "political relays" which made life even more "nerve-racking" for the operators in the Rankine Station. Relays are a specialized type of switch that can stop or allow the flow of power. They often work automatically. With wartime regulations, relays had to be set to prevent any power entering Rankine station from the United States. If one of Rankine's generators started to slow down and was on the verge or needing power from the U.S., the relay simply tripped and prevented any power from coming in. Operators tried to guard against this situation because it was difficult to restart a generator. Generators, relays, and other equipment could be damaged if anything went wrong. And when wartime enlistment in the forces left Canadian Niagara with a shortage of experienced personnel, it made the situation especially difficult.

The powerhouse had a sluiceway to prevent the ice from entering the penstocks, but it too could become jammed with ice. So in fall 1943, Canadian Niagara employees "cut an ice run through the triangular piece of concrete at the north end of the forebay, between the Power House and the sluiceway."[8] Given the wartime shortage of labour, outside contractors were simply not available, so the ice run was just one more job for overworked company employees. But management knew how to show its appreciation.

"On Monday evening, December 13th, Mr. Lawson and Mr. Bracken threw a party in the Chippawa Hotel for the men who had worked on the new ice run. A nice turkey dinner was served. Also, there was a bountiful supply of liquid refreshments on tap and the fellows all enjoyed it very much."[9] With wartime rationing limiting individual liquor purchases to 40 oz. per month, that bountiful supply must have been particularly welcome.

It was all part of keeping up output. Canadian Niagara was a war-essential service industry and demand kept increasing. For example, as Carborundum increased output, it needed more water and ran a new 10-inch water main from the Canadian Niagara standpipe beside the transformer house. Canadian Niagara had to build a new power line to Carborundum to supply additional power.[10]

Canadian Niagara also helped other utilities, particularly during bad storms. One fierce storm started on the night of December 30, 1942, and followed the all-too-familiar pattern of a big snowstorm which developed into sleet. Sagging and falling trees and limbs soon started the usual short circuits. Fort Erie was hit hard, but the situation was even worse in Cornwall. "The poles in that locality fell down like tenpins. They sent us an S.O.S. for help and Perce McAteer and Louie Rouse were dispatched 'Toot Sweet' to help repair the damage."[11] McAteer and Rouse were gone for more than two weeks.

The Rankine Generating Station generated power at 25 cycles per second or 25 hertz. But during the Second World War, and for a few years after, Canadian Niagara was also sending out power at what one operator called the "oddball" frequency of 66 2/3 cycle. The power went to the Ontario Hydro transformer station on Murray Street in Niagara Falls and then to customers needing that frequency. Operator Don Kelly recalls that the frequency converter housed in the transformer house was "temperamental." Wartime evidence suggests the same thing.

The frequency converter consisted of two interconnected units in the transformer house. An AC generator is the opposite of an AC motor, and they can be used somewhat interchangeably. Power from Rankine at 25 cycles turned the first unit as an electric motor, which then turned the second unit as a generator. The frequency

The annual Power Pioneers of Niagara dinner recognized the importance of current and retired employees. It was an exclusive club, open only to those with 25 or more years of service. In the spirit of the war and defence of Empire, the 1943 menu started with Empire Fruit Cocktail. Dr. Norman R. Gibson, president of Canadian Niagara, hinted at the future in his remarks: "The Engineering Department has locked up in its files complete designs for a unique plant which I can assure you will surpass anything that has yet been done on the Niagara River." The Famous Little Niagara Players ended the evening with a "History of Niagara Power's Early Days." In between, all were reminded of the importance of electricity and that "there is no war production activity more essential than the one in which you men and women are engaged." At this event, Ellis Holditch, who would later die on the job, celebrated 26 years of service. William N. Rickard, creator of the *CNP Tattler*, was in his 36th year at CNP. (Reproduction by permission of FortisOntario)

of the generator output depends on the speed at which it turns, so it was a matter of getting the proper speed in the motor turning the generator. As one report notes, "We had quite a fire in the motor of the frequency changer a couple of weeks ago. It sure made a mess of it and will require a new winding."[12] Given the manpower shortages, it took about four months to get the frequency changer back in operation.[13]

The fire probably wasn't the first and it definitely wasn't the last in the frequency converter. Don Kelly remembers a "scary sound, a high-pitched scream" when the frequency became a "runaway" and went into "overspeed." The fires he remembers were caused by a short circuit on lines feeding out from the 66 2/3 generator. With a short circuit, power feeds back to the source, but relay switches are supposed to trip to prevent the back-surge of power. They don't always trip in time. Also, with a short circuit, the current is always higher than in the normal circuit. When the higher current is too much for the generator to handle, it blows, destroys the insulation on the coils, and a fire starts. The fire might start small, but it soon grows because the rotating generator fans the flames. This generator was driven by a 25 cycles-per-second motor from another source that was not connected electrically to the generator, so it could keep driving the motor connected to the generator by a belt for quite some time. The fire damage could be serious. It was one more problem to deal with in a difficult period.

The Home Front

The war took a heavy toll in terms of money, materials, personnel, and disrupted lives. As each employee signed up, it meant one less person to keep Canadian Niagara generating power and sending it to industries that were running flat out to meet war production needs. At the same time, everything was in short supply — except red tape. It was a stressful time, and occasionally, people got a little testy. Sid Morris remembers that during the war, general manager John Lawson received a letter from the Buffalo [Niagara Mohawk] purchasing agent, asking Canadian Niagara's purchasing agent Lester Myers for the "serial numbers of all of Canadian Niagara's tires." Lawson tore the letter to shreds, saying, "I'm running this place, not Buffalo."

But there was more goodwill than testiness. As it had in the First World War, the Canadian government borrowed money from its citizens by selling Victory Loan Bonds. Quotas were set for various workplace units and there was fierce but friendly competition to see who could exceed them. In November 1942 Canadian Niagara "had another Victory Loan Drive in the plant recently and the quota for our employees was $5,875. We went over the top to the tune of $6,800, which was excellent, considering the high income tax deductions."

Canadian Niagara employees also held annual Red Cross Drives to raise money and employees made regular visits to the Blood Bank. By March 1944, Leonard Heximer appeared to be in the lead. He had donated "seven times, with a total of over 3000 c.c.s of blood."[14]

As the war went on, the Victory Loan quotas kept rising. So did the enthusiasm and resourcefulness of employees. In May 1943 Canadian Niagara held its Fourth Victory Loan Drive. "[We] have accomplished something which we at first thought impossible. They set us a quota of $13,600, which we thought was outrageous. However, all the fellows and girls got back of it and at we have at present oversubscribed our quota by $850.00. Some of us may be wearing patches and eating grass this summer, but we have done a good job and are proud of it. We had seven teams set up, headed by the following captains: Percy McAteer, Art Snider, Bill Hunter, Bart Atkinson, Art Clymer, Hugh Wilson and George Brown. McAteer's team went over their quota, but I am sorry to relate that he stooped to some pretty low and unethical business practices by giving a potato away with each Bond."[15]

We don't know if McAteer raised his own potatoes, but some Canadian Niagara employees helped the war effort by cultivating Victory Gardens that freed up farm-grown food to go overseas where it was needed so desperately. Growing food also made sense during a period of strict rationing. Liquor was in short supply too. A limit of 40 oz. per month seemed small, until it was reduced to 13 oz. in 1944. And by early 1943, gas shortages were making themselves felt. "There isn't much news about town these days, although there are a lot more people walking, now that gas is so hard to get." As one writer suggested, "Young couples who used to step on the gas will now have to stay home and gas on the steps."[16]

The Tattler was the lifeline between employees and families at home and those away in the services. With artwork and writing by Chief Operator William N. Rickard, it kept up a steady supply of news, cartoons, and jokes, such as the one in the June 1944 issue about the Texas-born captain of an all-Texas company in North Africa, who told his men: "Our job here is to promote good neighborliness, among other things. We've got to humor the natives. If they say Africa is bigger than Texas, agree with them." The April 1944 *Tattler* told readers that Earl Snider had a son whom he has not yet seen. The "Chins Up" section of *The Tattler* imagined what it would be like when bearded father and son finally met. (Reproduction by permission of FortisOntario)

The Tattler

At least good humour was plentiful. The best evidence of this is the wartime newsletter *The Tattler*, which kept staff at Canadian Niagara in touch with enlisted employees. The author, who hid behind the pen-name, The Scribbler, relayed news, commentary, and jokes month after month to those in the services and those at home. A two-page typescript of July 15, 1942, now in the National Archives of Canada, is either the first issue of the *Canadian Niagara Power Company Gossip Monthly News Letter* or simply the oldest to survive. The next month's issue was called *The C.N.P. Co. Tattler* and that name stuck.

The Scribbler was Chief Operator Bill Rickard. He never left any record of why he started the newsletter, but perhaps as the father of a son in the Navy, he was aware of the pain of separation from a loved one and the anxiety of not knowing if he would ever see him again.

CHINS UP

Major - "What is a manoeuver?"
Recruit - "Something you put on grass to make it green, sir?"

. . . . . .

EXPENSE REPORT
Period ending May 23, 1939.

Date		Amount
5 - 1	Advertising for Girl Stenographer	$ .50
5 - 5	Violets - new Stenographer	.65
5 - 8	Week's Salary for Stenographer	15.00
5 - 9	Roses for Stenographer	3.00
5 - 11	Candy for Wife	.75
5 - 13	Lunch with Stenographer	6.25
5 - 15	Week's Salary for Secretary	20.00
5 - 17	Picture Show Tickets - Self and Wife	.80
5 - 18	Theatre Tickets - Self and Secretary	7.50
5 - 19	Candy for Wife	.75
5 - 20	LILLIAN'S Salary	25.00
5 - 21	Theatre and Dinner with Lillian	21.75
5 - 22	Fur Coat for Wife	625.00
5 - 23	Advertising for Man Stenographer	.50

. . . . . .

"My family thinks there's something wrong with me," a woman complained to the psychoanalyst, "simply because I like buckwheat cakes."
"But there's nothing wrong about liking buckwheat cakes," the doctor murmured, puzzled. "I like them myself."
"Oh, do you?" The woman was delighted. "You must come up some day. I have seven trunks full."

. . . . . .

From the "Chins Up" section of the June 1944 *Tattler*. (Reproduction by permission of FortisOntario)

Although his son George was not a Canadian Niagara employee, enlisted men from Canadian Niagara asked about him in letters to the Scribbler. The Scribbler's son made his only *Tattler* appearance in June 1943. "Some of you fellows have asked about George Rickard who some of you met at the Fort Erie party. Well: he is at present home on a well-earned 28-day leave after a tough winter on the Atlantic."[17]

Rickard's style is nearly always cheerful and upbeat. When a generator blew, he noted, "The boys all took their coats off, rolled up their sleeves, spit on their hands, and went to work, singing the battle song 'Nothing Matters Now But Victory,' with the result that the job was completed by November 9th, and the unit was put on load on November 10th just two weeks after we shut it down. Pretty good, eh! Our hats are off to the boys who did the job."

He referred to the tugboat that managed the ice in the forebay as "Admiral Kent and his heap big canoe." "Incidentally, Jack Lamb is acting second in command of the Fleet and is able to take the ship in and out of port without knocking the power house down. Harvey McAteer is following in his dad's footsteps and has gone to sea in our Forebay, filling the various positions from Cabin Boy to First Mate." Harvey later joined the Norwegian Merchant Marine.

The Scribbler even managed to make fun of storms. "Just when we thought we were getting on so well, Old Man Winter went into a tantrum on February 11th [1944], and gave us a big kick in the pants, we had a big snow storm and a stiff east wind which brought a lot of slush ice down the river. We didn't get all of it, but we did get plenty with the usual results, however in a couple of days it was all gone and the old plant was once again running full blast."[18]

The *Tattler* brimmed with jokes about everything from rationing to the influx of women into the workforce.

"Six-Year-Old — I wonder why so many babies get borned at night?

Seven-Year-Old — Why, don't you know that? It is because they want to be sure of finding their mothers at home."

This one, reflecting the unfamiliar sight of women in trousers or smoking, might have come from the Fleet Aircraft plant in Fort Erie:

"Visitor (in defence plant) — Look at the youngster, the one with the cropped hair, the cigarette and trousers on. It's hard to tell whether it's a boy or a girl.

War Worker — She's a girl and she's my daughter.

Visitor — My dear sir, do forgive me. I would never have been so outspoken if I had known you were her father.

War Worker — I'm not her father, I'm her mother."[19]

Through it all, the Scribbler never let his readers forget why they were working so hard. "Every kilowatt is a nail in Hitler's coffin." "No. 8 is again turning out Kws to defeat the Axis." "Let us put that be-moustached paper hanger [Hitler] where he belongs." During one of the Victory Bond drives, the Scribbler put the money in perspective, "Just for something to do I figured out how long our $6,800 would pay for Canada's war effort, and would you believe it, it would last just about 45 seconds. It sure does cost money to run the war. But as the old lady said when she cried in the ocean, 'Every little bit helps.'" As an afterthought, he added, "Can't you fellows be just a little more economical!"[20]

Events big and small fill the pages of the *Tattler*. In April 1944 the Scribbler reported the "first fish story of the season. Our four-inch low-pressure water line on the thrust deck became plugged and when Bunny Frommert and Obs Heximer disconnected it, they found a 12 inch sucker in one of the fittings."[21]

These little details did an enormous amount to cheer the Canadian Niagara employees who enlisted. They responded, writing letters that were published in the *Tattler*, so they could stay in touch with those on the home front. From somewhere in Europe, Signalman Paul Bloomfield wrote to "Dear Scribbler and everybody.... This is very unusual for me to write a letter to so many people at once, but I believe it is the best way. Like a Scotsman would say, I'll be saving on the postage. Last night I received *The Tattler* from the good old CNP company. You know it is like a fresh breeze from home with all the bits of news from here and there of fellow workers."[22] CNP enlisted employees had only one complaint about *The Tattler*: "The Tattler improves with every issue. The only fault I can find is that it doesn't come often enough."[23]

For lonely men far from home, the *Tattler* was a lifeline. Jack Delaney confided, "It is pretty hard to realize I am so far from home, but I suppose eventually I will wake up to the fact, so do your best to keep me in touch with what is going on over there, and I will reciprocate."[24] The *Tattler* was a familiar friend for those who were far away.

Signing Up

Eleven Canadian Niagara employees had enlisted by July 1942. The Air Force seemed most popular, gaining Jim Bird, Jack Clark, Jack Delaney, Kenneth MacIntyre, Bud Reid, Omer Steele, and W. (Scotty) Sterling. George Hill was in the Navy, and three men joined the Army: Fred Caverley, Donald Fraser, and Joe Johnston.

More enlisted as time went on. In August, the month that Ed Staedt "received an invitation to live off the government for the duration [of the war], everything provided," the Scribbler opined, "There are not many young fellows left around town now. They certainly are picking them up fast." Early in 1943, Howard Shisler, Art Fretz, Sid Morris, Benny Benson, and Earl Snider "received that little slip of paper *requesting* them to go get a medical examination." Some wanted to sign up before the government called them, so they could choose which branch they went into. Don Kelly was trying "to get one jump ahead [by] trying to join the R.C.A.F." and it was rumoured "Earl (Poop Deck) Snider is seriously thinking of taking advantage of his early training on the Tug Boat, by joining the Navy."[25]

As the war progressed, individuals and companies had to deal with rationing for many things, including gasoline and tires, along with alcoholic beverages and foods such as butter and meat. These gasoline ration coupons were for one of the Canadian Niagara trucks.
(Reproduction by permission FortisOntario)

When Art Fretz joined the army, he was the 12th to sign up. By that time Joe Johnston was trying to get out of the Army and into the Navy. Then Ed Staedt joined the Army and Sid Morris, "who had the honour of purchasing the first Victory Bond sold at the plant," was next in line "just waiting for George the Sixth to invite him to his party."[26] By August 1943, 17 men from Canadian Niagara had signed up and the number continued to grow. Harvey McAteer joined the Norwegian Merchant Marine, where he served on a tanker. Then in March 1944 the Scribbler announced that "Bill Atkinson (Bart's son) who has been our electrical labourer for the past six months, has left us and has joined the R.C.A.F."

And in April 1944, *The Tattler* noted, "In future I will be unable to address you as 'Dear Fellers' as we now have a girl in the Services, Helen Danks, from our Fort Erie office, having joined the C.W.A.C's."[27]

Letters from those who enlisted provide a vivid picture of military life. Shortly after he joined up, Edward Clarke "got a haircut here at Manning depot [in Toronto] and you ought to see it. They cut and cut until all they have left is a little tuft of hair on top of your head. I look as though I have been partially scalped." Of course he wasn't going anywhere. "For the last 10 days I have been confined to barracks…because we have to be given a lot of injections with a lot of different kinds of serum. They cover nearly everything in the book: small pox, diphtheria, scarlet fever."[28] Along with the bad haircut and the injections, separation from family, friends and loved ones was the first thing that came with enlistment. Getting back to visit, or having wives visit, was a top priority. The luckiest ones were those whose wives could come to live with them, or visit them.

On September 24, 1943, Art Fretz wrote to the *Tattler*, "I'm in Newfoundland. The name of the place I'm stationed at is forbidden to be in a letter. It's not a bad place except for the weather, but I guess I'll have to get used to it. I am due for a furlough, but I guess I stand about as much chance of getting it as a snowball in hell. I've talked to some fellows who have been here for 22 months and they were just getting their furlough, so I guess at that rate I won't be home for the duration. Surely this damned war can't last 2 years more."[29] A year later, Fretz was still in Newfoundland, but under changed circumstances. "It sure puts a different slant on things with my wife here. It doesn't seem so much like the army, more like a job, but I'd sure trade it in a minute for my old one."[30]

In St. Thomas, Ontario, Jimmy Bird wrote to let everyone know that "Bud [Reid] has his family here now, so he is quite contented, does not care how long he stays here, and I am the same way. Getting those home cooked meals is a lot better than the hash they serve out here. Poor Jack is still living in at the barracks and has to make those weekend trips to Niagara to see his better half."[31]

Not all wives could join their husbands and not all husbands were in places where wives could join them. Then there was another problem: the cost of accommodations. "In Arnprior the Air Force is having a great time with the townspeople in the paying of rents. The people up here can rent a house for $13 a month. They turn around and rent 2 furnished rooms to an air force couple for $25 a month. Some profit, I'd say."

The company did its best to support enlisted employees. Canadian Niagara's Board of Directors decided that employees who enlisted would stay on the payroll at full pay for the first two months of their military service. Their generosity was much appreciated. In a letter to company management, Joe Johnston wrote, "My wife and I wish to thank the Canadian Niagara Power Company for the cheque she has been receiving. I understand that it is for 2 months. They sure have helped at home as the army won't send any money home until the end of the month."[32]

The management and board also made sure that those who enlisted had a proper send-off, if possible. On July 19, 1942, John E. Lawson, Vice President and General Manager, received a letter from Sgt. Scotty Sterling, who was taking his RCAF pilot training at Saguenay, Quebec. It read:

> "Dear Sir,
>
> I'll bet you didn't expect to get a letter from me so soon, but I had to write and thank you for taking me to the train. I was a little bit dizzy from the party but I knew everything that was going on about me and it was quite a thrill being taken to the train by my boss and having my girlfriend beside me.
>
> Mr. Lawson, Mr. Bracken, and all of the employees of the Canadian Niagara were really swell to me and I will always remember you and I will keep writing to you. That is the only way I can show my appreciation.
>
> I reached my new station and it is 10 miles from Baggotville.... My train trip here was a restless one but I blame that on the Scotch and soda, plus the other mixtures that I had on Saturday afternoon. There isn't much I have to say, Mr. Lawson, that is, right now, but I will let you know how I am getting on as I go along.
>
> Thanks again for everything. Hoping everything is in order and that everyone is keeping fine.
>
> Your friend,
> Scotty
>
> P.S. I didn't know a fellow could have such a headache after drinking like I had, but it was lots of fun."[33]

Mr. Lawson must have given quite a party, although the Scribbler decorously reported it as a "buffet lunch."

Scotty was one of six enlisted CNP men at the party on Saturday, July 11.[34] Lawson wrote personally to the five enlisted men who couldn't attend the party, talked about how "you were, nevertheless, in our thoughts during the proceedings," and said that he was mailing them a little memento, the same correspondence case given to those who had been present. He ended his letter, "I sincerely hope that this war will soon be over and that we will have you all back with us again."[35]

To let those in service know that they weren't forgotten back home, the operators pitched in to buy pen and pencil sets for men in the service. Jack Clark and Jack Delaney got theirs in person, others by mail. It was a tradition that would last, witnessed by the survival of smudged, pencilled lists with signatures and amounts of 15, 25, and 50 cents, beneath typed headings such as "Subscriptions for present for Bill Atkinson" or "Subscription list for presents for operators entering His Majesty's Service" or "Subscription list, to purchase a pen and pencil set, for Don Kelly who recently enlisted in the R.C.A.F."[36]

Over There

The Scribbler tracked the careers of those who enlisted. The earliest surviving issue of the *Canadian Niagara Power Company Gossip* reported that "Scotty Sterling is sporting a pair of wings now.... Let's hope that you can drop an egg right on Hitler's button." When Scotty arrived in Britain, the Scribbler expected to see "the Hun planes start falling like snow flakes!"

According to the Scribbler, Scotty's arrival in North Africa helped to explain a run of German bad luck. "I suppose you have all read of General Montgomery's drive in North Africa, and have wondered why he was able to chase [Rommel] the Desert Fox so fast. Well! I have the answer, but don't tell any one as it is a military secret, and this is it...Scotty Sterling is there to help him Yep! Scotty is pounding Hell out of Rommel."[36]

Life in North Africa was difficult and dangerous, but Scotty had time to relax as well. "Of late all I have been doing outside of flying is swimming and shooting. Our campsite is situated at the water's edge and it is just a matter of a few yards walk and you are in the water. We have 2 boats, one a rowboat and the other a kayak, which is similar to a canoe, and with these lots of time is spent sporting around the water and a few of the boys take the rowboat out and get a catch of fish.... We do quite a bit of shooting. Each fellow has a rifle, which was obtained from certain spoils, and lots of ammunition, and so we have a picked a spot and with tins and bottles as targets, we blast away for hours just as it suits one."[38]

Scotty never returned from the war. He was reported missing in action in August 1943, the only Canadian Niagara employee who died in service. Some of his friends cherished the hope that he would eventually be found, but as the months passed, the hope died.

The Scribbler abandoned his upbeat cheeriness for a sombre style in reporting the deaths and injuries of those in action, not just Scotty's, but also those of relatives and friends. In April 1944, there were two pieces of bad news. "Hector Smith's brother, Nelson, has been wounded for the second time in the fighting in Italy." Also, "George Hetherington's son has been wounded in Italy. We don't know just how badly, but are hoping that it is not serious. These are anxious times for we who have someone over there." In June 1944 *The Tattler* reported, "The tragedy of war has once again struck close to home. We learned with deep regret that our telephone operator, Josie Tilton's husband, Flying Officer Arthur Tilton, was killed in action overseas on May 16th, and was buried in the R.A.F. Cemetery at Chester, England."

The Scribbler, who had a son in the thick of the action in the navy, clearly felt the pressure of war at times, and at one point ended an issue of the *Tattler* with, "This letter may seem to be rather disjointed, but that is in keeping with the rest of the world these days. I have just rambled along and jotted down what came to my mind."[39]

Writing Home

Letters sent to Canadian Niagara, usually addressed to Bill Rickard, were often put on a circulation list and then posted on bulletin boards. Some of the letters now in the company papers in the National Archives of Canada still have the circulation slip attached, and you can still see the holes made by thumbtacks, indicating that the letters were posted on bulletin boards.

Food and drink were enduring themes of the letter writers. Reporting from Debert, Nova Scotia, about 60 miles from Halifax, Jimmy Bird told the gang back home, "There seems to be plenty of beer here, but the eastern beer is p — poor, even worse than the American brew."[40] Beer on the west coast didn't seem to be any better. Joe Johnston thought the Rocky Mountains were great but cautioned his friends, "You fellows back home had better not complain of the beer back there. This stuff they serve here is something like the old 4.4 beer we used to get, and worse than the Dixell's beer old Jake sold at the Chippawa Hotel."[41]

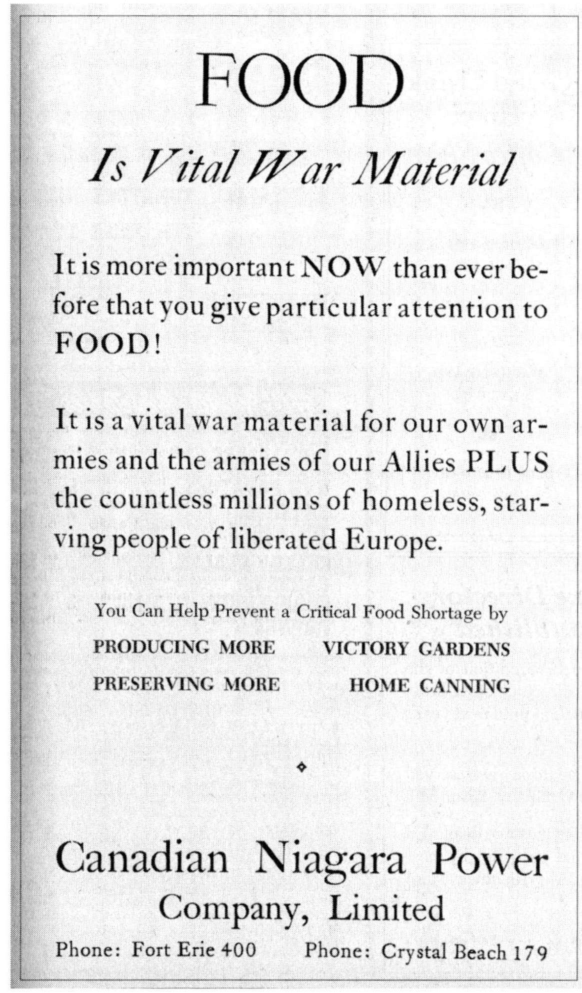

During the war, Canadian Niagara sponsored many newspaper ads to help keep up morale, sell Victory Bonds, and provide advice on how to get by during a difficult time. Many of their ads were aimed at homemakers, urging, for example, economy of electric power by defrosting refrigerators properly or how to make the best of food supplies by canning fruits and vegetables in season. This ad from the *Fort Erie Times-Review* of May 31, 1945, stressed the importance of food as a vital war material and that even when peace came, Europe would need a lot of Canadian food. (Reproduction by permission of Fort Erie Historical Museum, Ridgeway, Ontario)

When Earl Snider got overseas, he complained about English beer. "I think I could pay a pretty good price for a good old 'yellow cap' right now. The beer here is pretty expensive and we pay 1 shilling 1 pence, which is 22 cents, for one glass, and it isn't worth 2 cents."[42] In France, Don Fraser was about to head to Germany, "Just as I was beginning to be able to ask for wine in French, too. Oh well. Maybe when I get back Mr. Lawson will have his bar restocked." In parts of France, another drink was quite common, a distilled liquor made from apples. "They have a drink here called Calvados. About 3 shots of it and I could lift up a turbine bearing. Compared to it, Canadian liquor is like lemonade."[43]

But the letters were not just a litany of complaints. Many of those who enlisted were proud of learning new skills. Jack Delaney took the "trade test and passed it, so...I will go to St. Thomas to train as an aero-electrician.... When and if I get back to you, I hope I will be that much smarter."[44] Delaney went to England as a bomber squadron ground crew member, where he helped prepare the "big boys taking off quite regularly with their pretty presents for Hitler and his bunch. I hear that he doesn't like them very much, so we're going to put a few more ribbons on them for him."[45]

Bud Reid started his Air Force training in St. Thomas, Ontario. At first things moved a bit slowly. "The last 2 days they have been teaching me what a horseshoe magnet is. At the rate they're going, I think the war will be over before I leave here.

It's a good life, though, even if they have got us living in a crazy hospital [he was billeted in a former insane asylum]. I'm bringing my wife and kids to town so it will make it much better. I don't think I'm making much money, though. At noon today I had to buy a notebook and it cost 65 cents and my day's pay is 65 cents. There isn't much profit in that."[46]

When Reid left St. Thomas, he went to Winnipeg. "I am stationed at McDonald Brothers Aircraft Company. They build Avro Ansons for the RCAF. I'm inspecting everything from lumber to finished aircraft. It's really a swell job. We have nice hours: a seven-hour day, and live out. And we get lots of flying time at 75 cents extra per day. We have to go up with the test pilot and see if the damned things fly after we pass them. They have so far."[47]

In 1945 Jimmy Bird was at the Debert base, about 12 miles from Truro, Nova Scotia, where he had "been quite busy lately putting in new equipment. It sure is some job trying to keep up with the latest developments in training equipment. Some of this stuff becomes obsolete before we finish installing it."[48]

With D-Day and the invasion of France, the return to "good old Niagara" was finally in sight. One of those who survived the D-Day landings was Paul Bloomfield, who wrote a letter to the company in October 1944. "It seems like a long time ago since we landed on the shores of France and Normandy. Yes, landing barges and all....From the time we got our bearings in Normandy it marked the beginnings of our busy time and, believe me, it was busy. There were many a days in a row we plugged away from early morning until dark and we did not know what a day off meant.... We sure traveled a lot; sometimes we did not know where we were. It was very hard to keep track of us. Then the time came and we found ourselves in Belgium. The Belgians sure gave us a welcome, flowers and all.... I have been in and out of the Netherlands so many times that I lost count and some day, perhaps in the near future, hope to set foot on Jerry [German] soil."[49] The war was winding down.

"The Tomorrow of Our Yesterdays"

V-E Day (Victory in Europe Day) in Fort Erie was a time of "restrained rejoicing, not tempered by the riotous enthusiasm with which the news was received in many other cities."[50] A service of thanksgiving was held at the Mather Arch, and for many, their thoughts were "of those many Fort Erie boys whose names have appeared on the casualty lists and who will never come back again."[51]

In any case, the war wasn't fully over, and fighting continued in the Far East. On the front page of the newspaper describing V-E Day celebrations, there was a story about the youngest Fort Erie resident to buy a Victory Loan Bond (Rosemary Caroline Ellis, aged 3). Meanwhile, Canadian Niagara took out a full-page advertisement in the *Niagara Falls Evening Review* on May 9, 1945: "Say Thanks with Victory Bonds — our job is not finished." The ad stated, "The Boys Over There Have Done Their Part To Bring Peace in Europe ... We Must Continue To Do Our Part. Let's Finish the Job. Buy More and More, **Then Buy More Victory Bonds**."

On May 10, Lt.-Col. V. Currie, V.C., presented a Victory pennant to Canadian Niagara employees "as a reward for their excellent support of the 8th Victory Loan under the Payroll Savings Campaign. All the employees subscribed without exception and they achieved the impressive total of 132% of the objective set for them."[52]

Throughout the summer, Canadian Niagara placed ads in the Fort Erie newspaper to encourage people to buy victory bonds, plant victory gardens, and do home canning. Other ads informed readers that "Light Bulbs are Cheap and Plentiful," exhorted them to defrost their refrigerators regularly, recommended cooking whole meals in the oven to save food and power, provided safety instructions for electric ranges to "Save A Service Call," reminded them to oil fans and motors at least once a month, and explained the reasons for shortages of electrical equipment — "The Electric Equipment You would like is still part of the Bombers over Tokio [sic]."[53] Within a few days of the appearance of the last ad, however, conventional bombs over Tokyo had given way to atomic bombs on Hiroshima and Nagasaki, and the war ended.

Celebrations for V-J Day (Victory in Japan) in Fort Erie were also thoughtful rather than boisterous. Another ceremony was held in Mather Park, with a scripture reading, a hymn (O God, Our Help in Ages Past), prayers of thanksgiving, speeches by mayor H.A. Howe and the Reverend Canon Merry of St. Paul's Cathedral in Buffalo, and the national anthem. Canon Merry told his listeners that they had to look at the world differently, because the war had "clearly demonstrated that the world is just one big neighborhood and what affects any part vitally affects the whole."[54]

On August 23, 1945, an anonymous poem filled Canadian Niagara's weekly advertisement space in the *Fort Erie Times-Review*:

Peace

This is the Peace we have bought
With blood, sweat and tears
This is the harmony between
Persons and nations
This is the tranquility of mind
This is the to-morrow of our
Yesterdays.[55]

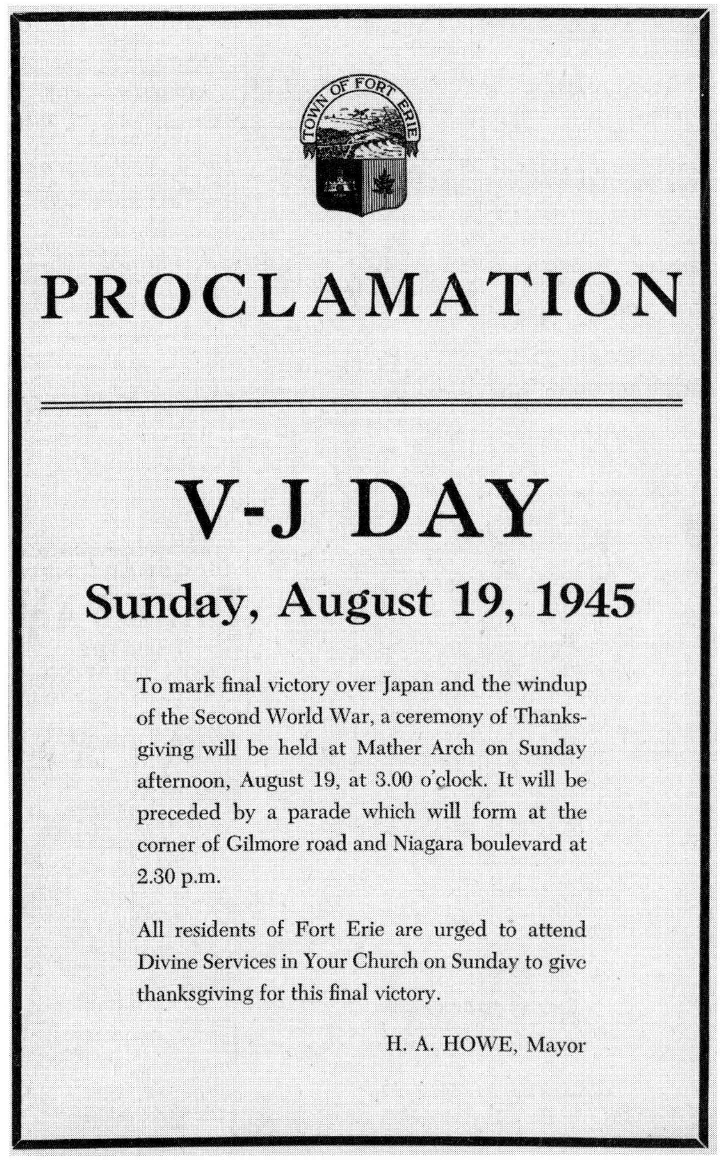

TOWN OF FORT ERIE

PROCLAMATION

V-J DAY

Sunday, August 19, 1945

To mark final victory over Japan and the windup of the Second World War, a ceremony of Thanksgiving will be held at Mather Arch on Sunday afternoon, August 19, at 3.00 o'clock. It will be preceded by a parade which will form at the corner of Gilmore road and Niagara boulevard at 2.30 p.m.

All residents of Fort Erie are urged to attend Divine Services in Your Church on Sunday to give thanksgiving for this final victory.

H. A. HOWE, Mayor

Fort Erie Times-Review, August 16, 1945. While V-E Day continues to attract more attention, V-J Day meant that the Second World War was completely over. At a time when so many were thinking of the need for peace and genuine international cooperation, it was fitting to hold the ceremony of thanksgiving at the Mather Arch. Alonzo Mather, an American businessman, had done much to promote Fort Erie and good relations between the two countries. As with so many important events in Fort Erie, it was shared with citizens of Buffalo.
(Reproduction by permission of Fort Erie Historical Museum, Ridgeway, Ontario)

In this photo of the inner forebay, taken by Hector Smith Sr. in February 1936, the water enters at the left and flows down the penstocks to the turbines. The widely spaced, ice-encrusted racks were designed to hold back large chunks of ice. The trash racks, *above*, with more closely spaced metal slats, have been raised to prevent them from freezing over entirely and blocking the flow of water. Sometimes so much ice accumulated in the inner forebay that only dynamite could loosen it. (Reproduction by permission of Mary Lou and Hector Smith Jr.)

Ice Problems in the Banana Belt CHAPTER 9

As a fruit-growing region associated with a mild climate, the Niagara Frontier is often called the Banana Belt of Ontario. Winter tourists, stranded motorists, and residents know better. Tourists came to Niagara Falls to see huge volumes of rushing water, but they were equally attracted to the frozen spectacles formed by ice. In a few rare years, the ice stopped the river dead. One of the earliest recorded times this happened was in 1848. Strong southwest winds drove enormous quantities of ice to the head of Lake Erie where it enters the Niagara River. Ice tends to clog up there and form a natural barrier to prevent large amounts of ice flowing into the river too suddenly. But this time, as the ice kept piling on top, the lower levels sank deeper and deeper, eventually settling on the bottom and completely blocking any water flow. The river bed was exposed from Buffalo and Fort Erie to the Falls, and there was no water to go over the Falls. The tourists loved it.[1]

In 1909, the situation was slightly different and far more destructive. One journalist wrote, "Twice this year…the voice of the river has been mute. The first time was late in February when, following a severe northerly blow, the falls ran dry, and now for the second time, the flood is frozen solid from bank to bank." By early April, the spring thaw was under way and "there was a large quantity of broken lumps of ice floating about the shores of Lake Erie." This was not unusual. But things started to go wrong on April 7, as the winds increased to a heavy gale. The storm that "tore out of the southwest…left ruin in its wake. The solid ice fields of Lake Erie were churned from end to end and piled in a huge conglomerate at the lower end of the lake."[2] The wind drove the ice into the Niagara River. The upper river can handle more ice than the river just above and below the falls. Above, the river is about a mile and a half wide, but just below the falls, and from there to Lake Ontario, the width narrows to less than half a mile. The difference in width meant that ice arrived at the bottom of the falls faster than it could float downstream.[3]

In April 1909, the entire Niagara River from the Falls to Lake Ontario was choked with shifting ice that kept rising and destroying whatever stood in its way. Postcards such as this one, which were made daily during the height of the destruction, flew off the shelves as visitors looked for views to send to friends or family or to keep as mementos. (Reproduction by permission of John Burtniak)

Accumulations of ice often formed ice bridges that attracted tourists. People clambered about on the ice — even though they were warned not to — and sometimes rested at illegally erected huts and saloons. Water going over the falls travelled under the ice bridge. Eventually it would break up and float downstream.

In 1909 an ice bridge had formed in the pool below the falls in the middle of the winter. The gale filled the river and falls with more ice. The mass of ice flowing down the river broke the bridge that then began to move down the river. Then the unimaginable happened. Before the surging mass of ice "could win freedom in the ample waters of Lake Ontario, the wind shifted again to come from the north. Instantly the moving floes packed at the mouth of the river. Each instant of cold congealed the pack more solidly and each hour brought added pressure from above." The north wind had blocked the exit for the ice that the southwest wind had driven into the river. By Good Friday, April 9, "the only portion of the river left open was about a mile through the surging whirlpool rapids." The rest "was one beautiful ribbon of glistening snow-covered ice." The ice rose far above the normal water level. The previous record had been 28 feet but this time it went higher, perhaps as much as 50 feet higher.[4]

In the gorge below the falls, the Ontario Power Company's power house was taken out of action. It had been built "at what all engineers thought to be a safe height above any and all possible dangers," but this was an unimaginably high river and "six monster ten thousand h.p. generators...with the numerous smaller electrical apparatus were submerged in 20 feet of water as well as the hydraulic machinery in the pit below the generator room." Fortunately, there was no permanent damage.

Further down the river, below the escarpment, the greatest damage occurred where the river banks were the lowest. "In the lower stretches of the river where the ice is packed the hardest, every boathouse, every fishing shack with its traps, often the sole capital of the owner, every private pumping station that supplied the summer homes that line the shores, every private landing and the piers of both the Niagara Navigation Company at Lewiston and Queenston are buried beneath thousands of tons of ice. At Niagara-on-the-Lake the wharf is also overwhelmed and all the bathing houses have been swept off their foundations." It might have been costly and destructive, it might shut off the power, but it was an amazing sight and the crowds turned out in droves. "All day long a constant stream of visitors poured...[in] to see a sight that is not likely to repeat itself within the lifetime of the present generation. Under a brilliant sun the river lay white and glistening to the horizon. And it was silent, absolutely voiceless for the first time within their memories."[5]

Virtually all of the ice in the river comes from Lake Erie. Lake Erie is the shallowest of the Great Lakes, averaging only 75 feet deep, and often freezes over completely. By contrast, very little ice forms in the fast-flowing Niagara River. The water enters the river at Buffalo and Fort Erie, right where the lake is its narrowest and shallowest. And the most common storm path in the area is from west to east, towards the Buffalo-Fort Erie end. It can be a cruel combination. The wind raises the water level in the east and lowers it in the west. Barometric pressure also has an effect. Winds generally blow from a region of higher to one of lower pressure. This condition can also cause changes in the water level. But the Lake Erie water level cannot build up indefinitely at one end. Gravity pulls the higher water back down while wind and barometric pressure differences push it up. As a result of these conflicting forces, the water in Lake Erie actually sloshes from end to end.

The sheer volume of Lake Erie ice could overwhelm the carrying capacity of the Niagara River. The normal spring ice load carried downstream by the river was about 200 square miles of ice. This is 2 percent of the normal amount of ice formed on Lake Erie in winter; the other 98 percent melted in the lake.[6] If more than 2 percent entered the river, it caused problems for hydro generating plants.

Early Ice Problems

As early as 1888, a study on river ice and its possible effects on hydroelectric power generation optimistically concluded that "ice troubles...should be easily observed and overcome."[7] However, as more generators were added, things changed dramatically. "In the winter of 1898-99, eight power units, 40,000 horsepower [total] capacity, were in operation [on the American side], and the water drawn more than doubled, but the velocity in the canal was still low enough to permit surface ice to form." Then "early in the morning of February 8, 1899, the water level in the canal suddenly fell several feet. The ice sheet, 18 inches thick, left without support, broke into great fragments, many of which were drawn flatwise across the rack openings, reducing still further the diminished supply of water to the turbines. Broken ice and slush packed the wheel-cases of the turbines and reduced their output, in some instances blocking them entirely. It clogged up the ports through which water pressure was transmitted to the so-called balance pistons by means of which the weight of the revolving parts of the turbines and generators were supported by hydrostatic pressure. Upon removal of this upward pressure, the entire weight was imposed on the groove and collar thrust bearings, designed to carry but a very small fraction of such a load. The thrust bearings of four units were burned out almost simultaneously."[8] After this disaster, the company made extensive modifications to water channels and booms, but problems persisted.

The engineers at the Niagara Falls [New York] Power Company had learned quickly and altered the design of the intake of Power-house Number Two using their experience with Power-house Number One. "The intake racks were enclosed within the power-house structure, water being admitted to them through submerged arches in the outer wall. The covered forebay was traversed by a small electric crane used for removing sections of rack specially designed to facilitate this operation."[9]

The Niagara Falls Power Company also had sluiceways to take ice out of the canal so it would not interfere with the turbines, but the "ice could seemingly not be induced to enter the sluiceways." The company tried a number of unsuccessful experiments, until "a tug was pressed into service. It was observed that when a stern line from the tug was made fast to a snubbing post while the propeller was driven at a moderate speed ahead, a powerful surface current was created that was most effective in driving the ice into the sluiceway." But this wasn't all. "The tug readily pushed through and broke up great masses of ice. With this assistance a canal full of ice could be disposed of in a very short time."[10]

Other measures, including booms to prevent floating ice from being swept into the plant's feeder canal, meant that by 1917, "the reduction of load at Power-houses Number One and Two by reason of ice have not been frequent or of importance."[11] Clearly the parent company had learned much about dealing with ice by time it started work on the Rankine station. As in Power-house Number Two, the bearing, shaft, and turbine system at Rankine was designed differently to support the system mass more effectively with the thrust bearing higher up. And Canadian Niagara also had an electric tug at least from the winter of 1907-08.

But Canadian Niagara had a less desirable location than its American counterpart. It was set back from the shoreline to let a road and street railway line run between it and the river. This was a sore point, particularly when one of Canadian

At right: This photograph shows the condition in February 1936, when the ice seemed to be winning. This was not a job for the tug, at least not without dynamite. Being out on the ice was particularly perilous. It was pack ice that had been driven in, rather than more stable ice that had frozen in place. Worse, the rapidly flowing water below, which was being drawn into the penstocks and turbines, eroded the undersurface until it was dangerously thin in places, even though on top it appeared that nothing had changed. The parallel poles joined at the top by a crossarm held up one end of the power line that fed electric power to the icebreaker through the cable that can be seen running up from the stern of the tug. (Reproduction by permission of Mary Lou and Hector Smith Jr.)

Below: 1938 was another very bad year for ice, but a great year for winter tourism. As the ice rose, so did the fears that the Falls View Bridge (the Upper Arch Bridge) would be pushed off its abutments. Those who were there at 4:20 p.m., January 27, were either thrilled or appalled when it actually happened. From his office in the Rankine Generating Station, Don Bracken could see conditions worsening by the day. He took pictures of the ice, and although he missed the moment of the bridge's collapse, the photographs tell a frightening tale. Before the bridge collapsed, the 12,000-volt cables that spanned the river on the underside of the bridge were disconnected. Any reports of bluish sparks were either from light standards on the bridge or hyperactive imaginations. The collapsed bridge settled onto the ice, which tore it apart slowly. (Reproduction by permission of Barry Bracken)

Above: A section of the collapsed bridge floats slowly by Niagara Mohawk's Schoellkopf Generating Station on the American side. Eighteen years later, in June 1956, this station too would fall into the river, when the cliff above and behind it collapsed. (Reproduction by permission of Barry Bracken)

Niagara's rivals with better political connections, the Electric Development Company (later known as the Toronto Power Company), was allowed to build right on the shore. The rival company, just a few hundred yards upstream, could draw its water directly from the river in the direction it flowed. It had far fewer problems with ice than Canadian Niagara, which had to draw water from the river and then make it turn roughly 90 degrees under the roadway bridge to enter the forebay.

At Rankine, "sluiceways were provided at the downstream end of the covered forebay and at the downstream end of the outer forebay, discharging into a channel leading back to the river at a point below the main intake. The head of the intake was guarded by a screen of steel bars mounted on stone piers. The screen was submerged to a depth of 4 or 5 feet below normal water level, but proved to be quite ineffective in excluding from the forebay either solid ice or slush. Ice entered it in quantities that could not be discharged through the sluiceways with the help of a powerful electrically driven tug installed in the outer forebay. Repeatedly each winter, wheel-cases and penstocks became clogged with ice and serous reductions in load had to be made for periods of many hours. As additional units were put into service, the necessity of remedying these conditions became more and more pressing, and in 1916, a radical reconstruction of the intake was undertaken," which was completed in 1918.[12]

The reconstruction involved relocating and enlarging the sluiceway in the inner forebay; increasing the depth of the outer intake from 12 feet to approximately 20 feet; and constructing an improved ice shield outside the original ice rack consisting of "a series of concrete piers with grooves for stop-logs to close the upper portions of the openings between piers, permitting water to be drawn from the bottom of the river, where it is comparatively free from ice."

But the severe winter of 1919-20 showed that the ice problems still had not been fully solved. "When the wind blows towards the Canadian shore, the ice from the large Canadian-owned power-plants above the location of the Canadian Niagara Power Company is thrust against the inlets, its forebay becomes filled and its waters clogged with ice, that in severe weather interferes frequently, although temporarily with the operation of the plant."

However, the American owners decided that further improvements were not worth the money. As John L. Harper, Vice-president and Chief Engineer, wrote, "At the present time [1921], no changes are contemplated in regard to this Canadian intake for the reason that any adequate ice disposal would require such an expenditure of money that the ordinary diminutions of power can be accepted with less loss than would be imposed upon the system in carrying continuously [paying the interest upon] the expense of an adequate ice disposal system for this plant."[13] Only the best was good enough for the American plant, but good enough was good enough for the Canadian plant. In fact, to make matters worse, as part of the improvements in 1918 on the American side, Harper had a boom constructed expressly "for the purpose of diverting ice from the American shore to the Canadian channel."[14]

The ice boom of 1918 was designed to address the fact that "under the influence of the prevailing southwesterly winds, most of the ice takes the channel on the American side of Grand Island and passes the power company's land in an ice stream varying in width up to several hundred feet. The ice sweeps along the two miles of river frontage owned by the Niagara Falls Power Company, with almost irresistible force...and always an impending danger to the industries dependent upon a contin-

The earliest known photograph of the Rankine ice breaker tug was taken in the forebay on September 24, 1906. That winter it would begin the first of many years of battling against the ice. It was later replaced by a heavier boat which also drew its power from the overhead trolley and sent it to an electric streetcar motor. (Reproduction by permission of FortisOntario)

uous supply of water as the primary source of their mechanical power."[15] During the worst ice conditions, the Niagara Falls Power Company's ice boom shifted most of the troublesome ice to Canada. This was the Niagara River's first significant ice diversion program. It has attracted very little historical attention, and appears to have been a purely American corporate decision with no political input and very one-sided benefits. What could Canadian Niagara do?

The year after Harper decided not to improve ice protection for Canadian Niagara, a Toronto *Globe* headline declared, "Big Chippawa Project 'Delivers' in Ontario's Sudden Hour of Need." The story was about the then-incomplete generating station now known as Sir Adam Beck No. 1. The ice had built up to the point at which both Canadian Niagara and its American parent had defaulted on delivering the quantity of power stipulated in their contracts with Ontario Hydro. To make up for the shortfall, "No. 1 generator of the Chippawa development at Queenston was placed under forced load" and the *Globe* reported that "for two days now this generator, rated at 55,000 horsepower capacity, has been stepping along merrily and delivering to the bus bars 60,000 horsepower. But for this, the many thousands of Hydro consumers on the Niagara system would have been without electricity."[16]

This was only a month after the generating station put out its first electrical power.[17] It was quite a publicity coup.

In November 1905, Canadian Niagara started keeping a "Comparative Statement Showing the cost of Protection against the Ice."[18] It is not clear how the figures were compiled, but the costs increased steadily, with some spikes for particularly bad years. In 1905, costs are shown only for November and December for a total of $90.43, but the following year, the company spent $606.22. The winter of 1909 was bad, January and February were the most expensive yet on record, and so was the $4,695.82 total. The following year, the costs fell; they did not exceed the cost of 1909 until 1920, which cost $6,949.54.

Canadian Niagara undertook a remedial project in 1936 and 1937. A 900-foot-long submerged weir was built at an angle to the intake and stretched upriver. It was designed to have both scenic and engineering benefits. It would increase the water flow, and hence scenic attractiveness, of the Horseshoe Falls close to the Canadian shore near Table Rock. The weir created its own rapids and added to the visible turbulence of the water. But the main intent was to restore the water level at the intake to the 1905 level. But even with the weir, ice remained a problem.

The Tug

From its earliest years of operation, the Rankine Generating Station needed special equipment to deal with ice. Running the ice breaker tug was a two-man job. One man worked at the controls, steering and setting motor speed and direction. The other looked after the electric cable that fed power from an overhead trolley that went across the forebay. Harry Saylor Sr. spent many shifts at the wheel of the tugboat in the 1950s. Ice could collect near the forebay entrance by the road bridge or right at the powerhouse walls. No matter where it was, the idea was to ram the ice with the tug or to ride up on top of it to break it into small pieces to get it moving and let water flow. As Harry explains, "You take a run, put the front end right up on top of the ice and you'd set there for a while. Maybe it would break loose, maybe it wouldn't. And then sometimes you'd have an awful time to get back. You really had to gun her to get backed off and pull the boat off." It was even harder than it sounds and there were lots of surprises.

One day, Harry and Sid Morris had been working the tug pretty hard. The ice was about four or five feet thick near the arches of the bridges. They been ramming and riding up on the ice and had gradually built a V-shaped opening towards the bridge. One of the bearings on the electric motor was running hot, so they stopped to let Joe, a machinist, do some repair work. It was near the end of the shift and Harry wasn't keen on having another run at the ice. The water was coming in quite well, and he was afraid that if they broke more ice, the entire flow would suddenly move into the forebay and cause even more trouble. But his boss told them to continue. "We got a pretty good speed up to ram into it, to break it. And that's when all hell broke loose."

While the tug was being repaired, the water had been flowing under the ice and had worn much of it away. Harry was expecting a jumbled mass of ice four or five feet thick. "So I hit it going full speed. And it was just like tissue paper. It cut right through. And if that bridge hadn't been there I'd've been on the other side of the high-

For many people, the story of ice at Canadian Niagara is the story of the ice breaker tugboat with the streetcar motor powered by an overhead trolley line that spanned the forebay between the road and the generating station. This was the ocean for what some called the Canadian Niagara Navy. This photograph, taken by Hector Smith Sr., in February 1936, doesn't have the same cabin shown in later photographs taken after Harry Saylor ran into the bridge and demolished the cabin. (Reproduction by permission of Mary Lou and Hector Smith Jr.)

way." The tugboat was halfway under the bridge, the motor and controls still worked, but the cab that hit the bridge was a wreck. Inside the boat, Harry looked at Joe the machinist who had been torpedoed during the war. "I said, 'Well, Joe, this is your second time to be torpedoed.' He said, 'Yes, but this time, I was a lot closer to shore than I was before.'" Sid Morris was working frantically to keep the electric cable under control. Somebody yelled at him to keep the power cable out of the water. As Harry remembers, "Sid was hollering back, 'I'm trying, I'm trying.'"

The company had not spent the money for a take-up reel for the cable. Instead the cable was just "in a coil in the back of the tugboat" lurking as an ever-present hazard. Usually the boat did not move quickly or unexpectedly, but there was always the worry about getting caught up in the coil. Because they had been doing repairs on the boat, there were four people on it instead of the usual two, and when the boat started moving faster and further than expected, the coil unwrapped faster than ever. After it was over, Sid marvelled at their good fortune. "We were lucky somebody didn't get tossed into the forebay, because the cable was in a coil at the back of the tugboat and if you got stuck in that when it was going, why, you'd be a goner." When the boat went further than it had ever gone before, the cable was stretched to its limit. "We were lucky we didn't get electrocuted. Good thing the breakers were in good shape and kicked out in the powerhouse." They didn't have any power, and that may have stopped an adventure from turning into a tragedy.

While Sid was hanging onto the cable and Joe was reliving his naval career, Harry recalls that back on shore "our boss was standing there laughing his head off." The boss thought it was too bad that the cab got smashed just after having work done on it, but as Harry explained, "That was the end of that. Nothing was ever said. We couldn't help it, we never knew anything like that was going to happen."

Hitting the bridge was a one-time adventure, but the tug had other kinds of collisions quite regularly. There were several reasons. Walter Moroz, who worked in the control room, points out that that most of the time, the tug was "along the [powerhouse] wall. There was a sluiceway there and the tug tried to keep the sluiceway open to keep the flow and ice going through there and shoving it out the sluiceway and back into the river." With the ice and the current, the boat bumped against the wall sometimes. There was another reason. When the tug was being made, the purse strings hadn't been opened very wide. The electric motor and controls had been salvaged from a streetcar. It was a great way to save money, but streetcars aren't designed to go from forward to reverse in a hurry. The motor would burn out if it wasn't brought to a full stop before changing direction. With the motor stopped, the current took control of the boat. And the closer one got to the powerhouse and forebay walls and the penstock intake, the faster the current. So it wasn't unusual for the boat to slam into the wall and be pinned there by the combination of the ice and strong current.

There was also no power steering. Turning the wheel and rudder was all manual labour. Harry Saylor remembers, "At night when you go home, you didn't have to bend over to untie your shoes because your arms were so long from trying to steer the darned thing. You knew it when you were on there for a shift. You knew you did a day's work, especially if the ice was running."

At times, the crew had to abandon ship. Low water was the worst, because it brought more ice in and made it harder to get the ice out. When the water level drops, ice can almost completely stop the flow of water into the forebay. The penstocks suck up a lot of water and with little or none coming in to replenish it, one day Sid Morris and Jack Lamb found themselves in a tugboat that suddenly shot towards the penstocks and grounded itself. Sid had had enough. When his boss suggested he go out and work the ice drag to try to clear things out at the river end, "I said, 'There's no use. We haven't got the water. You're just wasting your time and you're wearing the machines out.'"

The men in the tug weren't the only ones at Rankine who worried about the ice. The operators' job is to get as much power out of the water as possible, and under icy conditions this meant searching for an uneasy compromise.

The turbines can grind up a certain amount of ice and keep driving the generators, but with ice, one is always operating close to the edge. The ice may extend down several feet into the water. This means that the intakes can't draw much water from near the surface, or there will be too much ice in the turbine. With too much ice, it slows down and generates less power. If this is not corrected, the generator must be taken off-line. Once off-line, or even sometimes when on-line, there is the danger of even more

When the International Control Structure and the Ice Boom reduced the ice problem to the point at which the company no longer needed the tug, welder Frank Flake tried to buy it. He knew the boat and where to get an engine, and was good at repairing equipment. But a scrap dealer made an offer, too, which the company accepted. He would take the boat away to accurately calculate the scrap value, deduct removal costs, and give the company the difference. It turned out that the scrap value was exactly equal to the removal costs, so the company received no money at all. But it did get to keep the wheel, which is now housed in Rankine. Harry Saylor Jr. is shown in this picture taken by his father. (Reproduction by permission of Harry Saylor Sr.)

ice accumulating and making the situation worse. The worst thing is for the turbine to stop. When that happens, water and ice continue to enter. The water flows through the turbine, but the larger pieces of ice can't escape and they jam the turbine. However, an electric generator is the reverse of an electric motor. When the turbine turns, the generator generates electricity. The operator can reverse the process and send electric power into a generator to make it work as an electric motor. This turns the turbine to clear the ice and get the water flow back. At the same time, one can also raise or lower the head gate to adjust the water flow into the penstock and on to the turbine as needed. The trick is to have the gate far enough below the ice to let mainly water in, but high enough to allow water to generate enough power to bring the generator safely back on-line — easier said than done.

The timing is also tricky. Once the generator comes back on-line, it must be producing power at the same voltage as the rest and in phase with it in order to work as part of the complete system. If everything is not just right, the system tries to synchronize itself by either speeding up or slowing the generators down. With the enormous rotating mass of the system, this can cause damage to all of the rotating parts. The forces are so high that the entire building shakes. Operators use the synchroscope to judge just when to hit the switch to bring a generator on-line. If all goes well, it stays on-line and the ice has lost one battle in the ongoing war.

Ice can also interfere with lubrication. Much of the equipment in the generating station is lubricated by circulating water. If that water starts to freeze or if slush or frazil ice gets in the lines, the pressure drops, the flow of water slows down, and bearings and other parts can become dangerously hot. Walter Moroz remembers "going around and tapping in the lines on the thrust deck, to get the water flowing. Slush was the worst."

Fighting the ice was a normal part of life at Canadian Niagara, but everyone kept hoping it would get easier. With the company doing everything it could at the Rankine Generating Station, any significant improvement would have to come from developments elsewhere.

The International Control Structure and the Ice Boom

Under the headline "Power Is Affected by the Ice" a January 10, 1942 story reported that "Ice in the Niagara River has cut production of electric power at the Canadian Niagara Plant thirty-five per cent, John E. Lawson, Vice-President and General Manager said today, causing temporary suspension of the export of electric power to Niagara Falls, N.Y., and Buffalo." The weir, which was supposed to let ice flow over the top and send water into the forebay, acted as a trap when the water was so low that it exposed the top of the weir. As result, the water flow into the forebay was lower than usual and power output dropped, at a time when the war effort couldn't spare even a kilowatt. The problem was upriver, where too much ice was flowing into both American and Canadian channels. And with the American side being shallower, it jammed up earlier, sending even more than the usual share of ice floes to the Canadian side.[19]

In 1947, another newspaper article noted, "C.N.P. reduction is ten per cent; Water low, ice floes move in; part of Falls is dry." It was the same old story. "East

winds holding back the flow of water from the upper lakes into the Niagara River, lowered the river level three feet last night and the biggest ice jam of the season has developed in the upper river." The Ontario Hydro plant originally built downstream from the Horseshoe Falls by Ontario Power was unaffected, "with the Canadian Niagara Plant above the falls bearing the entire ice attack alone."[20]

Relief for Canadian Niagara could only come from reduced ice flow. And that could only come from international cooperation. Fortunately, Canadian Niagara wasn't alone in wanting to rethink the future of the Niagara River and its Falls.

In 1950 Canada and the United States signed the Niagara Diversion Treaty of 1950, one of the most important documents in the history of the Niagara River and Falls. By limiting how much water could be diverted, not in terms of what could be taken out but by how much had to be left to flow over the Falls, it set the conditions for new engineering works designed to harmonize scenic and power needs. During daylight hours of the tourist season, the flow over the Falls had to be no less than 100,000 cubic feet per second, and at least 50,000 at any other time. The same treaty charged the International Joint Commission trustees with preserving and enhancing the beauty of the Falls and River. The treaty opened the door to a new round of research, planning, and construction of a Remedial Works Program to deal with the recession of the Falls.

Thousands of years ago, the Falls were near what is now Lewiston,

This 1975 aerial view of the Rankine powerhouse has a number of clues to the story of ice at Rankine. The top of the photograph is south. The flat-roofed portion, along with the offices, cover the inner forebay to help protect the intakes from ice. This was a lesson learned from the open forebay at the Adams Station in Niagara Falls, New York. Two sets of two tall poles, one set at the south end and the other at the north end, supported the trolley that fed power to the tugboat. The bridge carrying the roadway stopped the boat from going out any farther, even when Harry Saylor Sr. ran at it at full speed. The structure farther out and parallel to the roadway was originally built on piers to hold the outer ice racks. These weren't enough, and a larger device that travelled on rails was added to carry the ice drag. The varying colour of the water's surface angling out from the north end shows the weir built in the 1930s to raise the water level flowing into the forebay. All these efforts to control the ice had only limited success. Niagara River ice is a formidable foe that can only be coped with, not beaten. Water that went into the forebay and didn't enter the penstocks left by the sluiceway that starts at the northeast (bottom left) corner of the building and joins the river near the bottom left corner of the photograph. (Reproduction by permission of FortisOntario)

New York. Erosion has kept pushing them upstream. This is generally referred to as the recession or migration of the Falls, or even the "suicide" of the Falls. One part of the Niagara Remedial Works Project aimed at slowing down the erosion by making the water flow more evenly along the crestline.

The construction of the Grass Island Pool Control Structure, more commonly called the International Control Structure, was the biggest part of the program. It was built about a mile above the Falls, where water is drawn off the river and conveyed by tunnel to power plants on both sides of the river downstream from the Falls. The concrete control structure, which extends about 1,500 feet out from the Canadian side, is divided into 13 sluices or openings, each with a movable 150-ton gate about 100 feet wide. Raising and lowering the gates controls the amount of water allowed to pass further downstream. Each gate is fully independent of the others. Gate operators decide which part of the river receives heavier or lighter water flows. As a consequence, the water level is not the same at all points.[21]

When the Control Structure opened in 1957, it was supposed to help Canadian Niagara's ice problems. It could direct more water close to shore, thereby raising the level at the company's intake. But it could also bring more ice or even other debris. When asked if the control structure made any difference to his work dealing with ice in the forebay, Sid Morris had a short answer: "I don't think there's been much difference."

It could even create new problems. After it opened, Walter Moroz and the other operators had to take some of their orders from the River Control people who worked there. Walter remembers one dark night when "you couldn't see and we were taking orders from river control. They told us to load up [that is, turn the generators on] and they loaded up [and sent more water down]. I told them I didn't think we should load it up. I can't see out there. At night you can't see in the water." He followed orders and soon "a whole bunch of logs and trees came down. Jammed everything up. Did a lot of damage."

One log came in under the head gate, but Walter knew it was too big "to get around the bend. It got stuck there" under the water. "It was a big job for the rigger." Some logs were small enough to go around the elbow in the penstock, but jammed the turbine, which had to be taken out of duty, opened up, and cleaned out. With a storm and a lot of debris coming down the river, the people at River Control "should have closed their inner gates, and instead of sending the logs through, sent them out around the end [of the control structure] and over the Falls."

The International Control Structure was based on considerable research, modelling, and creative engineering. "In the original design of the Remedial Works, power plant intakes, and ice escape channels, every effort was made to provide for satisfactory ice passage. It was recognized, however, that due to the magnitude and complexity of the ice problem at Niagara, a continuing study would be necessary to determine from experience whether additional measures were required."[22] A lot more was required.

The Power Authority of the State of New York (later renamed New York State Power Authority) and the Hydro-Electric Power Commission of Ontario (Ontario Hydro) operated two ice breakers in the Grass Island Pool area, and various other engineering changes were made, including extending the Control Structure, but it wasn't enough to deal with unusual conditions when huge quantities of ice move from Lake Erie into the Niagara River. The ice jams of the winter of 1963-64 were

When ice blocked the entrance to the forebay, the ice drag that ran along tracks on the weir pulled ice away, and sent it downriver and over the Falls. This concrete structure played a major role in film history, but not onstage. Canadian Niagara gave film-makers from Twentieth Century-Fox permission to mount cameras on the structure to take pictures of the violence of the river as it raced to the edge of the Falls. Back in Hollywood, editors superimposed on these shots an image of a runaway boat lodged against a rock midstream. A young lady climbs out of the boat onto the rock, and the boat, complete with villain, is swept over the Falls. The young lady is rescued by helicopter. It was not her first film, or her best, but *Niagara* catapulted Marilyn Monroe to fame. (Reproductions by permission of FortisOntario)

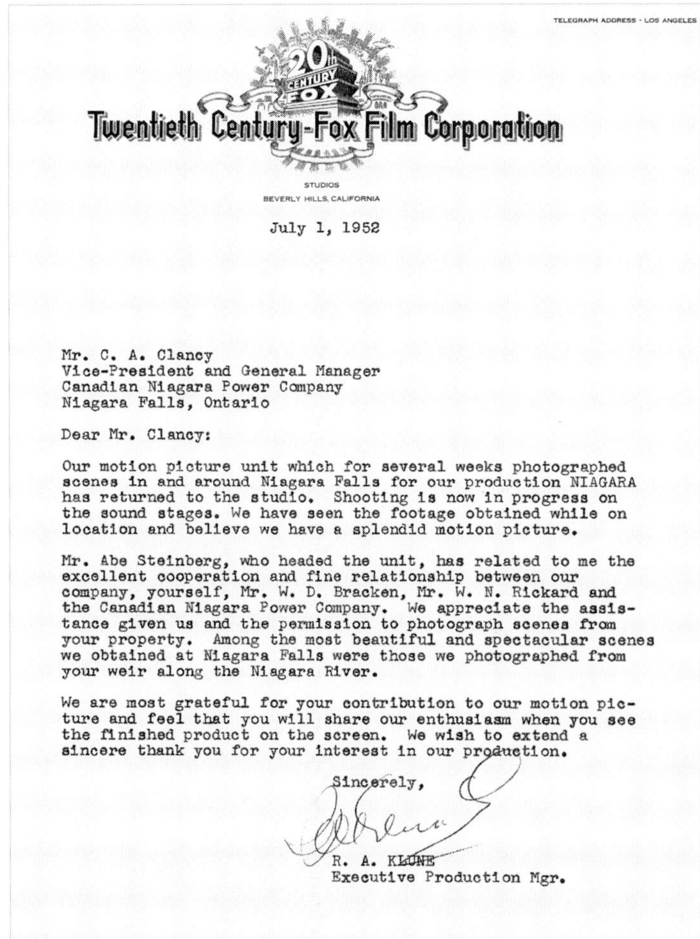

TELEGRAPH ADDRESS - LOS ANGELES

Twentieth Century-Fox Film Corporation

STUDIOS
BEVERLY HILLS, CALIFORNIA

July 1, 1952

Mr. C. A. Clancy
Vice-President and General Manager
Canadian Niagara Power Company
Niagara Falls, Ontario

Dear Mr. Clancy:

Our motion picture unit which for several weeks photographed scenes in and around Niagara Falls for our production NIAGARA has returned to the studio. Shooting is now in progress on the sound stages. We have seen the footage obtained while on location and believe we have a splendid motion picture.

Mr. Abe Steinberg, who headed the unit, has related to me the excellent cooperation and fine relationship between our company, yourself, Mr. W. D. Bracken, Mr. W. N. Rickard and the Canadian Niagara Power Company. We appreciate the assistance given us and the permission to photograph scenes from your property. Among the most beautiful and spectacular scenes we obtained at Niagara Falls were those we photographed from your weir along the Niagara River.

We are most grateful for your contribution to our motion picture and feel that you will share our enthusiasm when you see the finished product on the screen. We wish to extend a sincere thank you for your interest in our production.

Sincerely,

R. A. KLUNE
Executive Production Mgr.

particularly destructive in terms of flooding and costly reductions in the amount of power generated. The stage was set for the next stage of ice control.

The two power authorities applied to the International Joint Commission for permission "to operate the Lake Erie-Niagara River Ice Boom on a seasonal basis at the head of the Niagara River." This 8,800-foot boom extends almost from shore to shore between Fort Erie and Buffalo. The boom, which has been put in each year since the winter of 1964-65, has served the power industry very well. It has lessened ice runs to an acceptable level, but not eliminated them. Interventions of this scale invariably have opponents, who see it as causing harm in other areas. For years there were claims of adverse climatic effects such as later springs, poorer growing seasons, reduced fish population, and greater erosion. But in 1983, after 18 years of operation, a major study of the impact of the Ice Boom gave it a clean bill of health.[23]

Canadian Niagara no longer owns an ice breaker tugboat, but the memories linger. As for the Niagara River and Falls, the ice formations are still spectacular, but less so than before.

Chapter 10

Electric Meters and Meter Readers

Some people call electric meters the cash register of the electric utility industry. Most people don't think about them, unless they think their bill is too high. A few people try to steal electric power by altering or bypassing the meter. When that happens, the meter reader is usually the one to figure out that something suspicious is going on. Electric meters and those who install, repair, and read them are the unsung heroes of the electric utility industry.[1]

In 1925 the Sangamo Electric Company, a meter manufacturer, published a fascinating little book titled *Electric Meter History and Progress* by R.C. Lanphier. The author lamented that "The average man thinks of the great advances in the use and application of electricity in the past fifty years [that is, since 1875], principally in terms of great power houses with their huge turbines and generators. Or he thinks of the modern incandescent lamp, the electric motor, the manifold electrical household devices, or the radio. He does not realize that a great part of this development would have been impossible without accurate, simple and relatively cheap means of *measuring* electric energy and its factors."[2]

In the 80 years since that was written, we have come to rely on an almost infinite range of electric devices, from radar and microwaves to computers and DVD players. Much has changed, but not the way we take electric meters for granted: they are too reliable to attract much attention. Yet as Lanphier wrote in 1925, "Without the accurate meter of today — accurate initially and in service as well — the central stations of this country would be unable to go ahead with any prospects of continued successful earnings, and thus of continued ability to serve the public."[3]

Metering is at the heart of the business of selling electricity. Electrical engineer and consultant Dr. Jan Carr believes that the genius of Thomas Edison lay not merely in his inventions but also in his business model for the emerging electrical industry. At a time when economical transmission distances for electricity were measured in thousands of feet, Edison foresaw an industry based on electricity transmitted from central sources, rather than generated on customer premises as it often was in the early days, and paid for according to the amount consumed. He based the central-station, metered-supply electric industry model on the metered gas industry model.[4] "Central" in that the electricity came from a source that supplied many customers, and "metered" in that customers paid according to proven use. Sebastian Ziani de Ferranti, another of the great electrical pioneers, had a similar model. But Ferranti saw the future even more clearly. He advocated AC (alternating current) which could be transmitted long distances.

Before meters became the industry norm, some people paid by the number of light bulbs in their homes. This system made sense when consumption was regular and predictable; even today, most electric street lighting contracts are unmetered, with charges based on the number of lights and a defined daily lighting period. However, the arrangement had two drawbacks. First, it was unfair, because not every customer used the same amount of electricity: some kept the lights on longer than others. It could also be unfair to the electric company if customers left the lights on all day long. Second, it discouraged people from increasing the number of lights they used or, more important, from buying devices such as kitchen equipment, which extended the use of electricity beyond lighting. However, there is no evidence that Canadian Niagara customers were charged in this way; from the beginning, customers were apparently metered.

The standard electric meter was fair to both supplier and customer. However, because not all customers had the same needs, not all customers had the same type of meter. For example, some industrial customers needed guaranteed uninterrupted power. These were the customers whose processes could not be interrupted or reduced without significant financial loss and operational troubles. By comparison, residential and most commercial customers were not given an absolute guarantee that they would have all the power they needed whenever they needed it. Under most conditions there was enough power for all, but if not, some industrial customers had priority.

As we have seen, Carborundum was one of Canadian Niagara's biggest industrial customers. It needed a guaranteed supply for its electric furnaces. This meant that at any time, Canadian Niagara had to have enough electricity available to meet Carborundum's needs. Carborundum was therefore tying up part of Canadian Niagara's electricity supply, even when it was not using the maximum amount, and the billing system reflected this. Charges were based on the amount of electricity used, as with all customers, as well as other factors such as when the power was used and the peak, or largest power draw, during the billing period, usually a month.

This system was fundamentally fair. The Rankine Generating Station had cost a lot of money to build and took money to operate, whether demand for power was high or low. Rankine was legally restricted to producing a maximum of 100,000 electrical horsepower. And power could not be stored — if the water flowed by without generating electricity, it couldn't be called back later.

The Interesting Life of the Meter Reader

Some of Canadian Niagara's early industrial customers had designed their plants and offices to take advantage of electricity, but these were a distinct minority. Most customers occupied premises that had not been designed with electricity in mind. Everything had to be retrofitted. Early wiring would make a present-day building inspector shudder. Wires ran from the point of entry to wherever they were needed, using whatever route was available.

Nothing was more idiosyncratic than the location of the electric meter. Today it is taken for granted that the meter is either outside or in some easily accessible place. The opposite rule seemed to apply to many early meter installations: they were inside, in the most inconvenient places, imaginable just to get them out of the way. This made life interesting for meter reader and customer alike.

Retired meter reader Fred Barber enjoyed his job on the whole, even though the "bad weather was a pain in the arse" particularly during the great snow storm of 1977. It was a good job. "You were basically your own boss." Some days he'd feel like working fast and others a bit slower, but it didn't matter as long as he completed "one or two meter books a day, depending on the number of customers." And the company treated meter readers well, even buying their uniform of blue pants, grey shirts, and, depending on the season, jacket or parka. Moreover, for much of his career Fred had a company vehicle, although he couldn't keep it at home. He enjoyed meeting people, and most of the customers were pleasant. A few, however, gave trouble. When Fred came to cut the power off from customers who had not paid their bills, those "people got disturbed at me, even though they'd been warned enough."

Meter reader Jack McLeod liked his job too — the independence, working outside except in "real rotten weather," and most of the time, meeting customers. "They were mostly good with you, unless visiting to collect a bad account [or cut the power off]. That would rile them up a bit. I hated that part of the job." His wife, Ann, who worked at Canadian Niagara in billing and then in accounting, remembers how Jack and the other meter readers hated dealing with bad accounts. "It was so frustrating, because it was no fun to cut off the power. The company did everything possible to avoid cutting it off." But meter readers also were able to turn power on for new customers or for delinquent accounts who made some payment or had complained to the right person.

Meter readers soon got familiar with their routes, the people, and the odd locations of the meters. Homeowners liked the convenience of electricity, but didn't want to see the meter and the fuse box or panel. Until the late 1930s or 1940s, meters were often installed out of sight. Jack McLeod remembers finding meters in basements, attics, bathrooms, bedrooms, and other out-of-the-way spots. For example, bedroom meters were usually hidden away in the top corner of a closet. Canadian Niagara did not get to choose the meter location; homeowners or contractors did that.

Meter readers didn't simply walk in unannounced, even when the door was open or unlocked. Fred Barber would "always knock and never walk into a house unless someone answers the door…[it is] not a good idea to walk in; something might be missing and you'd get the blame." When the meter reader knocked and no one answered, it didn't always mean no one was home. Some of the vacationers or seasonal employees at the Amusement Park who rented cottages or small homes in places like Crystal Beach didn't think it was their job to let meter readers in, particularly before noon. Some cottage owners felt the same way. Then there were people who were casual about life and just hollered out "Come on in," whether or not they were up and dressed. Being a meter reader sometimes called for discretion and tact, and the ability to turn a blind eye at the right moment. As someone once said, "Meter readers are the gentlemen of the industry."

A Few Unscrupulous Customers

When no one answered the door, meter readers left a card that told the occupant how to read the meter and gave a phone number to call and leave the reading. Some people did and some didn't. And in some homes, people would phone in the reading but not let the meter reader in to see the meter. There was usually a good reason: they didn't want anyone from Canadian Niagara looking around. And that might mean they were using more power than they wanted the company to know about.

In the mid-20th century, as the use of electricity gradually became more widespread, the amount of current going to residential customers varied from one house to another. For example, when Jack McLeod married Ann, whom he had met at

Canadian Niagara's Queen Street office, they moved into a house with 30-ampere (amp) service, which was really intended only for lights. When they bought an electric stove and fridge from Canadian Niagara, they upgraded to 60 amps. As they added television and more appliances, they upgraded to 100 amps. In 1936, Leonard Heximer and his wife moved into a house on Fort Erie's Niagara Boulevard that had been built in 1924. It had an outside meter and 60-amp service. There were no problems hooking up one of their wedding presents: a new Moffat electric stove, bought from Canadian Niagara.

Stoves in particular need a lot of power. If you draw too much power from a circuit, the fuse blows: it is a safety device. When that happens, you are supposed to find out why your fuse blew and correct the situation. It might be a short circuit, but it usually means you are overloading the circuit. The fuse is designed to limit the amount of current in a circuit. If it is too high, the metal conducting strip inside the old-style screw-in fuse heats up, melts, and breaks the circuit. When a fuse blows repeatedly, some people put in higher amperage fuses, but for a serious overload, these too will melt and blow out.

A few people tried more creative measures. There were two common approaches. The first was simply to remove the fuses and replace them with copper. This was a potentially dangerous situation, because with no limiter on the circuits, the current flow could go far beyond the safe capacity it was designed for, overheat, and cause an electrical fire. The second was using extra electricity without the company knowing about it. This is called stealing electricity and the only question was whether it was intentional or unintentional.

Electricity goes from wires and poles on the street, inside the house to a main switch, then to the nearby meter, and then to the service panel with the fuses. Wires leave the service panel to various parts of the house or business. In newer installations, the meter would be outside and the main switch inside.

In those days, when home or business owners wanted to add capacity, they often found there was no place to start a new wire. Most people did the right thing, which was to add a new panel and fuse box with more circuits that allowed them to increase capacity from 30 to 60 amps or from 60 to 100 amps. But that meant hiring a contractor and paying for new equipment. Some people wanted a less expensive solution. They looked for a place to start a new line and tapped into the line between the meter and the panel or fuse box. Any electricity flowing through this line would be metered and paid for. Others found they could run a new line from the master switch where power entered the home. But this was "upstream" from the indoor meter. Any electricity running through this line would not be metered. This was stolen electricity.

Meter readers looked at the meter, panel box, and wiring as part of their job. They could spot irregularities quickly. Missing fuses and unauthorized wiring were the most common problems. There were others. For example, power running through a meter turns a metal disc, which in turn moves the dials that indicate the amount of power used. Occasionally someone would drill a hole in the meter casing and insert a wire to prevent the disc from moving for a few days or weeks, and thereby prevent the meter from registering power used during that period. They would remove the wire before the meter reader came. Other people would break the seal to get at the wires on the meter, rewire or bypass the meter altogether for a while, and then reconnect it. Meter readers could see that the seal was not intact.

Some people tried unusual ways to stop the discs inside the meter from turning. In one house where usually no one answered, Fred Barber was surprised to see that someone had "put a big magnet on the meter to try to stop the disc."

The job of meter readers was to help keep the system honest and safe. But they also had to deal with situations in which things were, as Jack McLeod puts it, "not always according to Hoyle." They didn't want open confrontation. They just wanted to see the game played fairly. If something was wrong, meter readers reported it to the office, and the company might follow up with a phone call or a visit to explain the rules. This educational role often fell to the meter shop, either the repairmen or the meter installers. The objective was to solve a problem, not to make enemies.

Meter tampering had to be dealt with immediately. Incorrect wiring, which amounted to theft of electricity, also had to be changed right away. The company was willing to accept the explanation that the homeowner hadn't understood that they were doing something wrong. However, as former meter repairman John Parisi observed, "The contractors should have known better." But this was not the kind of situation that customers tended to repeat, so just fixing the problem was enough.

Unsafe circuits were the hardest to deal with, particularly those where fuses had been taken out or bypassed. Here a little practical psychology went a long way. Parisi recalls, "We didn't say they had to get a new service. We just said what they had was unsafe." He found that the woman of the household was often the key to safer electrical practices. He would ask her if she slept in the house. When she said "Yes," he would reply, "Well, I wouldn't sleep in this house; it is unsafe." Then he'd let her do the rest. It usually worked.

Today, most meters are outside the house, and meter readers deal with fewer of these problems. As time allowed, the company assigned meter changers to change indoor meters to outdoor meters. The accounting department helped the meter department set priorities for this task, by identifying suspicious or uncooperative customers. Customers who wouldn't let the meter reader inside predictably refused to let the meter changer inside. But it didn't make any difference. The meter changer would simply install a new meter outside and leave the old one connected inside.

For the most difficult cases, as a last resort, the company informed Ontario Hydro, which was responsible for electrical inspections. If a customer did not cooperate with the inspector, the inspector told the company to cut off the power and the company had to comply. Things rarely got to that point, but the electrical inspector and Canadian Niagara had to take safety seriously.

Fortunately, most people were honest and wanted to do things safely. There were few unpleasant encounters, and when they happened, meter readers and other Canadian Niagara employees were deeply aware of the importance of customer relations. It was more than a case of normal courtesy. John Parisi explains, "We had to be extra careful, because we were a private company, an American company. We didn't want people saying, 'That American company is pushing us around,' so we had to be extra considerate."

Keeping the Meters Working

For its part, the company also had to ensure safety and fairness. Customers have the right to expect that meters work properly. When John Parisi started with Canadian Niagara in 1951, meters normally stayed in service for six years. At that point, a Canadian Niagara meter changer replaced it with a new one and the old one went into the meter shop to be serviced, calibrated, resealed, and checked by a federal government inspector. Only then could it go back into service. Parisi remembers that in the 1950s some of the meters dated back to the 1920s with patent dates from the 1890s.

As meter technology improved, the service period on new meters was extended to eight and then to twelve years. Bearings were the critical part. The meter disc ran on an upright shaft. The end of the shaft had a ball bearing that ran in a jewel cup like a watch movement. Wear or rust could slow the meter down. In the late 1960s and 1970s, Canadian Niagara started buying the newly available magnetic suspension meters. In these meters, the disc was supported on a magnetic field, and there was no need for thrust bearings which tended to limit service life. These meters needed only guide bearings to keep the shaft upright. Guide bearings lasted much longer, because they experienced far less force than thrust bearings, which had to carry the weight of the moving system. These meters were so reliable that a new testing system came in that was based on sampling instead of testing every meter. When Canadian Niagara bought 100 meters, only about 25 were tested. If these were fine, then the rest would be declared accurate too. Similarly, after eight years of service, a sample of that lot would be tested; if they passed, all of the others in the lot that were still in service were passed. Magnetic suspension meters have a normal life span of 20 to 25 years.

The meter is often called the cash register of the electric utility industry. It gives an accurate reading of how much power has been used. For some customers, it also records time of use and load, because different rates may be charged depending on time of use and peak load. Meters had to be checked, calibrated, and certified for accuracy. When meter testing shared quarters with appliance repair, it was often difficult to work under dusty conditions. For many years, the meter shop was across from the Queen Street office, in a former company house that had had the second floor taken off. When the interior was gutted and rebuilt, the meter shop, shown here with Harry Beer at work, was a clean, comfortable work place. The building is now used by St. John's Ambulance in Fort Erie. (Reproductions by permission of FortisOntario)

When the new Bertie Street Service Centre opened in 1973, the meter shop moved into completely up-to-date, dust-free quarters, shown here with Gary Bowman in the foreground. All meter work, including meter readings, is now contracted out. (Reproduction by permission of FortisOntario)

When John Parisi joined Canadian Niagara, the situation was typical of that in many companies at the time: some problems had been neglected during the Depression as well as the war years. The company traditionally bought meters from three companies: General Electric, Sangamo, and Westinghouse. The price was essentially the same, and the idea was to have more than one supplier, so if one company went on strike, there were two others with whom Canadian Niagara had a good working relationship. The company needed a reliable source for meters.

But many of the meters were getting old, "hard to repair or find parts for." Parisi, who had to work on them, called them "old junk." When Canadian Niagara customers changed from 25- to 60-cycle power, he hoped the company would scrap the meters and get new, easier-to-maintain meters. Instead, it had Ontario Hydro convert the old meters to 60-cycle. Parisi was disappointed and to this day is convinced that it would have been cheaper to get rid of all the old meters. Of course, the fact that he had to keep repairing the old meters might have influenced his views.

The Fort Erie meter shop was in a former substation near the Queen Street building. Meter repair shared the space with appliance repair. The appliance repairmen often did their work at customers' homes, but some jobs had to be done in the shop. The job of rebuilding refrigerator compressors left Parisi with some lasting memories. If the gaseous ammonia in the compressor started leaking, as it often did, everybody had to clear out until the gas dissipated. Then there was the general dirtiness associated with repairing appliances, which was not ideal in a space where others were trying to work on and calibrate delicate equipment.

Things improved when the company tore the second storey off the former manager's wooden house behind the substation and converted the ground floor into a meter shop with two separate spaces: a clean room for calibrating and a dirty room for cleaning and painting. When the Bertie Street Service Centre opened in the 1970s and the meter department moved there, the company closed the old meter shop. It is now the St. John's Ambulance building.

From Cheap Labour and Power to Electronic Technology

Carborundum's contract with Canadian Niagara gave it an assured uninterrupted supply of electricity, but it was based on a dual pricing system, which took into account total power consumption as well as the peak demand during a given period. At Carborundum, much of the power went into electric furnaces. Individual furnace operators, who needed to keep production up, did not know what the total power demand was at any given time. So the company assigned an employee the full-time job of watching the two meters, which printed out power consumption on tape every 15 minutes. When the demand meter approached the peak, which would have raised the electricity bill, this employee turned on a siren and furnace operators cut back on power. This was called "shaving the peak" and it was important enough to have one person per shift look after it. By the early 1980s, when wages and power costs were rising, an automatic peak load controller did the job.

But labour was not the only cheap commodity in the old days. Cheap power explains another situation that surprised Parisi when he started with Canadian Niagara. At that time, the National Energy Export Board limited the amount of power Canadian Niagara could export for sale to Niagara Mohawk, its parent company in the United States. The 25-cycle power sent there had to be metered. But as the number of companies generating 25-cycle power plummeted in the 1950s, the difficulty of finding replacement parts for 25-cycle equipment soared. When Parisi looked at the old 25-cycle meters in the Rankine Generating Station, he found one meter that actually registered "power going through it — even though the generating unit was shut off." Meter readers call this phenomenon "creep." It meant that the parent company in the United States was paying for power that wasn't actually being transmitted. Parisi couldn't believe that this situation had never been corrected, and attributes the oversight to low-cost power. It was probably cheaper to pay for unused power than to fix the old meter. Or maybe no one noticed. Whatever the reason, Parisi found it unacceptable and started looking around for better equipment. In the 1980s, he found it with a company making electronic meters.

Electronic meters have an indefinite life span, unless they are hit by lightning or subject to a power surge. They are read with a portable computing device, which downloads consumption information into a computer to calculate bills. The life of the meter reader would never be the same. It was more efficient, but it made for less contact between company and customer. Today, Canadian Niagara no longer employs meter readers; meter reading is contracted out.

CHAPTER 11

60-cycle Conversion — The Carnival and the Iceberg

In late 1957 and early 1958, Canadian Niagara switched all but a few of its customers from 25-cycle to 60-cycle power. This methodical process involved a great deal of planning and preparatory work before customers experienced "cut over day." On cut-over day in a given area, the company turned the 25-cycle power off, and replaced frequency-sensitive devices such as motors and fluorescent light ballasts. At the end of the day, the power was turned back on. Everything worked as before, but using 60-cycle power.

In many residential areas across Ontario, cut-over date was like a neighbourhood carnival. No one had seen anything like it. Trucks descended on the neighbourhood early in the morning and workmen swept through, house by house, street by street, with military precision. If kids were lucky enough to be on holidays, it was a day to remember. You could watch people working, trucks coming and going, and swap rumours about neighbours who had pulled a fast deal. These rumours were all much the same: since the power company promised to replace all motors that ran on 25-cycle power, some people saw an opportunity to get something for nothing. They would haul out a long-dead electric motor from the garage or basement, or even buy one from a junk store for next to nothing. The hydro guys would swap it for a brand-new motor. Of course, most people wouldn't dream of stealing from a power company, but there were always a few who took advantage of the situation.

During the 1950s, Canadian Niagara was just one of many utilities that changed from 25- to 60-cycle frequency. Why was this changeover necessary? Why had 25-cycle frequency been chosen in the first place?[1]

A Legacy Technology

Today the term "legacy technology" is usually reserved for out-of-date computer hardware or software. Y2K, the largest adventure in legacy technology conversion, came as the end of the 20th century approached, when it was feared that two-digit date entries for the year in which a transaction or event took place would cause malfunctions in the 21st century. But legacy technologies include all kinds of technologies from earlier eras.

When circumstances change, the answer to the question, What is the most appropriate technology? also changes. New technology enters the picture, but the older technology still works, and may still be the most appropriate technology for some uses, so the two coexist. So in the 1950s, 25-cycle power, which had been chosen as an appropriate technology early in the century, was largely, but not completely, replaced by 60-cycle power.

Canadian Niagara's original choice of 25-cycle power was the result of several factors. For one thing, the investors who backed the company saw the new primarily as a source of power for the American side of the Niagara Frontier. Therefore, power from the Canadian side of the river had to be compatible with that in the United States.

Another factor was the need to attract investment. At the time, it made better financial sense to invest in Canadian industry than to pay duty on products shipped to Canada. Canadian towns such as Niagara Falls, Fort Erie, and Bridgeburg looked to electricity from the Rankine Generating Station as a way to attract industry. Canadian Niagara's technology had to be attractive to American companies, such as Carborundum or Shredded Wheat, which already had 25-cycle plants on the American side. If Canada offered the same kind of power, such companies would be more inclined to invest in Canada because equipment could be used in either country without modification.[2]

For the Rankine powerhouse, 25-cycle power was therefore the only sensible option. However, at an even earlier date, the parent Niagara Falls Power Company had chosen 25 cycles from a variety of options. Why?

Edward Dean Adams, one of the founders of both companies, explained that they had been created "to develop power for use in manufacturing and transportation."[3] In other words, industrial uses were most important. Other uses, such as electric lights, were less important. Once the decision was made to use alternating current (AC) instead of direct current (DC), the scientists and engineers who investigated the frequency question found that the lower rate of alternation worked better for the large motors used in industry and in street railways. Higher frequencies were more appropriate for uses such as lighting, both arc lighting and incandescent lighting.

The question then became one of finding a frequency that suited heavy-duty power applications, but would not exclude lighting. In 1892, the Westinghouse Company had opted for "two standard frequencies of 60 cycles per second for lighting and 30 cycles per second where power in large units and rotary converters were to be used." Two frequencies meant expensive duplication of equipment. Consultants Dr. Coleman Sellers and Professor Henry Augustus Rowland, who did the original research for the Niagara Falls Power Company, worked to find out how low the frequency could go without "objectionable fluctuations" in light intensity with incandescents and "objectionable flickering" with arc lights. The lower limit, as dictated by the arc lights, was 25 cycles. They also considered 8 1/3 cycles or 16 2/3 cycles, which worked well for long-distance transmission, 41 2/3 cycles, which offered better performance for arc and incandescent lights, and 33 1/3, which might have been a good all-round compromise.

But there was another important consideration. The Niagara Falls Power Company was designing, building, and equipping its massive pioneering generating station in a process now known as "fast tracking." That is, the company was trying to move quickly by making major decisions even before the entire design was complete. Fast tracking allows a project to move ahead faster, but is riskier, because early decisions limit the options available later. Adams, who lived through it all and made key decisions, wrote, "The speed of the hydraulic turbines had already been fixed at 250 revolutions per minute. Professor [George] Forbes proposed that the generators should have 8 poles, which would produce 2,000 alternations per minute, or 16 2/3 cycles per second; the Westinghouse Company recommended 16 poles and 4,000 alternations or 33 1/3 cycles, which most nearly approximated the 30 cycles which it proposed as a standard as was possible to obtain at a speed of 250 revolutions [per minute]. It was finally agreed that there should be 12 poles and the result was 3,000 alternations or 25 cycles, this being the lowest frequency for which the Westinghouse Company would guarantee sound mechanical design." One couldn't resort to industry standards, because the industry was new and standards were only just emerging. But once the decision was made, it became its own standard. As Adams wrote, "This Niagara frequency was soon adopted by many other power plants."[4]

In addition to deciding on the frequency of the current it would generate, the Niagara Falls Power Company had to choose between single-, two-, or three-phase power. The three-phase system made it harder to design and build the generators (often called dynamos at the time). The company's faith in Nikola Tesla's alternating current induction motor was an argument for the two-phase system.[5] Then, just as the Niagara Falls Power Company was wrestling with the question of choosing between two- and three-phase, the Columbian Exposition of 1893 in Chicago, also known as the World's Fair, settled the argument.

> The Westinghouse Company showed at the World's Fair a complete polyphase system. A large two-phase induction motor, driven by current from the main generators, acted as the prime mover in driving the exhibit...with transformers for raising the voltage for transmission; a short transmission line; transformers for lowering the voltage; the operation of induction motors; a synchronous motor; and a rotary converter, which supplied direct current, which in turn operated a railway motor.... The apparatus was in units of fair commercial size and gave to the public a view of a universal power system in which, by polyphase current, power could be transmitted great distances, and then be utilized for various purposes, including the supply of direct current.... The Columbian Exposition of 1893...removed the last serious doubt of the usefulness to mankind of the polyphase alternating current. The conclusive demonstration at Niagara was yet to be made, but the World's Fair clinched the fact that it would be made, and so it marked an epoch in industrial history.[6]

In 1927 the Niagara Falls Power Company generating stations Number 1 and 2 were renamed the Edward Dean Adams Generating Station. The Adams plant on the American side used the two-phase system originally envisioned by Tesla and refined by the Westinghouse Company. Two-phase made sense in terms of generator design and powering motors. However, most of Canadian Niagara's power was destined for Buffalo, and three-phase was better for transmission and distribution. At that point, two- and three-phase power were incompatible. But advanced electrical technology was changing quickly — even more quickly than computer technology changed in the heady days of the 1990s. During the critical research phase of the Adams Generating Station, a brilliant technological innovation made it possible to link two-phase and three-phase current. Instead of choosing one or the other, electrical systems could have both.

Charles F. Scott, a consulting electrical engineer for the Niagara Falls Power Company, proposed and developed what is now known as the Scott connected transformer, a device that allowed for the transformation of two-phase current from the generators to three-phase current for transmission to Buffalo. Generation and transmission were two different operations, with different phase requirements for maximum efficiency. Scott's invention made it possible for each operation to work efficiently without the needs of one dragging down the efficiency of the other. This innovation reduced the cost of generating electric power at Niagara Falls and delivering it to Buffalo, and gave the designers of the Rankine system an extra degree of freedom.

Finally, the need to safely bring generators on- and off-line was another factor that influenced the decision to adopt 25-cycle electricity. Each generator had to feed into a common stream of electricity. This meant they had to operate in parallel. Parallel connection was a difficult task that revolutionized electrical generation. As technological historian Robert Barnett puts it, "Connecting generators in parallel is far from a trivial task. The machines [generators] must be brought up to the same speed, and then their phases must be quite precisely matched before the switch is thrown to connect them. In the days before synchroscopes, the matching was done by eye, comparing the brightness of incandescent bulbs connected across the generators' terminals. The lower the frequency, the easier to make the comparison. Since throwing the switch at the wrong time could wreck a generator, ease of paralleling was a powerful argument for keeping the frequency as low as possible."

Barnett notes, "in the first two decades of the twentieth century, seven large hydraulic plants were built on the Niagara River. By the beginning of World War II, that area boasted the largest concentration of hydroelectric generators in North America, all of them producing 25 Hz [that is, 25-cycle] power."[7] It was an extraordinary legacy of the power pioneers of the Niagara River. But by the 1950s, it was a legacy that was looking less and less attractive, both technologically and financially.

The Perils and Rewards of Increased Demand for Electricity

Early electrical generators tended to be single-purpose and single-customer installations designed for a specific load. For example, some very wealthy families in New York had steam-driven electric generators solely to light their own homes and stables. These were expensive, required a full-time attendant, and tended to irritate the neighbours, who complained about dirt and noise.[8] Industrial customers, such as aluminum plants and paper mills, also had their own generating stations. They would choose a frequency that suited their purposes, and design a generator and other equipment such as motors to fit the specific load.

The Rankine and Adams generating systems had been designed to meet a customer load dominated by industry, from traditional saw and planing mills, furniture factories, flour mills, and steel manufacturers to newer industries, such as electrochemical and electrometallurgical plants. Homes and offices were not an important part of the market for electricity. But as the decades passed, more homes and offices were electrified and electricity found more and more uses in these places. The demand for a higher frequency power, better suited to lighting and appliances, began to rise.

The fluorescent light tube also demanded power at more than 25 cycles. The idea of a fluorescent light dates back to the mid-19th century; at the Columbian Exposition in Chicago in 1893, Nikola Tesla displayed one he had designed.

The Queen Street Service Centre, shown in this photograph from July 1946, included garages, storage for parts and equipment, and offices for billing, drafting, and planning, as well as the sales area and demonstration kitchen. During the conversion to 60-cycle power, this was Action Central for the company. (Reproduction by permission of FortisOntario)

However, the modern fluorescent tube light is based on the 1926 patent of Edmund Germer, which formed the basis of the light that General Electric brought into wide commercial use in 1938. In the years following the Second World War, particularly in the 1950s, fluorescent lights became one of the symbols of modernity and stylishness.

With 25-cycle power, the current ranges from zero to maximum positive to zero to maximum negative and back to zero 25 times per second. The element in an incandescent bulb maintains its glow long enough at zero or low current that the alternations in current were almost imperceptible, and the light seemed to shine steadily. Not so with fluorescent lights, where light intensity from the gas in the discharge tube clearly rose and fell with the changing electrical current passing through it. As a result, the lights had an irritating flicker in places with low-frequency power.

But it took more than fluorescent lights to propel the change. Homes, offices, and industries were finding more and more uses for electric power. This had several important implications. For one thing, with the growing dependence on electricity, people found power outages more and more unacceptable. Outages couldn't be eliminated completely, but greater interconnection — the emergence of the grid — provided a more reliable supply of power. The grid, in turn, called for a uniform type of power. It wasn't possible to connect power at different frequencies.

The existence of different frequencies also presented a challenge for appliance and electrical device manufacturers. Some manufacturers of frequency-sensitive domestic devices, such as refrigerators, made two lines, one for 25-cycle and the other for 60-cycle markets. Others made dual-frequency devices, which could work on either 25 or 60-cycle power, depending on how the wiring was hooked up. Dual-frequency devices included sealed-unit refrigerators, oil-furnace motors, controls, and transformers, as well as sump pumps and fluorescent lights. These devices cost more to manufacture, but their creators had an eye to the future. Many Canadian Niagara customers, also with an eye to the future, were beginning to buy dual-frequency equipment.

Meanwhile, the invention and improvement of the small fractional horsepower motor revolutionized home and office. Early electric motors, such as those used in streetcars or machinery such as lathes and milling machines, had been large and powerful. In factories and shops, one motor often drove many machines through lineshafts, belts and pulleys. Small fractional horsepower motors, by comparison, were embedded in each machine and carried out small tasks. They were everywhere — at the office in the duplicating machine, the electric postal meter, the electric typewriter, the electric adding machine. In the home the list was even longer — blenders, egg beaters, mixers, refrigerators, vacuum cleaners, floor polishers, clocks, and record players. New uses for electricity were constantly being added: can openers, carving knives, toothbrushes, hair dryers, hair curlers, and razors. The basic electric stove didn't need an electric motor, but who wanted basic? Most people wanted a stove with an oven timer and perhaps a rotisserie. More motors. Electric clothes washers and dryers were becoming more popular and dishwashers were entering the market. More and more motors.

After the Second World War ended, consumers wanted to get on with normal life. For some, this meant replacing older appliances or buying ones that they could not buy during the Depression and war. In this ad in the *Fort Erie Times-Review*, Canadian Niagara asked customers to be patient. The new ranges would be available soon. What no one knew then is that in the late 1950s, they would all have to be replaced or rewired for 60-cycle power. (Reproduction by permission of Fort Erie Historical Museum, Ridgeway, Ontario)

In the basement or garage workshop, electric motors proliferated. The popular electric drill was soon accompanied by electric jigsaws, table saws, band saws, grinding and buffing wheels, spray painters, even electric concrete mixers and air compressors. Large coal or wood furnaces had been converted to forced air oil or gas furnaces, which needed motors. And the basement sump pump had an electric motor, to keep the basement from flooding.

One symbol of the new electric age was introduced in 1954: the TV dinner. You kept it in your electric freezer; you heated it in your electric stove; you ate in front of the electric TV. And the three-compartment tray? Made of aluminum, the metal at the heart of the electrometallurgical industry.

As manufacturers produced more appliances and gadgets that ran on 60-cycle power, customers with 25-cycle power faced decreasing choice and higher costs for equipment that used an electric motor.

Despite its huge investment in 25-cycle equipment, the electrical industry began to recognize a need to convert to 60-cycle power. Existing transmission and distributions systems could deliver more power at 60 cycles than at 25. The common phrase was "more power with the same amount of copper," which really meant a lot more power with a bit more copper.

It was a daunting task, but it had to be done, and soon. Every year of delay meant more devices to convert and that translated into more expense for the company.

Preparing for Frequency Changeover

When Ontario Hydro started its frequency conversion program in the 1950s, it was inevitable that Canadian Niagara would follow. The company did not stop generating 25-cycle power, it added 60-cycle power. Therefore its first task was to make sure that it had a reliable source and infrastructure for 60-cycle power. This meant making the necessary changes at the Rankine Generating Station and transformer house, in the substations, and with the pole-top and other transformers that stepped the power down to the voltages at which the customers used it.

John Damato started working for Canadian Niagara in 1956 as an electrician, helping to get the transmission and distribution system ready for upcoming 60-cycle conversion. His first job was installing lines from the Rankine transformer house to the new heavier feeders that went to Fort Erie and then on to Buffalo.

Overall system growth had started to tax the Canadian Niagara transmission and distribution system, and the company needed to increase its capacity before changing over to 60-cycle power. In addition to replacing the frequency-sensitive transformers with ones that were "much bigger in size and wattage," they replaced breakers and completely rewired transformer substations. The 60-cycle conversion was therefore accompanied by an extensive upgrading and makeover of the transmission and distribution network.

John also helped tie the Rankine transformer station into a 115-kilovolt, 60-cycle power source from Ontario Hydro. As part of this project, Ontario Hydro built and fully equipped a 115-kilovolt, double-circuit transmission line that ran along seven towers from the Niagara Falls Murray Street transformer station to Canadian Niagara's Rankine transformer station. To accommodate these changes, Ontario Hydro installed an airbreak switch and tower near the Allanburg station, and changed the layout in the already congested area near the Murray Street station. This work added \$35,144 to the \$110,435 needed for the new 115-kilovolt line, bringing the total to \$145,579 for new 60-cycle transmission lines to the Rankine transformer station.[9]

In the Rankine station, new or rewound transformers had to be installed to accommodate the new frequency. Heavier lines, switches, and relays were also needed. Power left the Rankine station at 34 kilovolts and travelled to neighbourhood substations, which stepped it down to 13.2 kilovolts. From there it flowed to pole-top transformers, where the voltage was stepped down even further, and finally it entered houses and businesses. Some larger businesses had power delivered at 13.2 kilovolts to their own vaults or transformer rooms.

Like the Rankine station and the substations, pole-top transformers had to be converted to 60 cycles. Keeping up with these transformers kept linemen busy just as transformer stations kept the electricians busy.

Finally, in every single house and business that used electricity, all frequency-sensitive devices had to be changed. The trick was to carry out the entire conversion from 25 cycles to 60 cycles along the path from the Rankine station to the individual customer in a single day. Everyone had to work together. The first scheduled cut was November 26, 1957, and work continued five days a week until December 23. It stopped for the holidays and restarted on December 30, continuing on a regular five-day-a-week schedule until May.[10] There were 219 regularly scheduled cut-over dates, the last scheduled for May 22, 1958, in Crystal Beach.

Even with outside contractors and Ontario Hydro doing much of the work, 60-cycle conversion kept Canadian Niagara employees extremely busy. Linemen were essential. Progress through the ranks to journeyman lineman was usually a combination of training and experience. In order for each line crew to maintain the right balance between apprentices and journeymen, new linemen had to wait for someone to retire or move to a different job and a position opened up. However, in the years before conversion, Canadian Niagara accelerated the training process.

Babe Douglas, who joined Canadian Niagara in 1952, was told he "wouldn't be a qualified lineman for at least 10 years," but when Tod Docker and Bob Hanna joined a few years later, they were rushed through the process. As Tod recalls, he became "next to qualified." He could work with and do everything a journeyman lineman could do, as "long as there was somebody within hollering distance" but "you didn't get paid for it."

Canadian Niagara has long had a strong tradition of flexibility: employees do whatever needs to be done whenever it needs doing. This approach helped keep the conversion process moving. As an electrician, John Damato normally "looked after anything electrical in the powerhouse," but during conversion the electricians did whatever was needed to free up the lineman for things that only linemen could do. Linemen climbed poles and electricians climbed structures.

Linemen did the heavy work replacing pole-top transformers. First, the 25-cycle transformer was disconnected and lowered to the ground. Then the lineman had to raise the 60-cycle transformer up and connect it. Easier said than done. Tod Docker helped to put up transformers and he recalls there were "no bucket trucks.... It was very limited equipment. We had a huge line truck and they stuffed us in the back of that to go out to the jobs. Everything was hand climbing. Everything. We had a derrick on the back of the truck and you put it together for hoisting." Some things could be hoisted with a power take-off unit called a "redhead."[11]

Syd Freeman remembers that the wooden legs on the three-legged derrick were about 20 or 25 feet long. They would assemble it by hand, attach a three-pulley block and tackle top and bottom, and raise the derrick up by hand. The linemen on the pole waited for the transformer to rise as those below pulled on the rope. When the transformer was close to where the linemen wanted it on the crossarm, the linemen would "horse it around to get it where you wanted." There were various transformer mounts, but the linemen's favourite "had a curved hanger that dropped over the crossarm." The transformer was bolted onto the hanger, which went through the holes in the crossarm that had been measured and drilled in advance.

Some poles were too tall for the derrick. Then the linemen had to haul up a "gin pole" that they chained to the top of the hydro pole. The gin pole was an ages-old hoisting device. It had a movable angled boom with pulleys at the end. A three-pulley block at the top and bottom gave good mechanical advantage. The crew members on the ground either pulled on the ropes or used a power take-up reel sometimes called a capstan. It was a simple device, a powered rotating horizontal drum about 10 inches in diameter. The rope from the transformer ran through the top pulley block, down to the bottom pulley block, and around the capstan drum two or three times. The drum rotated, but if the rope was left slack, it slipped off. If it was pulled tight, the rope gripped the capstan drum and the transformer started to move. As Syd Freeman explains, "If you wanted it up a little bit, you just nudged it."

For decades, Canadian Niagara sold appliances directly to their customers and allowed them to make time payments on their electricity bills. It was an important convenience at a time when appliances were not sold locally. When this situation changed, Canadian Niagara did not want to compete with local merchants. Former employee Chris Wren took over Canadian Niagara's appliance sales and service business under his own name. He continued to use the Queen Street Service Centre windows for displays and perhaps some of the interior floor space as well. The display above the stoves on the left is for Westinghouse coloured light bulbs. In the 1950s, more customers realized that 25-cycle power meant less choice in a world dominated by 60-cycle power. When Canadian Niagara switched to 60-cycle power, customers could use a wider range of electric appliances. (Reproduction by permission of FortisOntario)

There was enough friction to keep everything in position. But slacken the rope off the capstan too much, and the transformer would start to come down. The job looked easy, but required a lot of skill. Strength and stamina helped too. And this work went on day after day, street after street, pole after pole, on every one of the 219 scheduled cut-over dates.

Fortunately, as with other utilities that changed from 25- to 60-cycle power, Ontario Hydro, largely through contractors, looked after the customer's end of the conversion. For residential and commercial customers, Canadian Niagara paid all the conversion costs; the customer did not have to make any of the arrangements. These customers used single-phase three-wire (230/115 volts) or two-wire (115 volts) electric service. In preparation for the conversion, Ontario Hydro had for some years subsidized the manufacture of dual-frequency equipment to reduce the eventual changeover costs.[12] This made things go more smoothly, but didn't save the company much money because, at conversion time, Ontario Hydro recovered the subsidy costs from Canadian Niagara. Dual-frequency appliances also cost more to buy initially, but less to convert, and customers having them got a rebate from Canadian Niagara.

Industrial customers were defined as "a customer of the Company who purchases three-phase power at the Company's standard or contract voltages to serve his plant or the needs of his business."[13] Industrial customers had more complex equipment and greater variety of installations, and paid most of the costs. For these customers, the work was done by an independent contractor who had to have an appropriately experienced professional engineer on staff and "the tools, rewind facilities and sufficient staff to do the job within the required time." The contract was between contractor and customer, but had to be approved by Canadian Niagara. Industrial customers also tended to make certain improvements at the same time since the power was going to be off anyway.

Canadian Niagara did not want its industrial customers to suffer a loss of efficiency as a result of 60-cycle conversion. Therefore it agreed "to pay for the necessary capacitors to restore the Customer's power factor to the same value as it existed when operating at 25 cycles." As planning for conversion proceeded, companies often looked more closely at the efficiency of their power use and identified ways in which they could be more efficient. To encourage this process, Canadian Niagara agreed to a cost-sharing agreement when "the 25-cycle power factor was less than 90% and the Customer wishes to raise his 60-cycle power factor to 90%."[14]

In late March 1957, Charles "Chick" Clancy, vice president and general manager, sent a letter to all Canadian Niagara customers to tell them about the coming changeover. Most people had probably heard about frequency changeovers, because Ontario Hydro had been doing this work elsewhere, and the topic was in the news. However, Clancy wanted to make sure people understood exactly what was going on.

The letter stated, "The change-over from 25 to 60 cycles is being done to standardize and improve the electric frequency in this area. You will find when this change-over is completed that your lighting will be improved, and that your electric appliances will operate more efficiently and will be available in the standard 60-cycle models at more economical prices than present 25-cycle appliances.... Commercial and residential customers will not be asked to pay any costs involved in changing over their existing appliances to the new and improved frequency. This will be done by the Company." Someone had written on the copy in the company files that about 50 industrial customers would have to pay approximately 20 to 30 percent of their changeover costs.[15]

The Twenty-week Countdown

A company pamphlet sent to customers explained what would happen, when, and why. It was a detailed, carefully planned process.

Twenty weeks before the cut-over date, every customer had to be visited personally to get a rough idea of what would be involved. Each was given a letter describing the process.

At the 17-week mark, the company mailed another letter to each customer to let each know that a technician would be coming to list all frequency-sensitive devices. The company allowed 12 weeks for this process, because it also included deciding how "to engineer the conversions based on information obtained by the inventory technicians," procuring materials, assigning contractors, and, for unusual situations, negotiating special arrangements.

Four weeks before the cut-over date, a company representative made another visit to customers to check that everything had been included and that nothing had changed since the original inventory.

Three weeks before the cut-over, the company mailed customers a notice of the exact date when the changeover would take place. This notice also asked customer what they wanted to do with 25-cycle clocks and fans, which were simply replaced rather than rewired.

On the day, changeover "commences about 8:30 a.m. The Company's Contractor supervises all the work of cutover, and deals directly with the Customers. Power interruption and customer inconvenience is kept to a minimum.

Work continues throughout the day and should be completed by 5:00 p.m."[16]

Dave Renshaw, who went on to become the Director of Public Works for the Town of Fort Erie, was a student at Fort Erie High School in 1957. He remembers that someone from Canadian Niagara even came to his high school to talk to the students about 60-cycle conversion and assure them that electricity would be more efficient and appliances would be cheaper after the conversion. Talking to students was a good way to get messages to parents. There were six children in the Renshaw family and not much money to spare. Dave remembers that "father got really exercised about" 60-cycle conversion. His parents and their friends gathered around the kitchen table talking about who was going to pay for this. They had one question about 60-cycle conversion: "Will this cost us more money?"

In Ridgeway, another teenager, Bill Johnston, lived on South Mill Street, next door to his grandfather, Chester Johnston, a cement contractor. Chester's employees made cement blocks in the barn behind the house. With the help of a huge electric motor, about two feet by two feet by three feet long, three men (or two men and a teenager) produced about 1,600 cement blocks a day, one at a time. A long leather belt joined the electric motor to a line shaft and pulleys mounted just below the ceiling. The line shaft, pulleys and more belts drove the cement mixer and the conveyor belt that took the mixed cement to an electric-powered block-forming machine.

When the conversion was over, the men had a motor that measured about 16 inches by 16 inches by 2 feet long. No one believed it could do the work of the big old brute. But sure enough, it worked fine. Chester, his employees, and his grandson Bill concluded that there really was something to Canadian Niagara's talk about greater efficiency with 60-cycle motors. Moreover, although some commercial customers paid part of the conversion costs, Canadian Niagara didn't pass on any of the $20 it cost for that new motor.

Bill's mother Gladys thought 60-cycle conversion was wonderful. "You gave them half worn-out clocks and fans and they gave you brand-new stuff, and even let you pick the design you wanted." People also got new radios for old, but on some domestic appliances such as wringer washing machines, the company just replaced the motor. "But it was the same deal, you got a new motor for a half worn-out motor." The replacement motors were generally reconditioned motors that had been rewound for 60-cycle power. The Johnstons even got a new motor for the pump that ran the filtration system for their in-ground swimming pool.

The conversion program also accommodated anyone who didn't want a new motor with a "Cash-In-Lieu of Conversion" offer. Willoughby Tool & Die opted for $227.47 instead of a new motor. Wilkinson Plumbing said goodbye to a pipe threader and hello to $126.58, while the owner of a health belt (an electric device that supposedly helped one lose weight) took home $18. One W. Wilkinson got $36 for not converting his electric player piano.[17] The list of other unconverted devices included floor polishers, record players, humidifiers, fans, jigsaws, circular saws, refrigerators, and paint sprayers, not to mention to the coin-operated phonograph at Eddie's Lunch and a $15 "bug killer."[18]

Canadian Niagara destroyed the clocks that it took in on exchange. As a result, those people who chose not to convert their clocks ended up owning a scarce commodity. Today antique dealers and collectors are very pleased when they find one of those rare 25-cycle clocks or radios.

Adding up the Costs

Dave Renshaw's parents weren't the only ones who wondered what the whole conversion program was going to cost and who was going to pay. The $4,668.38 that Canadian Niagara paid out in January 1958 to customers who didn't want to convert their 25-cycle devices was pocket change. Most of the costs were huge, but then so was the job.

After a preliminary survey in summer and fall 1957, Canadian Niagara planned on converting 35 three-phase industrial power customers and 9,182 domestic and commercial customers. This number was based on the number of electric meters that the company's meter readers checked routinely. The final count, however, came to 9,426 domestic and commercial conversions. There were more customers than meters because in some cases, two or more dwelling units were on the same meter. To save money, Canadian Niagara decided to convert the meters instead of installing new ones, even though this would have been a good time to move to more modern meters.

After elaborate negotiations, Canadian Niagara and Ontario Hydro signed a cost-plus contract that spelled out who did what. Of course, even Ontario Hydro did not have the resources to do the job alone and had to depend on subcontractors. It was a seller's market and contractors did well during the process of conversion. St. Catharines electrician Moe Edwards was too young to remember the conversion, but when he was a young electrician's apprentice, the older electricians loved to talk about the "glory days" of 60-cycle conversion and how much money they made. Whatever they did, they apparently earned their money fairly, since customer dissatisfaction was almost unheard-of. On a job that cost over $3 million, Canadian Niagara paid out slightly less than $1,500 in claims for loss or damage.[19]

Still, three kinds of "surprises" added to costs.

First, on cut-over days, Canadian Niagara's residential and commercial customers turned out to have, in total, 3,700 more 25-cycle items for conversion than they had had on inventory day. The junkyards and second-hand shops must have been empty. Extra items meant ordering more parts, more installation work, and more time and money to do the job.

The second category of surprises resulted from "domestic customers changing equipment after final inventory without notification to the Company. Such changes in equipment cause a tremendous inflation in costs which might not be readily appreciated when studying statistics." A customer might think an equipment change was too insignificant to report to the company. After all a Frigidaire was a Frigidaire, wasn't it? But it made a different to Canadian Niagara. One customer had a Frigidaire refrigerator that was listed as Model M-6-42 in the inventory. On cut-over day, the workmen found a Model M-4-38. But it wasn't just a matter of telling the customer that they had the wrong parts and the customer was out of luck. Customer service was too important. Instead, the company devised a routine that increased its own costs.

"Action would immediately be taken to: — (a) Install a loan refrigerator. (b) Requisition material for converting the previously unknown refrigerator. (c) Return conversion material procured for the inventoried refrigerator. (d) Cancel the original Contractual Order to convert as part of an organized Cut-over. (e) Instruct a Contractor to convert the M-4-38 refrigerator as an individual job. (f) Co-ordinate the customer, contractor and material. (g) Remove loan refrigerator." Conscientiously living up to that little phrase "co-ordinate the customer, contractor and material" turned out to be surprisingly costly.

The "no-shows" were the final type of surprise. On May 22, 1958, when the contractors finished their last scheduled cut-over, there were still 822 unconverted customers with 25-cycle equipment. This was almost 1 in 10 of the customers, who had to be treated as expensive special cases. Most were American cottage and summer home owners who couldn't be reached.

By the end of June 1958, only "six to ten seasonal customers" had not yet returned to their residences and consequently had not been converted. Nabisco, by that time owner of the Shredded Wheat plant, was the only industrial client still to be completed. The program would end on July 18, 1958, and "customers in good standing who have claims for conversion after this date would be given a cash-in-lieu allowance and [have to] secure the services of their own contractor to make the necessary changes."[20]

In 1957 Canadian Niagara and Ontario Hydro had estimated the job at $2.7 million, but that was before the surprises. By September 1, 1958, the costs stood at $3,250,351, an increase of 20 percent over the initial estimate. Meters and meter work cost an additional $90,000. By the end of 1958, interest on borrowed money added another $44,000, bringing the grand total to over $3,384,000. That figure included $877,389 that was Canadian Niagara's portion of the shared costs for 104 industrial customers. Commercial customers had cost the company $490,443.77[21] and the residential customers averaged about $230.00 each.[22]

What Had Been Achieved and What Did the Future Hold?

All in all, though, the conversion had run remarkably smoothly. On average, about six frequency-sensitive items were changed for each commercial and domestic account. The table below provides an interesting picture of life in the areas served by Canadian Niagara. There were still quite a few houses without electric refrigerators or electric washing machines. Less than one in three homes had a record player. The relatively low number of frequency-sensitive items that had to be changed showed the wisdom of changing frequency as a pre-emptive strike before the use of electric appliances and gadgets soared.

Item	**No. per customer**[23]
Clocks and fans	1.13
Refrigerators	0.984
Washers	0.680
Heating equipment	0.581
Fluorescent lights	0.551
Miscellaneous	0.502
Hobby motors	0.491
Record players	0.305
Timing mechanisms	0.280
Pumps and allied equipment	0.249
Fans — field conversion	0.232
Average items per customer	**5.98**

Canadian Niagara's customers could now buy less expensive and more efficient 60-cycle devices, as people did elsewhere in North America. And as of April 1, 1957, electricity rates went up. It was the first increase in more than 20 years. Pricing was based on zones and use. Even so, Canadian Niagara customers still paid less than those who bought power from other utilities in the area.

Two industrial accounts opted to stay at 25 cycles: Carborundum and the Niagara, St. Catharines and Toronto Railway. Both had invested heavily in 25-cycle electrical equipment. In addition, the Railway faced an uncertain future. If 25-cycle power worked, it made sense to keep it just to avoid unnecessary expenditure. Carborundum was not in financial difficulties, but its electric furnaces and other heavy equipment had been designed to run at 25 cycles. Conversion would have meant lower efficiency and greater electricity costs.

Canadian Niagara continued generating power at 25 cycles and could supply these customers. Before conversion, the company had looked to the future and found it would be prohibitively expensive to convert its 11 generators to 60-cycle. One of the conversion expenditures was $144,579.15 to modify Ontario Hydro equipment at the Murray Street transformer station in Niagara Falls and to construct a 115-kilovolt double-circuit transmission line from there to the Rankine Generating Station.

The 60-cycle feeder line from Ontario Hydro was more than a simple capital expenditure; it heralded a new relationship with Hydro, which would take the 25-cycle power Canadian Niagara generated but didn't need. When electrician John Damato was putting in heavier lines, switches, and substation transformers, he saw the emergence of more than just a higher-capacity system. He and some of his co-workers realized that Canadian Niagara was starting the journey from a generating and distribution company to a distribution company only. They might not like it, but they could see that staying in generation would not be an option.

After 60-cycle conversion, Canadian Niagara supplied 60-cycle power to all but two of its customers. Ontario Hydro supplied the 60-cycle power and took the 25-cycle power that Canadian Niagara could not sell elsewhere. Ontario Hydro either sold it as 25-cycle power or converted it to 60 cycles. Under these circumstances, questions began to emerge. Did anyone really need the Rankine Generating Station? What was the best use of the water that Rankine used? The same water could have been directed further downstream to Ontario Hydro's Sir Adam Beck Generating Station, which had a higher head of water and therefore generated more power from the same volume of water. Did Canadian Niagara make sense as a separate entity or should it be absorbed into the province-wide Ontario Hydro system? There were a lot of questions that no one wanted to ask openly or even think about.

Chapter 12

In Praise of the Women Who Got the Bills Out

Many employees regarded Canadian Niagara as family. And as in most families, some sibling rivalry lurked beneath the harmonious family image. Frank Flake worked at Rankine, where some of the "older fellows" felt a little superior to the folks in Fort Erie. "We were making all the money in the Falls, because we generate the power. Those guys [the linemen] in Fort Erie just put up poles." The attitude in Fort Erie was "We're making all the money, because this is where the customers are and where they pay their bills. If it wasn't for all these bill payers, where would you [people in Niagara Falls] be?"[1] In truth, all the jobs were important. And the women who made up and got the bills out worked as hard as anyone.

A Cumbersome System

The Fort Erie Historical Museum in Ridgeway has several electric power bills sent to customers in 1925 and 1931. Unlike today's bills, they are postcards on which parts have been filled in by hand. They are also easier to understand than most computer-generated bills. The bill from 1931 has more categories than the one from 1925. Both have been filled in by hand on preprinted forms, with only the billing date period typed. This combination of hand and mechanical billing continued for decades until computerized billing arrived.

Shirley McAteer remembers the system well. She started in the billing department on Queen Street in 1965. She met her husband there — Harvey McAteer, a second-generation employee who had taken leave from the company to serve in the Norwegian Merchant Marine during the Second World War. He was line superintendent when they met. After they married, Shirley stayed at Canadian Niagara until 1974, when Jason, now a Canadian Niagara lineman, was born.

Shirley later moved to accounting and then customer service. These jobs gave her a complete understanding of the billing and collection cycle. It started with the meter readers, who followed predetermined routes to various houses, businesses, and institutions, visiting each in order and writing the reading in the meter book. Each customer had a special account number. Each route had its own hard-covered meter book. Each book had its own place back in the Queen Street office.

When the meter book for any route came back to the office, the women started to prepare bills. The process was intended to produce an accurate bill as well as accurate records in the office about the power the customer had consumed, the rate, and the total owing. Meter readers recorded only raw data — the actual reading — in their books. The women had to do the rest. They subtracted the previous meter reading from the most recent reading to give the power consumption in kilowatt-hours. They also calculated correct consumption amount for readings from meters that required a multiplier. That is, businesses that used a lot of power had meters that recorded only a fraction of the power used. The amount written in the meter book as consumption had to be multiplied by 10, 20, 30, or as much as 400 to give the actual power consumption. Even some residential customers had multiplier meters. Once the correct consumption amount had been written by hand in the meter readers' books, the women had to calculate the money owing. Again, it was not simple.

Shirley Athoe explains, "At one time, we had something like 15 different rates," depending on whether it was a residential or commercial account, or a seasonal or year-round residence or business, and according to the customer's location, such as Crystal Beach, Bertie Township, or Fort Erie. Calculating the cost of electricity was so time-consuming that Shirley made up large charts that showed how much a given amount of power consumption would cost for each of the rates. The amount could be read off the chart quickly and entered into the meter book. But some calculations were more complicated. Certain industrial customers had a demand meter, which indicated peak demand, even if only for a minute or so, during the thirty-day period. Bills for these customers were determined by peak demand, plus the costs of power used in blocks, each of which had a different rate.

Once power consumption, rate, and total cost had been determined and written into the meter book, it was time to transfer the data to the bill. The bill the customer received was only part of the original form that had been filled out in the office. Names were printed by a foot-powered Addressograph machine, which pressed inked metal printing plates against the billing form. Each embossed plate was made in the office by a Graphotype machine, which stamped the customer's name, address, and account number on a metal address plate. The sound of metal type hitting the metal address plate made the Graphotype machine very noisy and the work was tiring. When Fort Erie no longer needed these machines, they were sent to Rankine where Annette Brinda says the Graphotype was so noisy it had to be put in a separate room.

Between July 2 and August 4, 1931, the Ridgeway Public Library used 18 kilowatt-hours of electric power at 5 cents each. There was a lower rate for consumption over a set limit which the library did not reach. Prompt payment yielded a healthy 23 percent discount. The other side of the bill, which was mailed to customers, warned of loss of service for failure to pay, and a $2.00 reconnection fee. The perforated edge indicates where the part of the bill was detached when payment was made. (Reproduction by permission of Fort Erie Historical Museum, Ridgeway, Ontario)

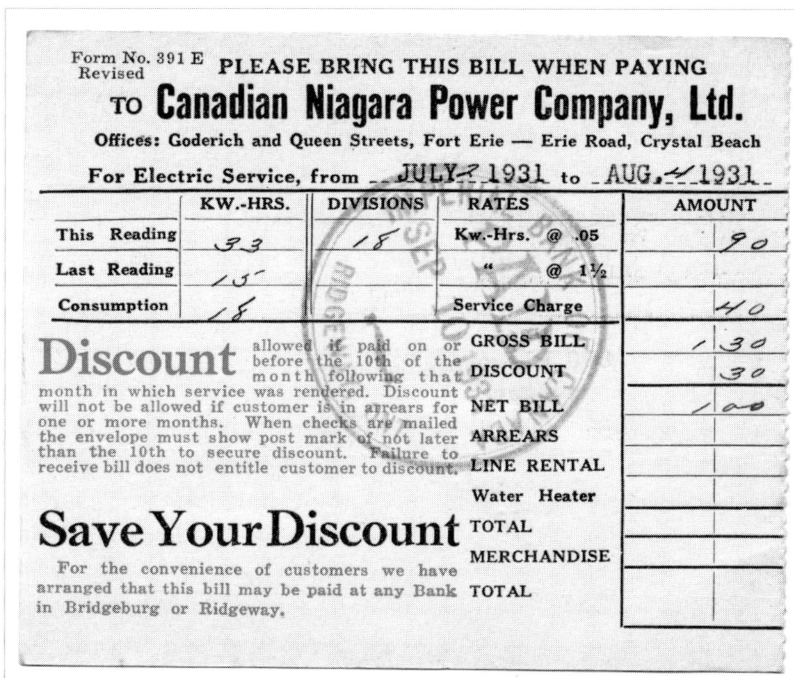

Form No. 391 E Revised PLEASE BRING THIS BILL WHEN PAYING

TO **Canadian Niagara Power Company, Ltd.**

Offices: Goderich and Queen Streets, Fort Erie — Erie Road, Crystal Beach

For Electric Service, from JULY 2 1931 to AUG. 4 1931

	KW.-HRS.	DIVISIONS	RATES	AMOUNT
This Reading	33	18	Kw.-Hrs. @ .05	90
Last Reading	15		" @ 1½	
Consumption	18		Service Charge	40
			GROSS BILL	1 30
			DISCOUNT	30
			NET BILL	1 00
			ARREARS	
			LINE RENTAL	
			Water Heater	
			TOTAL	
			MERCHANDISE	
			TOTAL	

Discount allowed if paid on or before the 10th of the month following that month in which service was rendered. Discount will not be allowed if customer is in arrears for one or more months. When checks are mailed the envelope must show post mark of not later than the 10th to secure discount. Failure to receive bill does not entitle customer to discount.

Save Your Discount

For the convenience of customers we have arranged that this bill may be paid at any Bank in Bridgeburg or Ridgeway.

There were thousands of address plates, one for each customer. Shirley McAteer explains that to keep them organized, "We had this huge cabinet, and drawers and drawers of these metal plates. Each one would correspond with the book that the meter readers were reading." The address plates were supposed to be in the same order as the meter book entries.

When it was time to prepare bills based on a particular meter book, printing the billing forms was the first step. One of the women in the billing department inserted a drawer full of Graphotype address plates in the Addressograph machine and started to work.

The form was a piece of cardboard about 4 inches by about 10 or 12 inches. First, it had to be addressed in two places — on the part that would eventually be

The telephone switchboard was the usual place for women to start with Canadian Niagara, just as men usually started in janitorial or cleaning work. This photo graph from March 1950 shows supervisor Stella Learn overseeing the work of four telephone switchboard operators at the offices in the Rankine Generating Station. (Reproduction by permission of FortisOntario)

mailed out and on the part that would stay in the office. After the form was put in the Addressograph machine, the woman started to work it with both legs. She put the put the bill form in, and "pushed a foot pedal, a big thing came down with a roller on it and rolled over the name plate" which printed it on the bill form. When she pushed again with the same leg, the name was printed again farther down the card. Remove the bill form, put in another, and push the foot pedal with the other leg. Repeat — endlessly. This had to be done for every account every month.

Each card was then inserted into a billing machine. The order was crucial, and the women checked to make sure that the name on the bill was the same as that in the book from which they were taking consumption and charges. Later, with a different machine, they had to enter the data only once to have it printed on three separate parts of the billing form. Shirley McAteer explains that the National Cash Register billing machine she used worked somewhat automatically. "Once you put in the consumption and the meter readings, then you put in the amount of the bill — say it was between $10 and $12, which was a feasible amount back then. It would automatically print it three times. First on the section that the customer got to keep after he paid it. Then there was another little stub on it. When they came in and paid it, we took [the second stub] off. And then a third part, which had the amount on it again that you kept in the office at all times. Then when the bill was sent out and the customer paid, we would take the stub and match it to the one that was in the file...[so] we'd know it was paid." Inputting once and having it printed three times was progress — of sorts. But even then the bills still weren't ready to go out.

The books had to balance. The total power used and the total amount billed had to match the amounts in the pile of bills. If they didn't, it was back to comparing meter book figures with figures on the bills until the mistake was found. Shirley Athoe recalls, "You really had to concentrate on what you were doing." The women in the billing department had to make sure they didn't put the figures for one bill on somebody else's bill, because then "you'd get all off-cog." And they often had to do this while they were physically tired, because of the legwork needed to print up the billing forms. Sometimes, she remembers, her legs would be so tired that they shook.

Once the billing forms came out of the billing machines, they went to someone in the billing department, who checked each account for arrears. Any unpaid amounts were added to the bill manually. Finally, the forms were ready to leave the billing department, but they were not yet ready for the customer. They had to go to accounting on the main floor of the Queen Street building. There each pile of bills went to a "great big paper cutter," says Shirley McAteer. "And I'd flip through them and cut them and then I'd file away the section that the company kept. The other section got mailed out to the customer."

Even mailing out bills had its challenges. During Shirley McAteer's early years, one "poor girl" had to run the mailing machine. She had to put in each bill by hand, one at a time, and "turn a handle with each bill. And in order to make the handle turn you had to push this little button. So here was this poor girl — she might have 5,000 bills that had to go out at the end of the month. And she'd have to put each one in, push the button, turn the handle, and then take it out. She must have been dead by the end of the day. Fortunately, I only had to use that machine a couple of times." Better mailing machines were available, but the office manager at the time was a little tight-fisted. He "didn't want to requisition anything. I guess he figured that if he didn't spend too much money, he'd be thought more of."

When Shirley McAteer "moved downstairs to do the accounting and the collecting of the bills," she learned about new ways of getting tired. "The bills were all due on the 16th of the month.... You could pay them before...but most people came in on the last day or the second to last day. So I'd be standing up for a long time. And of course I'd have to open all the mail." She still had the same office manager, so this process was not automated either. She had to open every letter with a letter opener.

That office manager probably never understood why some women were so happy on his retirement day. "The next guy who came in got me [an automatic letter opener] that you just stuck the letter in and pushed a little handle and it cut it right across which made it a lot easier." And the poor girl using the old feed-the-letters-in-one-at-a-time mailing machine? She eventually got one of the new ones where she "sat the pile of bills in it and it whipped them through by themselves."

Before leaving in 1974, Shirley McAteer spent the last five or six years of her career at Canadian Niagara in the customer service department. Part of her job included keeping track of customers who moved. When she had been in the billing

A new bill-paying arrangement to improve our service to you

On January 1, 1950, the Canadian Niagara Power Company, Limited, will inaugurate an important service to its customers in the Crystal Beach area.

Arrangements have been made with Mr. P. R. Dyke, Real Estate Office, 21 Erie Road, Crystal Beach, by which you may pay your monthly electric service bill in their office on any week day from 9:00 a.m. to 5:00 p.m. The old part-time bill-paying office on Erie Road is being closed, effective December 31, 1949.

A new 24-hour service has been established to handle trouble calls. If you have power trouble at any time of the day or night, simply telephone Zenith 41000 and our despatcher will take your call. There is no charge for a Zenith call, which will be listed opposite the Power Company's name in your telephone directory.

CANADIAN NIAGARA POWER COMPANY, LIMITED.

Canadian Niagara's first postwar task was to meet the pent-up demand for appliances. Then came improved service and customer convenience, with innovations such as radio-dispatched trucks and 24-hour toll-free telephone service. The company also made it easier for people to pay electricity bills in Crystal Beach. This small form on blue paper appears to have been mailed out to customers. (Reproduction by permission of FortisOntario) department, someone from customer service would bring her a "slip of paper" with the new information, and her job was to go to the noisy Graphotype machine to type up a new metal printing plate. Then the new plate had to be inserted into the right spot in the right drawer to be ready for printing with the Addressograph machine. When she moved to customer service, Shirley was the one who sent up the paper so someone else could put up with the noise of the Graphotype machine.

The billing and collections system remained fundamentally unchanged for years, except for minor equipment purchases such as letter openers and mailing machines. Shirley Athoe remembers that before computerization, there were no radical changes, only minor improvements such as a billing machine that "might work a little faster" or "newer calculators." It didn't take much to make an improvement. For example, "The original calculator that we worked with was unbelievable... . It had so many buttons on it...rows and rows and rows" and it couldn't even print. You just put in your figures and the answer "would just pop up in a little slot almost like a slot machine where the numbers come up at the top." These numbers would then have to be entered manually into the billing machine.

One of the most important changes, however, had nothing to do with machinery. For many years, one of the most peculiar features of this labour-intensive billing system was that all bills came out on the same day and all had a common due date. The work flow was anything but smooth and evenly paced. Then, Shirley Athoe remembers, "after a while, somebody's logic set in and there were 20 something [billing and] due days instead of one." One wonders if such an obviously beneficial innovation would have been so many decades in coming if those who set policy had done the billing themselves. It might also have had something to do with the fact that there were no women on the Board of Directors or in senior management positions.

The Beginning of Dramatic Changes

The first computers at Canadian Niagara were used for office work at the Queen Street building. Shirley Athoe took readily to the new equipment and soon found herself acting as a liaison between the computer consulting company and Canadian Niagara to make sure the women got "what we needed and what we wanted." Once the programming was over and the equipment installed, it was time to input customer data and for that "we hired extra girls.... We all worked 16 hours a day just to get all the information on computers." That lasted a few months and it turned out one computer wasn't enough. The company needed two computers with big tape drives on reels. These tapes had to be backed up every night and stored securely. The women took turns at this job. "We'd come in 15 minutes late [in the morning] and stay 15 minutes [extra in the afternoon]...to back up the day's work" and then put the tapes in the vault in the basement. "That was pretty scary at night. Not so bad in the summer, but in the winter, it's dark at 4:30 and in those days you worked right to 5 o'clock."

The early computers still required manual input of power consumption data and billing still required as many people as before. But it did mean that the Queen Street office could get rid of the old equipment, which went to Niagara Falls, where it was adapted for payroll work and later donated to a local church.

Throughout this period, billing, accounting, and customer service were separate functions. Employees stayed in one area for many years and got to know both the job and the customers very well. Employees had the time and interest to train the next generation to the same high standards. It was more than a job, it was a way of life.

Other Office Equipment in the Lives of Women

When Annette Brinda started work at Rankine in 1974, she worked in the attic above the control room. The company was "rewriting the procedure books for the operating department and each page had to be typed up onto one of those Gestetner wax sheets." She typed on multi-layered stencil sheets without a ribbon, to cut through the waxy top sheet. "If you made a mistake with a letter, you had a little bottle of waxy stuff that you put over it and you would blow on it. Hopefully it would dry soon, and then you could type the correct letter. When the page was finished, you took the stencil to the Gestetner machine."

The Gestetner duplicating machine was once a standard part of every office. "You had to put your stencil on a drum and make sure it had ink. It was messy. The machine had an arm you would turn by hand and it would pick up the ink. And you'd keep turning to make your 50 copies or whatever you needed. I did that for months and months." One nice thing about the Gestetner machine was that it printed on regular paper.

"The photocopier, too, was unbelievable. There was a pink film that came on a roll that you put into the photocopier" which transferred the image onto a waxy paper that soon turned brown. It was, as she says, "definitely pre-Xerox."

Perhaps the most exciting innovation before computers was the IBM Selectric typewriter with the dancing "golf ball" that printed the letters. "Everything was nice about that machine. I was amazed. I was happy with it." It even had an optional self-correcting feature but, as she laughingly recalls, "only the General Manager's secretary had the one with the self-correcting tape."

The clean-up and repair work for the Great Ice Storm of 1976 went on for weeks after the ice had melted. The area was in such a mess that some days it was hard to know where to start. Each day, employees set out to clear debris and restore power, hoping not to encounter too many unpleasant surprises along the way. (Reproduction by permission of FortisOntario)

The Great Ice Storm of 1976 CHAPTER 13

Monday, March 1, 1976

Monday's snowfall looked as if it would be just another photogenic storm with nothing more newsworthy than a few fender-benders on slippery roads. Ridgeway's electrical supply went out before 9 a.m., but that small problem was soon fixed. The *Niagara Falls Review* carried the usual spring snow storm photo: an attractive young lady kneeling down to look at plants covered with two inches of snow. The caption looked friendly, "A return to higher temperatures is forecast Tuesday." Other front-page items seemed more interesting: UFOs; proposed gun control legislation; and Richard Nixon returning to the U.S. from a trip to China.[1]

But as night fell and the temperature rose, snow gave way to freezing rain, ice and sleet. Electric utilities dread this kind of weather. The ice coats everything — trees, transformers, switches, electric lines, and cables. Even after relatively light ice storms, tree branches that can no longer bear the weight of the ice may bend down to touch electric lines and cause a short circuit. Sometimes the power goes out when the weight of the accumulated ice drags one cable down on top of another. More often, the weight of the accumulated ice breaks branches or drags entire trees down, bringing electric lines with them.

Monday afternoon started a new week of four-to-midnight shifts for Bob Douglas at the Bertie Street Service Centre. Bob was a former operator who had changed jobs when the installation of monitoring equipment at the Rankine Generating Station reduced the number of operators needed. Now, as caretaker switchman, his duties included taking phone calls after the regular operators had gone for the day. As the shift wore on, the number of calls about power outages rose so dramatically he started to phone around to find some linemen to work on an emergency basis.[2]

Point Abino and Crystal Beach got the first glimpse of how bad things were going to get. Bob Douglas remembers lineman "Babe" Douglas — no relation — calling from Crystal Beach and saying over and over again, "You won't believe this. You won't believe this." In fact, some people didn't believe early accounts of the severity of the storm. Bob Douglas lived in Niagara Falls and the storm never touched Niagara Falls. When he tried to tell friends and neighbours in his city about how bad things were, they didn't believe him.

But in Crystal Beach, Babe Douglas and his crew found "ice build-up was so heavy that secondary lines were coming down. We had to put them back up. Unfortunately, the company only owned three bucket trucks, none of them large. The small one was strictly for streetlight replacement; it didn't even have stabilizing outriggers. The other two were medium-sized." Babe had to put on his spurs and climb up poles. "On the side the wind was blowing from, I had to use a hammer to chip the ice off the pole and the cross arms." As he was driving along Erie Road towards Crystal Beach, he radioed in to report what was happening. The sound was what hit him most. Nearly 30 years later, he remembers it. "All I could think of was the sounds of a helicopter crashing down through trees. That's what the ice, wind and breaking branches sounded like." But there wasn't much time for reflection. "We just kept getting calls over the radio. I didn't even sleep the first night."

Still, it just seemed to be a particularly bad day and things would soon be back to normal. City and regional road workers started at 9:30 p.m. and worked through the night. The region had 21 ploughs, 5 graders, and 30 sanders working, and by the next morning, most roads were centre bare. The thaw during the day also helped get rid of both ice and snow.

On Monday night, temperature dropped from a daytime high of 2°C to an overnight low of –5°. During the night, the sleet and freezing rain had started to take its toll and by 1 a.m., Canadian Niagara Power line crews were out trying to repair the damage.

Tuesday, March 2

Tuesday morning at 8 a.m., the temperature was –2°C.[3] But during the night, "ice and snow knocked out two electric power mains, leaving just a single main to serve residential and small business users. The power flow, however, was considerably below industrial needs."[4] Because of the resulting voltage drop, Canadian Niagara asked radio station CJRN to start broadcasting messages to ask industries to drop all unnecessary load by turning off electric motors and reducing lighting levels. Homeowners were asked to do the same.[5] As a result, more than a thousand workers were sent home without pay. Some employers "among them, Harber Manufacturing and Hooker Chemical, kept their workers at the plants and had them doing clean-up and other chores that did not require heavy machinery."

The power was off completely from 8:30 to 9:50 a.m. in Fort Erie. But even when it came back on, "factory equipment did not get adequate power until a second main was put back in to operation in mid afternoon." Later in the day, some of the employees who had been sent home were called back to work.[6]

At that point, it still looked as if the area was in for just another bad sleet and ice storm. Traffic lights had gone out, but there were no major accidents. And the fire department radio communications system fared well, thanks to a company set up by a former Canadian Niagara employee, Wren Communications. The company's back-up power generator kicked in when the power went out. The back-up power system at Douglas Memorial Hospital also performed flawlessly.[7]

On Tuesday, vice president and general manager William J. Hunter sized up the situation with the guarded comment, "We could have some trouble."[8] Babe Douglas and his crew kept working, and Babe even got some sleep on Tuesday night "on a table at the Bertie Street Service Centre" in Fort Erie.

The Niagara District weather office called for freezing rain overnight followed by above-freezing temperatures and more rain on Wednesday. During the evening, a critical line went out, putting the Rose Hill pumping station out of service.[9]

As Tuesday night turned to Wednesday, the growing weight of the accumulating ice dragged down more tree branches and feeder lines.

Wednesday, March 3

Early on Wednesday morning, an unidentified Fort Erie woman phoned her neighbours for help "because the accumulation of solid ice outside her doors had locked her in." She was lucky. Her phone worked. On Gilmore Road Jackie Bateman and her three children woke up to find ice so thick they couldn't force the door open. Their phone wasn't working, but her seven year old managed to get out a window and go for help. The electric heaters had gone out when the power lines went down and the house was getting colder. A St. John's Ambulance crew took them to a friend's home in Niagara Falls. The drive through Fort Erie might have been the most frightening part of the whole ordeal, because "it was like an obstacle course trying to drive around the downed power lines."[10]

At this point, the town of Fort Erie, including Crystal Beach and Ridgeway, "was in a complete blackout; this included the hospital, industries, pumping stations, schools and residents." Canadian Niagara again contacted CJRN to broadcast warnings that the blackout was going to be a long one. After a quick tour to Ridgeway and Crystal Beach, Bill Hunter could see things that were even worse there than in Fort Erie. He went to the Mayor's office and emphasized the seriousness of the fact that every pumping station was without power and that not all of the fallen lines would be dead.

Some homes and businesses had now been without power for 24 hours. Lines were down everywhere and short-circuited railway crossing signals had been ringing all night. Phone lines were down in many areas and could not be replaced until hydro lines had been repaired. The Fort Erie fire department communications centre was knocked out between 7:30 and 9:30 a.m. — fortunately, there were no fires during that two-hour period. "The Fort Erie detachment of the regional police was using a gas lantern to light their offices and because the power was off, they were unable to operate their own private gasoline pumps, and were forced to start a temporary account at a gas station in Stevensville."[11]

Mayor John (Jack) Teal declared the town a disaster area at 9 a.m. He went on the air at radio station CJRN to warn listeners of the danger of falling wires and tree branches, and to appeal to area residents to use as little water as possible. Although the power loss had silenced conventional radios, people listened to the message on battery-operated transistor radios. As the editor of the *Times-Review* pointed out, "throughout the three-day ordeal, CJRN was like a voice in the dark. For many, the reports on the radio were the only contact with the outside world."[12] Radio broadcasts reduced the number of phone calls to Canadian Niagara, but hard-pressed staff still handled 8,500 calls.

Civic employees set up an emergency headquarters in the town hall. At noon, Niagara Region's disaster plan went into effect. Heads of regional departments, regional and Ontario Provincial police, fire departments and other emergency services met to discuss the plan, which included evacuating people from homes without heat and handling requests for information. Many telephones were out, but police, fire, and some city vehicles had two-way radios, as did Canadian Niagara. This two-way network linked the various organizations. The situation was chaotic and there was naturally some confusion, but patience and good will prevailed.

Remarkably, no accidents or injuries were attributed to the storm in Canadian

As Line Superintendent, Harvey McAteer, shown here talking to Alvin "Slim" Putnam, was at the centre of the storm. There were conference calls with Niagara Mohawk, endless calls and reports from the field, and major safety concerns, particularly with so many additional people flooding into the area to help out, people who were not familiar with the Canadian Niagara system. For ten days, Harvey didn't have time to go home, but in that era, at least you could smoke at your desk. (Reproduction by permission of FortisOntario)

Niagara's service area. That was more than good luck. Few people realize that even when the lights are out in houses, fallen power lines may still be live, so the company launched a public safety campaign about the danger from power lines. Radio announcements were made and notices appeared in the newspaper. But that still wasn't enough. Police cruisers and fire trucks with loudspeakers cruised the streets, warning all residents to stay indoors and not to touch any downed power lines.

Unfortunately, staying at home was not an option for some people. Gas and oil furnaces need electricity. That came as a surprise to many people, who had never thought about the electric fan that forces the hot air through the duct work or the pump that sends hot water through pipes to radiators. Those who used small gas heaters were simply delaying the inevitable: it just took a little longer for a house or apartment to become uncomfortably, or even dangerously, cold. And not everyone had neighbours, friends or family to take them in.

Other communities reached out to help. Although the storm had spared Niagara Falls, it touched the hearts of many in that city and elsewhere. In the 1970s, winter tourism was not a big part of Niagara Falls's economy and the Sheraton-Foxhead hotel was closed for the season. Management reopened it, recalled staff, and declared the rooms available for free to anyone without heat and power. The Pilgrim Motor Inn did the same thing. ABC Taxi offered free transportation to hotels. Restaurants offered free food for those in hotels or for people who were still at home, but couldn't cook. Social services staff set up a shelter at the East End Arena, and spent four hours stocking up on food along with borrowing beds and bedding for the expected onslaught of cold, hungry citizens. But with free hotels and restaurant meals in Niagara Falls, only 12 residents spent the night in the arena.

Bill Hunter had realized by this time that "the situation was beyond our capabilities and outside assistance had to be obtained."[13] Only a few of the town's 23,000 residents had electricity.

Help started to arrive the same day. The Niagara Falls Public Utility Commission sent four bucket trucks with eight men, the Region of Niagara and the St. Catharines Parks and Recreation Department sent three bucket trucks and nine men, and Canadian Niagara hired a local company, Kent Tree Service, to provide one bucket truck and four men. More were on the way from elsewhere in Ontario, but Bill Hunter called Dick Torrey, the President of Canadian Niagara, to ask for help from the American parent company, the Niagara Mohawk Power Corporation. He also contacted Charles Davidson, Manager of Customs at the Peace Bridge and A. Highfield, the officer in charge of immigration. They agreed to let equipment and line crews enter Canada to help repair the damage, and get power back to Canadian Niagara customers as fast as possible.[14]

As men and equipment poured in, a young draftsman in the Bertie Street Service Centre found himself busier than ever. Blaine Desrosiers remembers, "I spent five days cooped up, sleeping on a table top.... My job was to coordinate [message] slips...as the calls were coming in." But with outsiders outnumbering Canadian Niagara employees, Blaine also had to keep producing circuit diagrams and road maps that were detailed enough to guide people who knew neither the roads in the area nor the electrical system layout.

Wendy Spry, left, Sharon Bews, and Shirley Athoe lost count of how many sandwiches they made during the aftermath of the Ice Storm, while linemen, riggers, and other Canadian Niagara employees were working flat out to restore power. (Reproduction by permission of FortisOntario)

There weren't enough people to go around, but wherever possible, Canadian Niagara wanted one of their employees keeping an eye on outside crews. The Niagara Mohawk crews from Olean and their supervisor were sent to the Crystal Beach area to work with Babe Douglas and his crew. Babe instantly recognized the supervisor as a former neighbour who had lived in Fort Erie and commuted to Buffalo to work for Niagara Mohawk before he was transferred to Olean. They got along just fine. The outside crew helped clear the branches, downed wires, and other debris out of the way and put in new lines or poles, but they did not work on live wires. Only Canadian Niagara linemen could do live connections; they were familiar with the system and had been trained on it.

Restoring power to the water system was one of the highest priorities. On Wednesday, all pumping stations were without power, including the Fort Erie South and the Bertie Water Treatment plants. Water in elevated storage tanks was rapidly being used up. The old Bridgeburg plant on Niagara Boulevard could be called on if the water pressure in Fort Erie, Stevensville, Crescent Park, and Douglastown dropped too low, but the plant hadn't been operated in three years. Fortunately, it wasn't needed. Late in the day, Canadian Niagara restored power to the Rose Hill Pumping Station.[15]

By the end of Wednesday, about 6,000 customers were still without power, 17 major units, 9 support vehicles, and 57 men were out in the field, and 16 people, including 6 supervisors, were at Canadian Niagara facilities, trying to undo the damage.

Thursday, March 4

By Thursday, the worst of the storm was over, but the work continued. For Canadian Niagara employees, the days had started to blur into each other. Babe Douglas recalls that they ate most of their meals in a restaurant in Ridgeway, "usually just one big main meal a day." Volunteers at the Ridgeway and Crystal Beach fire halls also kept them going. "They were tremendous, always providing sandwiches and coffee." At Canadian Niagara, the office staff pitched in too. Wendy Spry remembers, "When they weren't managing the phones, they were down in the basement, cooking meals, so that when the line crews came in, hot food was there on the table."

The kitchen had been built for lunches and parties. When the storm hit and the power went out, the cupboards were bare. But "the community rallied around.... Rich's Food Products sent cases of dairy products and chocolate éclairs and anything we needed to keep the fellows going." Everyone remembers the éclairs. For something more substantial, "the restaurants in town provided food and the grocery stores opened up and let us go in, and get the food that we needed to keep people going."

When they weren't feeding hungry men, Wendy Spry remembers, the women in the office were answering phones. The phone calls were different from those they'd received during other storms. In lesser storms, "you rely on the public to tell you which areas are out, and whether a transformer has blown up — to give you advice on where you're needed most." But this was different. "The girls spent 16 hours a day on the phone, calming the elderly and sending people around to take food to anybody who needed help.... We did what needed to be done. We were out of power for 10 days. That's a huge amount of time." She remembers that Harvey McAteer "literally slept in his office on the bookcase for 10 days." At one time there were up to 20 men sleeping on the floor in the Service Centre.

As power gradually returned to the area, progress started to slow. On Thursday, Bill Hunter explained, "We're just not going to make that much progress from here on in [because] there's just too much debris around in the areas we haven't gotten to yet." Crystal Beach and Ridgeway were the worst off. Hunter didn't offer any time estimates. All he could say is that Canadian Niagara would just keep on working. "Right now, all we want to do is get the services hooked up again."[16] Fort Erie civic employees figured they had to get rid of 1,000 truckloads of branches.[17] Bill Johnston, the Bertie Township road superintendent who lived in Ridgeway, told reporters, "It will take 25 men and 15 trucks two 40-hour weeks to collect the debris, which is being taken to the town dump."[18] One optimistic reporter suggested "a wealth in firewood from fallen trees awaits those who want to take the trouble to cut it up."[19] In fact most of it would be hauled to the dump or burned. A Ministry of Environment official said the Ministry would turn a blind eye to bonfires, unless there were complaints of undue smoke pollution.[20]

Smoke came from other areas too. Some people tried to cook meals in fireplaces — often with disastrous results. Firefighters were called to at least ten storm-related fires.[21] Fred Fontana, Chief of the Ridgeway Fire Company, cautioned people about relying on fireplaces during the power outage. Most fireplaces are not properly insulated for continuous service. "When fires are kept burning for hours, or days...the chimney overheats and the adjoining wood begins to smoulder, eventually breaking out into flames."[22]

By the end of Thursday, another 1,000 customers had had their power restored.

Friday, March 5

By Friday morning, when 11 Niagara Mohawk Eastern Division men arrived from Warrensburg, New York, 4,500 customers still had no power. Canadian Niagara now had 20 major units — bucket trucks or diggers — plus 11 support vehicles, 69 men working outside, and 15 inside.

Some of the crews hurriedly sent from the United States seemed not to have packed everything they needed, so Zellers opened up. Everything was charged to the company. Getting the power back on was the first priority.

Meanwhile, the *Niagara Falls Review* reported, "Merchants suffer losses as frozen foods go soft." Brunton's Store on Dominion Road in Crescent Park spent nearly 40 hours without electricity, and lost an estimated $3,000 in frozen food, meat, and ice cream. Other stores such as the Derby Food Market, the A & P, and Michener's Food Market, all in Crystal Beach, suffered similar losses. In Crystal Beach, firefighters went to Quinn Trucking in St. Catharines, borrowed a freezer truck and drove it back. Grateful residents put frozen food in large green garbage bags, labelled them, and put them in the truck until the power came back on. Some Ridgeway-Crystal Beach Kinsmen drove to Hamilton and brought back a 2,500-pound truckload of dry ice, which kept 60 freezers cold for three or four days in the most badly damaged area.[23]

About 2,500 homes were still without electricity and the weather forecast called for 45-mile-an-hour winds. Lightning lit up the evening skies, but that poked only a temporary hole in the fog. Felled trees, branches, and wires turned roads into potentially deadly obstacle courses and caused further outages in places where power had previously been restored. Many of the broken branches remained tangled in the branches of standing trees. On Friday afternoon, high winds tore many of these limbs loose and as they fell they "knocked down some feeder lines in...Ridgeway and crews had to be diverted from Crystal Beach," which was the area most in need of repairs.[24]

When the storm closed Fleet Industries, most of the employees were sent home. Security guards like Heinz Frick of 76 Albert Street had to stay on the job. His wife and children were at home with "no heat and no light." Mrs. Frick said they "were really petrified with the big trees crashing," particularly when a large branch fell on the roof, went through the attic, and stuck through the bedroom ceiling. When the lightning storm started about 10 p.m., she and the children left the house and headed to one of the hotels in Niagara Falls.[25]

In terms of getting customers' power back on, the electrical system could be divided into two parts. Canadian Niagara was responsible for everything from the generating station to the wires on the poles that ran along the streets, and customers were responsible for the line that ran to the house from the hydro pole. Throughout the storm area, but particularly in the Ridgeway–Crystal Beach area, many of the stacks, or private service connections, on the outside of homes had been torn off. Under normal conditions, customers would have to hire an electrician to repair a downed line, and wait for it to be inspected before service could be restored, but there were too many downed stacks and too few electrical contractors. The rules had to be bent.

The Ontario Hydro Electrical Inspection Department gave the company permission to make short-term, uninspected temporary hookups. Canadian Niagara drew up an easy-to-understand legal form for customers to sign. It indicated that they understood the hookup had "the temporary approval of the Ontario Hydro Electrical Inspection Department" only, and that "The connection is made with the full understanding that my electric energy use will be estimated, based on previous use in similar months, and that I shall have a duly authorized facility installed by the Ontario Hydro Electrical Inspection Department as soon as possible and in no event longer than fifteen days" from the signing date.[26] At least one local company, Weinmann Electric Ltd., diverted crews from normal construction work to repairing house service lines.[27] The company earned extra money and homeowners had secure hookups.

Saturday, March 6

The lights were going back on everywhere, and only 60 people remained at the Sheraton-Foxhead.[28] But just as people were getting their power restored or returning home from a hotel or a relative's house, they had to face a new problem: flooded basements. Saturday's newspaper showed a picture of a fireman in Crystal Beach using an auxiliary pump to suck water from a basement flooded by melting snow from the storm.

In some parts of the Fort Erie area, most notably in Crescent Park, Crystal Beach and Ridgeway, the underlying shale does not drain well. This difference was not very important when houses had cellars with floors made of earth, stone, or brick. No one expected a cellar to be completely dry year round. Cellars were for storage, not part of the house. Some even had outside trapdoor entrances. But over time, the cellar became the basement, part of the house's living space. Basements were expected to stay dry and smell clean. External water was drained away through weeping tiles. This posed few problems in areas with underlying gravel or soil. But in areas of shale, such as Crescent Park, and parts of Ridgeway and Crystal Beach, poor drainage was a problem. Basements needed sump pumps.

In some parts of Canada, basement sump pumps are virtually unknown, but in the Fort Erie and Niagara area they are common, particularly in low-lying areas with a high water table. Builders would dig a pit, or sump hole, below the floor in the lowest corner of the basement, and put a removable wood cover on it. As the water table rose around the house, so did the water level in the sump hole, which had an automatic electric pump in it. The pump came on as soon as the water reached a certain level. Few homes had a backup with a gasoline-powered electric generator and few of the pumps were designed to operate underwater. When the power went out, the water level rose, but the pump had no power. Once the sump hole overflowed and the basement flooded, the problem could rectified only by bringing in a portable generator and portable pump.

In Crystal Beach alone, firefighters operated on shifts with 10 machines, pumping out flooded basements. "Ridgeway firefighters were receiving an average of 20 calls an hour, a spokesman said, and had been to some houses two or three times." No matter how hard the crews worked, many basements were unsalvageable. Fred Rail, Deputy-Chief of the East Bertie fire company sighed, "It really makes you sick to see some of the basements that have expensive furniture and paneling in them."[29]

Sunday, March 7

A pencilled note in the company archives, written at 9:30 on Sunday morning, reads in part: "Harvey, Slim and Warren toured the [Crystal] Beach area — situation worse than previous reports." They added "This a.m. 2,100 customers still out, including 500 services down." Three and a half hours later, there was another penciled note "About 3,500 is revised est. of outages."

Meanwhile, men and equipment continued to pour in, including the first contingent from the Cleveland Illuminating Company: 22 men and 10 vehicles — four bucket trucks, four service trucks, and two cars. When American crews started to arrive, Wendy Spry remembers the buzz in the office. Word spread that the customs officers at the Peace Bridge had "just let the convoy come over; you know, there was no paperwork, no nothing, they just opened the Bridge and let them go, because we just needed help so badly." In fact, behind the scenes, Bill Hunter had cleared the way for the convoys that seemed to cross without paperwork.

This photograph summarizes a typical day's work in the aftermath of the Ice Storm. Remove branches, remove downed wires, reattach service connections. Repeat. The two men on the ground are from Niagara Mohawk, and the one on the left has the name "Sam" on his hard hat. (Reproduction by permission of FortisOntario)

The Cleveland Illuminating Company crews left a lasting impression on Harvey McAteer, who went to the bridge to meet the convoy. "I'd never seen anything like this in my life." Harvey knew that when you travel to help another utility, "you load your truck with every piece of material you can get on it, hoping you can use it in the other place." With all that equipment, he had expected that getting through customs might be difficult. So he arrived early and was talking to the customs officers, when they spotted "a parade of lights coming down the Peace Bridge." The trucks were following a station wagon that "pulled up to the customs booth. The customs guy waved them through to the front" of the line.

A man got out of the station wagon, walked over to Harvey, and asked "Who are you?" Harvey explained, and the man said, "Good." The customs officer asked, "What have you got on board?" to which the man responded, "Personal belongings and materials." The officer asked, "Would you say it was worth a lot of money?" The Cleveland man replied, "Well, it depends on what you're going to do with it." To Harvey's amazement, that answer seemed to be good enough. The customs officer said, "That's all right." Customs had been cleared.

The Cleveland man might have been there to work for Harvey and Canadian Niagara, but he wanted to establish the ground rules right away. First he asked if the hotel rooms were ready. They were. Then he started a little speech that he had undoubtedly used many times. "Okay, I'll tell you something. If you got anything you want to know, or you want to tell these guys something, or if there's any reprimanding, I'll be the one to do it. You come to me, not them." Harvey said, "Okay."

Then the Cleveland man insisted on seeing the rooms. The trucks weren't going to budge until he had. "So we went out and looked at them at the Garrison Inn. 'Fine,' he said." Back at the Peace Bridge, he took command again. "He said, 'Follow me out.'

So I did. The trucks followed all out in a line and they lined up all in front of the motel. Not a door opened on any truck. He put up his hand, with four fingers. First four trucks, the doors opened, the guys got out, got their bags in at the registry desk. As soon as they leave, he goes out the front door [and signals again]. Four more."

These crews specialized in construction and storm rescue work all over the States. They were on the road all the time. Their odd life involved an almost military discipline, combined with a certain wildness when they were off duty.

In the morning when the trucks were gassing up, the Cleveland crews "didn't get out of the trucks. They all sat in there — the other guys were out, milling around. Not them." They didn't even eat anything that hadn't been authorized. Harvey remembers that one Cleveland crew member was having a coffee and an éclair in the Canadian Niagara service centre when his boss walked in. "Out! Put it down!" The crew member obeyed instantly. As Harvey reported, "There was no back talk, nothing." Harvey "found this fellow hard to believe, that he would deny them a cup of coffee and an éclair." It was a different style of management from Canadian Niagara's.

At the same time, Babe remembers that the Cleveland men were "kind of wild — worked about 18 hours a day, drank for about two hours, they didn't eat until they had a belly full, then they slept for about two hours." Everyone recognized they were good at what they did, but they were also light-fingered. "Those people were professional thieves." In Babe's eyes, the biggest loss was "two or three 40-foot pogo sticks." These were long-handled poles for pulling wires or limbs; they were only six feet long when collapsed, but could be extended to 40 feet.

It was not that they lacked equipment. Dave Durward remembers, "They came with everything. When their foreman signalled them out, it was kind of like Wagons West." Babe Douglas recalls that the Cleveland crews had a "stake truck with all the tools in the world in it. They did this work all over the country and had quite a unique set up; it was a machine shop on wheels. They even had a hoist to take a truck motor out, rebuild it and put it back in." But nothing seemed to make a bigger impression than their ability to fix hydraulic equipment — the heart and soul of bucket trucks. "They could put a new end on a hydraulic hose in a heartbeat."

Monday, March 8

The Monday newspaper carried the gloomy headline, "Long list of power users still waiting for hookups." Bill Hunter told the newspaper that 125 workers in 39 vehicles were working during daylight hours to restore power.

That day, somebody handed the phone receiver to Harvey McAteer with one word: "Buffalo." It was Dick Torrey. "He was CEO of this outfit, but he also had a position over there [with Niagara Mohawk]. He called me up and said, 'How soon are you going to be done?' And I said, 'Geez, it'll be three or four days yet.' 'No,' he said, 'I don't think so.' I said, 'What do you mean?' He said, 'We're going to be done in two days.' I said, 'Can't do it.' He said, 'What are you doing this afternoon?' I said, 'What do you think?' He said, 'I'll be over to pick you up.'

"Well, he came over with a limousine and picked me up. I said, 'Where are we going?' He says, 'We'll take a little ride around.' So we rode around. He said, 'I think you could get done in two days, couldn't you?' I said, 'Well, we'll try, but I don't think so.' 'He said, 'We're going to be done in two days. I'll talk to you tomorrow. If you can't be done within two days, I'll send a crew over for every goddamned pole you've got here. So we're going to be done.' "

And "in two days we were done, whether we were done or not. We announced to the papers and everything." Was it actually all done? "No, not over here. I don't think over there either. The majority was done. It was a public relations thing. The radio, TV, and all that stuff." For a lineman, the job is done when it's really done and going to stay done. For someone who is constantly fielding questions about restoring service, the job is done when you can stop saying "Pretty soon." Both views make sense.

Tuesday, March 9

By Tuesday morning, the number of people without power had been reduced to 1,900. That was further reduced to 600 by 5:30 pm.

Wednesday, March 10

This was the day on which Canadian Niagara announced that the job was done as of 5:30 p.m. It was almost true. Many cottages in Crystal Beach that remained empty for the winter still had no power, but they were a low priority. And there was a big difference between getting the power back on and getting it back on so that it was likely to stay on safely. Bill Hunter explained to reporters that Canadian Niagara and contractors had "to replace temporary lines put up in the last week," and that it would "take quite a while to get things back to a secure state."[30]

Until most of the damage had been repaired, there was little time to think about what the total bill would be, but everyone knew it would be big. The company just had to keep on working and spending until the power was back for everybody. March 10 was the peak day for outside help, with 83 vehicles on the job. Of these, only 15 belonged to Canadian Niagara. The remaining 68 came from outside and 57 were expensive heavy trucks. On the same day, 215 people were working on storm recovery, 39 of them Canadian Niagara employees; the remaining 176 were hired from outside.[31]

Thursday, March 11

The Immigration Department agreed to let the 11 Niagara Mohawk crews continue working in Canada until Friday, but the Cleveland crews left Thursday morning. Harvey McAteer went to the Garrison Inn to look over the bills and clear up any loose ends. The crews weren't supposed to charge any alcohol to the account, but as Harvey described it, "These guys are professionals. They worked this circuit a lot and they knew how to deal with expense accounts." All things considered, Harvey thought the bill "wasn't too bad," but when the clerk asked about the "lady" who remained in one of the rooms, Harvey said she couldn't stay and Canadian Niagara wasn't about to foot the bill for her.

Friday, March 12

It was time to start taking stock, and saying thank you. During and after the storm, Margaret Tolan, Bill Hunter's private secretary, typed dozens of letters to thank various people and groups for their assistance. In addition to thanking crews for their dedicated work for "long hours under much less than favourable circumstances," many of these letters portrayed the bucket truck as hero: "The bucket truck was an invaluable piece of equipment and continued in this mass power outage to prove itself as the line vehicle for all occasions."[32] Canadian Niagara didn't have enough bucket trucks to handle the situation, and Hunter thought that something might be done about the condition and number of trucks that the company owned.

It was also time to start thinking about long-term needs again. Canadian Niagara hired Bernard and Grey Construction Company to work on permanent repairs to damaged lines. It was a big job, with 12 men, one radial arm derrick, two double buckets and one single bucket, along with four pickup trucks, working from March 15 to March 26. But even by the end of the month, the lines still needed more work. The *Niagara Falls Review* reported that four weeks after the two-day storm, "Canadian Niagara crews are still working to repair street lights and power lines" as well as to make temporary repairs more secure. In Ridgeway and Crystal Beach, some "street lights still do not operate."[33]

Post-Mortem

A few days after the crisis was over, Hunter met with his managers to review the events of the storm. The Bertie Street Service Centre had played a key role. "The whole operation was directed from the Service Centre and we were extremely fortunate that this emergency did not occur while we were at the Queen Street location. The large yard and stores facilities enabled all trucks to be loaded with a minimum of confusion." Only one aspect of the Service Centre let them down. "The one area which could be improved on is the operating room. It was too small for the efficient operation of the personnel for an emergency this large. If the wall between the operations room and the clerk's office were removed, this would be an improvement."

At the time of the storm, Canadian Niagara had employees at the Rankine Generating Station in Niagara Falls and two Fort Erie locations — the Queen Street office and the Bertie Street Service Centre. After the opening of the Bertie Street Service Centre, Queen Street served largely as the company's financial and administrative centre. But during the storm, "Queen Street personnel were at the Service Centre handling telephone calls and assisting with lunches and food for the men as it was needed." The advantages of having all services in one place were clear.

In the early hours of the crisis, the company also felt it had been hampered by two things. One was simply a "lack of CNP trucks and manpower; we were unable to make a proper survey of the possible extent of the overall situation in the early hours of the disaster." It had also been difficult to locate people, including the mayor, and the management team "recommended that a local telephone directory of CNP personnel and other civic employees be updated and from time to time, possibly every six months."

Line Superintendent Harvey McAteer mentioned the expenses run up by some of the workers from the United States and the need to find or make maps to provide directions for out-of-town crews. The management team recommended that in future, "a local area person be assigned as a runner for each out-of-town crew or group of crews."

Canadian Niagara wasn't the only organization carrying out a post-mortem on the storm. The Region and the Emergency Measures Organization learned quite a few hard lessons during the emergency. Mike Taylor, a councillor from Fort Erie, concluded, "We had too many chiefs.... If there are improvements in the disaster plan needed, perhaps it should be in the area of coordination of efforts."[34] While there was no lack of commitment and enthusiasm, efforts were sometimes hampered and tempers flared unnecessarily because of shortcomings in coordination, communications, and knowledge of supposedly accepted procedures.

Con Eidt, Metro [or Region of] Niagara Director of Engineering, lamented the fact that people tended to act on their own without informing others. Apparently an unnamed senior government official had called the army for help at one point. The Region narrowly averted a major expense when someone from the army thought to contact the Emergency Measures Organization before setting out. The army stayed put.[35]

At one point, the Metro Niagara public works committee had offered help to Fort Erie and Port Colborne, but city officials in the latter town turned the offer down without telling anyone. "The region was inundated on Thursday [March 4] with calls from irate Port Colborne people who had not been offered shelter and food." When Fort Erie Mayor Jack Teal asked the region for graders to plough ice, the message got through to St. Catharines, but not to the right people in Fort Erie. Four snow ploughs arrived in Fort Erie, where the local "road superintendent told the drivers to go back to St. Catharines," but that mix-up was resolved. An uncoordinated appeal for food sent hot meals to Niagara Falls when "the hungry people were in Fort Erie."[36] The lapses were more irritating than life-threatening, but local councils would have to pay more attention to regional disaster procedures.

Adding up the Costs

It took a long time for all of the invoices to arrive and be explained to the point at which Canadian Niagara could pay for them. However, by May 1, 1976, a fairly clear picture had emerged. Canadian Niagara's normal payroll of $4,165.68 had jumped to $58,193.30, an increase of 1,297 per cent — and that wasn't counting benefits or supervisory time. But this bill was dwarfed by the cost of outside contract help, which came in at $247,397.30, without counting about 1,700 meals, 523 motel room nights, and other expenses, which added $29,504.11. The costs just kept mounting. Mileage and a pole digger for 236 hours cost $7,536.97.

Wendy Spry, who had joined the company two years earlier, remembers that the storm brought her "the biggest pay cheque I'd ever had in my life." One line worker's monthly pay cheque ballooned from $876.80 to $3,635.98, thanks to $1,093.26 earned at time and a half and $1,665.95 at double time. Another employee, whose regular pay was $397.90, earned $1,812.18 during the storm.[37] The total estimate came to $501,770.29, of which lost revenue was a seemingly insignificant $25,000 or 5 per cent of total storm costs.[38] The final cost exceeded $650,000.

Canadian Niagara and Niagara Mohawk shared a $1-million-deductible insurance policy that covered some — but not all — of the costs. Canadian Niagara's insurable expenses came to $308,192. Niagara Mohawk had much higher insurable losses and therefore got a bigger piece of the insurance payout, but paid out a greater portion of the deductible portion. The insurance company paid Canadian Niagara $218,192, and Niagara Mohawk $2,818,186.[39] When everything was over, Canadian Niagara was out of pocket more than $400,000 in 1976 dollars.

As for the residents of the area, insurance agents and adjusters had to explain that standard homeowner policies did not cover flooded basements, including damage from sewer backups or sump pump failures. Insurance was no help when it came to damages to paneling, furniture, clothing, or books stored in a basement. Nor were the "wind and hailstorm" clauses any help, because they excluded "damage caused by water, waterborne objects, ice or waves." However, damage to houses from falling trees and poles was generally covered, and in some cases "broken power service stacks and lines leading from street poles to houses may be covered."[40] Homeowners ended up paying more than they thought they should have, but so did the municipality.

The Town of Fort Erie was able to get the Region of Niagara to pay for some of the damages, but it was hard hit nonetheless. The Ontario government had ended financial support for the Emergency Measures Office the year before. Up to January 1 of that year, the EMO had been funded 75 per cent by the federal government, 15 per cent by the province, and 10 per cent by the municipality. Starting in 1976, the federal government would pay 61 per cent and the municipality 39 per cent.[41] Subsequent efforts to get any kind of disaster relief from the province failed in the legislature.

The choice of a single word determined the financial help that Fort Erie received. Had Fort Erie come through an "emergency" or a "disaster"? As the *Fort Erie Review* reported, the distinction was important. "Because last week's ice storm was officially called a disaster, Metro Niagara public works committee voted...to pick up the bill for all work done by regional forces. Had the situation been called an emergency, all costs would have been charged to Fort Erie."[42] Maybe regional government wasn't so bad after all.

The Scorecard

Perhaps the greatest lessons were those listed in the "Scorecard on a Storm" that was published in newspapers and in a letter that went out to all employees. Niagara Mohawk had turned to the public relations firm of Marlow/Johnson in Fayetteville, New York, to prepare the Scorecard.

It was an effective document and Canadian Niagara paid to have its own Scorecard on a Storm published locally. The Scorecard explained that the 41 Canadian Niagara employees working to deal with the storm were able to accomplish what they did because they had had so much help. It shows the tremendous variety of people, services, and occupations involved and thanked the many people "WHO, ALL TOGETHER, GOT THE POWER BACK ON."[43] The Canadian Niagara family was bigger than anyone had imagined.

SCORECARD ON A STORM:

THE TOLL

- Ice, wind, sleet damage covered 57.6 square miles.
- 181 miles of primary circuits out of service by falling trees and limbs.
- 8,500 telephone calls received and handled.
- 900 meters shorted out.
- ALL CANADIAN NIAGARA POWER CUSTOMERS AFFECTED

THE COST

- 10,000 area customers knocked out of service.
- Damage: more than $650,000 to electrical power equipment and facilities.
- 16,000 manhours spent on emergency repair service.
- In a continuing emergency situation, 41 Canadian Niagara employees combined with 173 men of various other utilities and contractors who came to their aid from Cleveland, various parts of New York State and neighboring utilities in the Niagara Region.

THANKS: to Canadian Niagara customers who suffered the full brunt of the storm damage for their reasonable patience while experiencing inconvenience and discomfort.

THANKS: to police and fire departments and personnel, Regional and Town officials and employees, public works personnel, telephone company employees, Provincial Gas employees, private contractors, Canadian Customs and Immigration officials - all of whom participated in a necessary co-operation, or in repairing the damage that speeded the return to normalcy following the worst storm in terms of power damage in the history of Southern Ontario.

THANKS: to suppliers and transportation facility operators for racing to us vital materials and replacement parts essential for repairing the widespread damage.

THANKS: to newspaper and broadcast officials and personnel for their publicized warnings on the dangers of downed wires and tree limbs as well as other storm-related information which effectively backed up the men and women in the field working to restore power.
THE FACT THAT NO ONE IN THE GENERAL PUBLIC WAS INJURED PROVED THE GREAT VALUE OF SUCH REPETITIVE WARNINGS.

THANKS: to restaurant and motel operators and their personnel for caring for our crew members who were forced to eat and to sleep when and if they could, wherever they happened to be.

THANKS: and to the Canadian Niagara men and women who worked throughout the unprecedented situation, 16 hours and more at a stretch - the trained and willing specialists who responded to the 8,500 calls; the emergency crew members from Regional Municipality of Niagara, Niagara Falls, Ontario, Niagara-on-the-Lake, St. Catharines, Thorold, Chippawa, Buffalo, N.Y., Niagara Falls, N.Y., Olean, N.Y., Cleveland, Ohio.

THANKS: to all the others who were part of the largest emergency team effort ever put together in the Niagara Region; the vehicle mechanics, service personnel, engineers, dispatchers, hardware suppliers and supervisors and managers who organized the eight day effort and
WHO, ALL TOGETHER, GOT THE POWER BACK ON.

CANADIAN NIAGARA POWER COMPANY LIMITED

Goderich & Queen Street, Fort Erie, Ontario

The "Scorecard on a Storm" summarized the results of the disastrous storm of March 2, 1976. It was the largest emergency team effort to date in the Niagara Region and thanks were owed to many individuals and organizations. (Reproduction by permission of Fort Erie Historical Museum, Ridgeway, Ontario)

This undated photo is typical of how linemen and riggers worked for decades before stringent safety regulations were introduced. None of these men on a high tower is wearing a safety belt. Their fedoras protected them only from the sun. These men probably worked for the Niagara Falls Power Company or Niagara Mohawk on the American side of the Niagara River. (Reproduction by permission of the estate of Daniel Dumych)

Dangerous Work

CHAPTER 14

The Old Folks' Home

Sometimes when you hear the same joke over and over again, it gets pretty tiring.[1] Back in the 1960s, Earl Snider and Harry Saylor Sr. used to meet their friends from Chippawa Hydro for a drink at the Chippawa Hotel. As Harry recalls, "Every time we'd see one or the other of them, they'd say, 'How are you fellows doing at the old folks' home?'" That's what a lot of people called the Rankine Generating Station. It looked peaceful on a sunny day when Canadian Niagara staff were having lunch on a bench in the park.

Things looked very different down in the lower levels of the generating station. One night, when Earl and Harry were working overtime below the turbine deck, Earl decided to "invite a couple of those loud-mouthed bastards from Chippawa" to see what work in the Rankine Station was all about. Earl and Harry had been working hanging upside down over the turbine tailrace. It's a terrifying place, and that's what they wanted their guests to see.

As the Chippawa workers descended ever deeper into the generating station, they became less and less cocky. By the time they got to the turbine level, they were clearly rattled. To Earl and Harry's satisfaction, one of the guests suddenly shouted, "The hell with you! We're not going any further than here." And from that day on, the Chippawa Hydro crowd "never said another word" about the "old folks' home."

Earl and Harry faced danger high above the ground too, although they didn't get much respect for that from their Chippawa friends. One night in the early 1970s, during a sleet storm, the men heard that there was a problem on the 90-foot high tower at Chippawa Creek. Ice was forming on the tower and a loose guy wire was whipping about dangerously in the high winds, threatening to short-circuit the other lines. As Harry recalls, "I had to climb this damn tower. It was covered with ice so thick I had to break it off in chunks so I'd have something to hang onto." Part way up, the rungs stopped, "because you're getting into the live wire." He had to get to the other side of the tower, but it was too slippery to stand, so he just sat down and started sliding across, hanging on to the cross braces. Then he climbed up the other set of steps to reach the guy wire and cut it. The timing had to be just right; Harry had to cut it when it was over clear ground, not over the Chippawa Hydro lines lower down. Earl and Harry watched it whip back and forth. At a moment when the wind was blowing away from the lines, "Earl said, 'Cut her.' So I cut her and just as I cut her she went flying." Unfortunately, something happened, maybe the wind shifted at that moment, and the line landed on the Chippawa Hydro lines.

Harry and Earl saw "a big flash." About half the residents of Chippawa lost their power. It stayed off for some time, because a power transformer had to be replaced. The Chippawa men didn't congratulate Earl and Harry for their nerve in going up the tower in the first place. They'd just ask, "How are the high-wire guys tonight? Gonna cut any more wires?"

Coping with Stormy Weather

Many of the dangers faced by Canadian Niagara linemen were the result of bad weather. During one particularly bad winter windstorm in the 1970s, Point Abino, where many of the area's biggest summer estates and mansions were located, was badly flooded. At that time of year, most of the residents were gone, except for the elderly Mr. and Mrs. Mackie. They lived on the Baird estate, in a house that had been built for them by their employers when Mr. Mackie had become bedridden.

Babe Douglas got a call that the place was completely flooded and the power was out. The only way to the property was a twisty private road. Babe assumed that if he could stay on the road, he'd be all right. "I figured the truck was high enough that the water wouldn't be that deep. I get half way in, I'm flooded. The engine was flooded. The water is coming in the cab.... My radio was still alive. It eventually died, but I called out and said, 'You guys know where you can get a boat? I'm stranded.'"

Jim Nigh, who had been on another job, drove to Point Abino, borrowed a small aluminum boat, and started rowing out to rescue Babe. Meanwhile, Babe, who could see that the fuse had blown in the pole-top transformer, had realized that it was a simple job to fix it if anybody could get there with the right tools and a fuse. When the boat arrived, Babe stepped in and Jim rowed around to the tool bed of the truck. Babe found some fuses, a set of spurs, and a belt. But as Babe recounts, "I was supposed to be the boss," so Jim, who is now Line Superintendent, "strapped on the tools, rowed there, tied the boat to the pole, got out of the boat, climbed the pole to the transformer, fused it in, came down, and rowed to the house."

There they found that the basement had been partially flooded, so even when they got the power back on, the furnace wouldn't work. However, the couple had some electric heaters, a gas stove, and a good supply of food. Babe and Jim made sure they were all right before they left.

Snow also provided plenty of drama. In 1977, the worst winter snow storm in decades hit the Niagara Frontier. Some cottages on the lakeshore were completely buried in drifting snow that blew in from the frozen Lake Erie. Workers located some houses only by digging down from the television aerial. Frank Snitchen, owner of Garrison Power Products and a Skidoo Club member, was one of the volunteers who used to take Canadian Niagara workers out on calls. Frank drove and Babe used his "glow detector" to see if the overhead lines were live. The snow was so high that poles were buried. He stood up on the snowmobile to work on the lines. They were heading towards the lake when all of a sudden Babe felt a pain in his leg. "We went back and investigated. My leg had got hit by the chimney of a cottage!"

Storms create emergencies, and emergencies sometimes call for improvisation, rather than strict adherence to rules and standard operating procedures. During the Great Ice Storm of 1976, lineman Syd Freeman came across a place "in, I believe, Crescent Park, where the people were living in their kitchen with the gas stove on [as their only source of heat]. There was a newborn baby, home just three days from the hospital." Syd quickly sized up the extent of the disaster. The service stack was bent horizontal, the line from the pole top was drooping dangerously over the roadway, and the "leads for the meter all busted." There was so much damage that it couldn't be fixed properly for days. According to the rule book, Syd should have left and told the family to find somewhere else to stay. But the baby was starting to turn blue with cold.

Syd had a chat with the mother. "I said to her, 'It's either go back to the hospital with your baby, or I'll see if I can do something, if you don't mind losing a pane of glass in your cellar window.' " He attached the line from the pole to the side of the house at the roof line to give it road clearance. He let the line drop down past the broken meter and "to the ground in front of the cellar window." Then, as Syd puts it, I "put my toe through a cellar window and put cardboard over it, taped it up, cleared the top side of the fuse box, hooked the line right into their box and gave them power. There were no fuses, no meter, nothing. I said, 'If you have a short or anything at all, yank them out.'"

He did it "so their furnace would go on, because the baby looked almost blue to me. It was so cold, and I thought, 'It's either that or go back to the hospital with your baby.' And I thought, 'Well, they're here, I'm here, just do what you can.' " When the furnace came on, Syd recalls, "The lady was crying. She was quite happy with me."

Syd had thought that situation through carefully. When asked if he'd made any other temporary fixes, he said, firmly, "No. That's the only one. My heart went out to them."

Rescues and Roof Tiles

The history of Canadian Niagara is filled with one-of-a-kind stories. For example, Sid Morris remembers that one time, "a dog got stranded in the forebay on the ice and we couldn't get him close enough. Hector [Smith] got on the tugboat and put a rope out. He just wiggled it a little bit and the dog got kind of friendly. The dog got close enough to him, he grabbed the dog, put him in the tugboat and they called the SPCA. They came along and picked the dog up. Hector's a very good fella."

However, some difficult jobs were not part of the job description, as Bob Douglas found out soon after joining Canadian Niagara in 1947. He started on the four-to-midnight shift at the powerhouse. "I started as a janitor. Everybody did." The next step up was electrical janitor, but by that time, he was quite an experienced hand at keeping Rankine spotless. On the day shift he cleaned windows, a lot of windows. "You had so many of those big bays to clean every day, or every week. You'd do a certain amount and then the guy coming after that would do the next bunch. That job was rather unsafe. You used a makeshift type ladder that sometimes tipped over. It was a wheeled thing. Then you had to go up on the crane and do the crane track windows. That was kind of interesting."

Shooting pigeons inside the generating station with a pellet gun was another of his duties. Pigeons were an unavoidable part of life in the powerhouse. The huge windows provided necessary cooling and ventilation for generators that throw off a lot of heat. With 10 or 11 generators working, the place would have been impossibly hot if the windows hadn't opened. But windows also let pigeons in. Bob explained, "I don't like to shoot things, but it was just a case of self-preservation you might say. They were dirty birds and they never left — at least, not unless they went out feet first." The place was a lot easier to keep clean without the pigeons.

The windows also created problems with the roof tiles. The generating station is on a north-south axis with large banks of windows on the east and west sides. Breezes that blew in windows on one side, and out the windows on the other side, helped cool the building, summer and winter. However, in stormy weather, gusty winds that blew in the windows occasionally pushed up roof tiles — which were attached to the roof ribs, rather than to a solid underlay — and then sucked them down again. The men would run to close the windows. If they didn't make it in time, tiles that had been cracked and worked loose rained down onto the plant floor. You could see open sky where the tiles had been and the rain poured in. During the July 27, 2005 celebration at the Rankine Generating Station to honour one hundred years of commercial electric power production, John Crotinger quietly remarked that "some people who never worked here might think it was romantic. I'll bet they weren't here during a wind storm when tiles were crashing down on the floor like big rocks."

The following day, someone would have to replace the tiles. Getting up on the roof was generally rigger's work. It wasn't easy. Harry Saylor Sr. remembers that walking on the "tiles, especially if they were a little dry and had a film of dust, would be just like walking on a bunch of ball bearings, slippery as anything. You had to have a rope on you when you were working on that. You couldn't stay there by yourself unless somebody was holding you on the top."

Two men were needed on the job, "One on the peak with the rope holding the guy that's putting the tile in." The other man on the roof worked at the hole, and hauled tiles up, hand over hand with a rope, until he had enough to fix the hole.

In the late 1950s, the entire roof was re-tiled. The roof crew was made up of "riggers, electricians, machinists, carpenters, everybody." Orin Steele, the machine shop foreman at the time, and later general manager of the Fort Erie office, designed a platform that went up to ceiling height so the men could reach up and fasten the tiles down from the inside. This made the job much easier.

Some hard, potentially dangerous jobs were responses to problems. Others were scheduled in advance. Every year for many years, the company completely overhauled one of the eleven turbine and generator units. It was a big job with lots of overtime, and there was always a sense of urgency, because a unit that wasn't working wasn't making money. In addition to the electrical crew, the company pulled in extra workers, such as the machine shop crew and carpenters. This meant working with people who weren't familiar with the powerhouse. One man was regarded as a particular hazard because of a combination of overenthusiasm and bad timing.

It was difficult getting the turbines apart at the bottom of the pit so the parts could be hauled up by crane. Huge nuts had to be loosened by putting a wing nut on top of the nut to be loosened and then hitting the wings with a sledgehammer. Two men would stand facing each other on either side of the nut and swing in unison. Timing was crucial. Both sledgehammers had to hit each of the two wings at exactly the same time. If one hit first, the other wing would start to rebound just as the second sledgehammer was hitting it and "it would just about break your arm, it would bounce back so much."

The job demanded skill and a sense of rhythm. As Harry Saylor Sr. explains, "It doesn't matter how hard you hit, unless the two of you hit together to loosen the nut up" it won't loosen. A carpenter hauled in to help at disassembly time "couldn't wait to pound things." Unfortunately, "he had no rhythm at all. One guy refused to work with him, [complaining] 'That guy is going to kill me. He's swinging twice as many times as I am. I'll swing a sledgehammer with anybody else, but not him.' "

In addition to routinely scheduled maintenance, there were a few once-in-a-lifetime jobs. John Damato joined Canadian Niagara in 1956. Within a period of only a few years, the company changed over to 60-cycle power, temporarily shut down the tail race tunnel to the face of the gorge for inspection and maintenance, and completely re-tiled the Rankine Generating Station roof. John worked on all three jobs. As he explains, you didn't bring an outside contractor in "if you could do it yourself. Everybody busted their fanny to do it, to make sure that things are operating properly, so things were fixed up."

Everyone pitching in was an important part of how Canadian Niagara did things. The work force was talented and flexible and so were schedules and tasks. You went where you were needed. As John explains, during the roof re-tiling, "I'd be up there pinning the tiles, and I'd be called, 'Hey, there's a problem there, go down and fix it.' When I got that job finished, I'd be back up pinning tiles." Flexibility and having a little slack built into the system meant that a lot of one-of-a-kind or unusual jobs could be handled economically without outside help. And it was not as chaotic as it might seem. The men knew the powerhouse and they knew each other. Both kinds of knowledge added immeasurably to safety.

Having a Blast

Attitudes towards safety have changed over the years. There are more regulations than in the past and safety is approached more formally. Jim Nigh is the Canadian Niagara line superintendent. Today, he cannot allow his workers to do many of the things he did when he was a young lineman. "In the 1970s, everything was freewheeling and in the 1980s [safety rules and procedures were] getting a little tougher."

Consider dynamite. As a young lineman, Jim worked putting up new poles and lines beside the Queen Elizabeth Way while traffic whizzed by. They drilled holes for dynamite and then "we'd go over to the quarry to the dynamite shack. They'd say, 'Yeah, here's the dynamite, here's the [blasting] caps.' You got it all in the same box. They'd send the bill to the power company.... We didn't even have plastic mats." Plastic or rubber mats were often used to absorb the shock moving upwards and to prevent dangerous debris from flying in the air.

Instead of mats, Jim says "When I started, we took a bunch of butts [sections] from a pole, laid them over the top and chained them all together." This approach had its problems. One day, "somebody was jackassing around and put in way too much dynamite...and this stuff [went] flying all over." When the crew was blasting out old pylons near the Peace Bridge, they "smashed a guy's car. I forget what happened there."

Riggers didn't climb poles, but they often worked with line crews on various jobs such as drilling or blasting holes and putting up poles. So it was nothing out of the ordinary when a line foreman asked Harry Saylor Sr. and another rigger to help put up poles in Ridgeway. However, it did seem a bit unusual when the line crew foreman showed them where he wanted to put up a pole, told them to drill the holes for the explosives and go ahead with the blasting, and then left. Harry and his partner had helped with blasting pole holes, but when it came to deciding how much dynamite was needed, they were inexperienced. They put a lot of explosives in, wired them up, covered the area with a mat made of rubber tires, and put some timbers on top, to try to hold everything down.

It didn't quite work as planned. "We got a big bang. Rubber mats and everything went up in the air," some even up into the telephone wires that, fortunately, did not come down. When the line boss returned, he was furious. "What the hell are you doing? We're going to put in one pole, not a substation. It's big enough to put a foundation in for a house." Harry admitted they were lucky "nobody got hurt." He had often used dynamite for clearing ice, but had overestimated the amount he would need for the hole. He concluded that the blasting shouldn't have been left to "two clowns" who "knew nothing about dynamite, really."

Sid Morris, however, had considerable experience with blasting. Neither he nor his foreman, Warren Kent, expected any surprises when Sid, along with two Fort Erie men, were preparing holes close to some cottages near Prospect Point Road. Sid recalls that as Warren left, "He gave me the [blasting] caps and said, 'Be careful.' I said, 'I'll watch it.' The first hole was good. The second hole was all right." But the third one picked up a buried water line and left it standing up in the air in a U-shape. It was bent, but not broken. It wasn't even leaking. Sid figured there was only one thing to do. "We dug a trench and set it back in." The men figured nobody needed to know, other than the boss. When the foreman came back, "I said to Warren, 'I moved the waterline.' He said, 'Did you break it?' I said 'No.' He says, 'Well, we won't worry about it,'" and laughed. They both knew that the one-inch pipe was part of a cobbled-together private water system for cottages. Nobody knew where these lines were.

Dynamite also helped break up ice. As described in Chapter 9, the ice breaker tugboat in the forebay was supposed to disperse the accumulated ice so it could float out through the ice sluiceway and over the falls. But sometimes the tug got stuck and couldn't move, particularly if it was driven against the wall of the generating station and held firmly by the strong current and the ice.

There was no designated tugboat operator. The job went to whoever was around. One day, when line foreman Warren Kent was operator and Sid Morris the helper, "the tugboat was stuck by the powerhouse on an angle," in danger of capsizing and sinking. Warren called for dynamite. "We dug a hole [in the ice] beside the tugboat, put the dynamite in, put the wires up to the controls, pushed the controls in," and "it just picked the boat up and set her straight."

When the ice was bad and dynamite was needed, it was needed in a hurry. The men wanted to keep some handy. But it wasn't safe to keep dynamite inside an electric generating station. "We used to hide it in the bushes along the sluiceway. You couldn't do that today. Today you've got to have a special place, and you've got to have a permit and everything. We never had a permit." The bushes were very thick and the dynamite was well hidden from all but those who were in on the secret.

Percy McAteer lived with his family in one of the company houses on the top of the escarpment near the transformer house. It was one of the perks that came with being on call around the clock. In the 1930s, a weir had been built out into the river to divert ice away from the forebay, but sometimes pack ice would jam up close to the mainland above the weir. Percy and Sid had to deal with it. One day, it was Sid's turn to get out on the ice to plant some dynamite. The problem was jammed-up chunks of ice rather than solid ice. Crawling out on it meant spreading one's weight, and the downward force, as evenly as possible on a plank.

"We put a plank out, put a rope around you [someone on shore would hang onto the rope], and you'd crawl out on the ice, dig a hole and put the dynamite in" and crawl back. The dynamite loosened the ice. If the water level was high enough, the ice floated over the weir and over the falls. But if the water level was low because ice jams further up river were reducing the volume of water draining out of Lake Erie, then you might as well put your dynamite away and stay inside where it was warm.

Linemen, riggers, and other Canadian Niagara employees don't blast any more. It is a job for specialists with licenses and permits. Jim Nigh "can't remember the last time we hired one."

Getting a Hit

Driving rain and lightning storms bring out linemen when others stay indoors or in their cars. Lightning arrestors and fuses prevent pole-top transformers burning up, but the fuses have to be replaced after a lightning hit, a short circuit, or a strong power surge. Sometimes a connection simply shakes loose and has to be reconnected. The tasks are varied, but one thing is constant: linemen have to be outside climbing up and down poles in bad weather. Babe Douglas remembers one of these days very vividly.

"I grabbed the [transformer] lead, which was hot, with my right hand, and there I was hung up." He'd had a hit — 2,400 volts surged through his right arm. He couldn't pull it free. He reacted quickly. "I was lucky enough that I had my other hand loose. I drew it back and swung it across as hard as I could and banged my forearm and broke myself off it. Otherwise I would have been...." The words are unspoken. "You know, sometimes it crosses over and you're a goner. I was lucky this didn't get through my feet. It stayed on one side of my body and that's because I broke myself free very quickly. Had I not, I'd have been fried probably or very seriously hurt. I ended up with just a little mark, which went away."

Babe knows why the accident happened. In those days, the construction group had seven-man crews with four men riding in the back cab of the truck. When a lightning storm occurred, the truck would go out to do numerous small jobs to keep the power flowing. There would be a lot of short stops. A lineman would go up a pole quickly, fuse or tighten a lead on a transformer, come down, get back in the truck, and move on to the next job. It was crowded in the truck, "so it as very hard to remove your sleeves and remove your gloves each time. So we used to test them [the gloves] prior to starting and we'd put everything on. And usually for the first ten jobs, you probably left everything on, because you took turns getting out." The men wore leather gloves over insulated rubber gloves; the leather was intended to protect the rubber from nicks or cuts.

"The leather gloves were soft leather, and they would get really, really soggy. [That day] my gloves were soggy. I had the equipment [rubber gloves, leather gloves and rubber sleeves] on for maybe two hours and we stopped at a pole. There was a loose connection on the pole. I had to go right into the live side of the transformer to fix it. It was one of the drop leads, so I upped the pole and unbeknown to me, on the way up I caught a nail."

Something sharp had gone through the leather and put a small hole in Babe's rubber glove. He was no longer insulated. Electric current could now pass through wet leather gloves, through the rubber and onto his skin. Babe was an accident waiting to happen. Fortunately, although he didn't know about the cut in his glove, he followed the rules. "One important rule whenever you're going to grab a hold of anything that is alive, or can become alive, is: *belt in first*. You don't go up and grab something and stand there with only one hand."

Babe has thought about that close call many times over the years. "I don't think you could call that carelessness. It was just an unfortunate thing that my glove caught a nail on the way up. It wasn't anything that I'd done completely wrong. How could you know?"

Normally, before each climb, he, like every other lineman, was supposed to take each rubber glove off, cup it to his mouth, and blow in to see if air leaked out. But even that would not have helped; the glove had torn on the way up. It was a rare occurrence.

Even rarer, Babe got a second hit. This time, he says, it "was somewhat my fault and I was the first to admit it." It happened during another ice storm. "We were replacing some wires that got pulled down" by the ice. The circuit was dead, so he was working without rubber gloves. No safety violation here. The line had broken in two and he was going to splice it together, "I'm in the ditch, standing in water, I had good rubber insulated boots on, and I'm putting it together." That's when a miscommunication occurred. The men on the pole thought Babe had finished the splice and had let go of the line. They also thought the other line from the same cross arm was dead. But the lines touched and the dead line wasn't dead after all.

"They pulled that wire up over the cross arm, and the other side of the cross arm was alive. They pulled it up while I still had a hold of it. It touched the wire on the other side. They realized what they had done, quickly removed it, but it threw me. I was thrown maybe ten feet and pretty stunned, but I went back to work the next morning." After the hit Babe was "so thirsty all I wanted to do at the Doctor's office was get a cold drink. I couldn't get enough to drink. My gums were so dry." The shock loosened some teeth which he eventually lost.

Syd Freeman has a scar on one thumb where he got a hit when there was not enough separation between the dead line he was working on and a nearby live line. "We were putting in a new line beside a hot line down by McAvery Cemetery, near Garrison Boulevard." Linemen don't wear rubber gloves when working dead lines, but then the wind picked up and the lines started blowing around. Syd realized the potential for danger and went to get his rubber gloves. The wind blew the new line straight at his face and instinctively he put his hand up to ward it off. What he didn't know was that the wind had also blown the new line into contact with the old live line. Both lines were now live.

"I couldn't pull my hand away. Your muscles tense right up. I can remember the guys saying 'Let go,' and I said, 'I can't.' And then it swung free. It sort of threw me backwards across the ditch [and about 8 or 10 feet] onto the gravel road. I was on the ground. You've had the sensation where you've maybe come in out of the cold weather and had a nice hot meal and you'll sit there and you want to read the paper and you're dozing? That's exactly the sensation I had. I wanted to go to sleep. When I was on the ground and the guys were doing resuscitation on me. Lance Taylor was saying, 'Wake up. Syd! Wake up!' And I said, 'I'm not asleep.' Then all of a sudden, 'You just got zapped. You just got zapped, Syd!' I started to get up, and Lance said, 'Relax for a minute.' "

"I pulled my glove off and there was the hole — you could put a pencil in there to the bone — no blood, because it was cauterized. I went to the hospital and they put me on oxygen for three or four days and measured me for water intake and water output, to see if any of my juices were cooked. The doctor refused to let me go back to work for a month. I wanted to go back the next day, put a band-aid on it, let me go work. But he refused. He said, 'You could go back into shock and you could be in trouble.'"

Dr. Derbyshire's explanation was good enough for Syd, but not for the line foreman, who thought Syd should go back to work. The doctor is reported to have said "'You're a lineman. You do your line work. I'm the doctor, let me do the doctor work. If he goes into shock and dies, his family could sue you people. He's off until I tell him to come back to work.'"

Not everyone was as fortunate as Babe and Syd. On Saturday, December 16, 1911, three electrical fitters at Canadian Niagara were standing on the top of a live 12,000-volt transformer. One of them, A.H. Baldwin, saw "a sudden flash and then darkness, and found himself on the floor." A cable had come loose and hit them. Baldwin and another man were severely burned; the third man, Melvin C. Brown, was electrocuted. Brown's pallbearers were all from Canadian Niagara. At the coroner's inquest, assistant superintendent A.D. Robb told Dr. G.T. Kellam that he had considered it unnecessary to make the transformer dead before the men went to work above it, because all the equipment was inspected and the insulation was considered satisfactory. Unimpressed, the coroner found the company negligent and ordered that from then on, men would not be allowed to work on live transformers. It was a new industry and an important lesson.[2]

Pylons or towers have to be painted regularly to keep the rust at bay. But in order to discourage unauthorized climbers, they were not designed for easy climbing. Painters scrambled up and down with their brushes, paint buckets and safety belts. The painter in the centre here is Jack Lamb; his safety belt is not attached to anything other than his waist. There was always the danger of nearby live wires, and in 1935, a 27-year-old Canadian Niagara employee named Melvin Ernest Higgins was electrocuted while painting a transmission tower when he got too close to the live circuit. He was thrown to the ground. (Reproduction by permission of Sid Morris)

Melvin Ernest Higgins was a maintenance man and painter who worked out of the Rankine Generating Station. On September 16, 1935, he was part of a crew that had been working for three months painting the two lines of towers that that carried transmission cables from the Rankine transformer station to Fort Erie. The towers had to be painted periodically. To balance safety and the need to maintain the power supply, painters did one side at a time and the live side was marked with a red flag.

Higgins was electrocuted that day, but no one could explain why. The coroner, Dr. John M. McGarry, questioned the members of the painting gang, the foreman and the assistant superintendent Don Bracken. No one knew why Higgins had got too close the live wire and been thrown to the ground. Hector Smith Sr., who started with the company in 1931, remembered that day for the rest of his life. In 2003 when he looked back, all he could say was, "He might have slipped a little bit and grabbed something. I don't know." (His son, Hector Smith Jr., who also worked for Canadian Niagara, married Mary Lou Higgins, second cousin to the late Higgins.)

Keeping the Public from Getting a Hit

One of Canadian Niagara Power's earliest accidents involving a member of the public occurred in 1903. As the *Niagara Falls Daily Record* described it, "With his wife and daughter Mr. [Hiram] Stanton was walking along the Military Road in Stamford township when he touched a guy wire leading to the ground from a pole carrying electric wires to Mr. Acheson's residence in the Lane. In a flash Mr. Stanton was drawn to the wire and convulsed with the full force of electric current....His wife and a friend of Mr. Stanton's raised an alarm after trying in vain to release the victim. Mr. C. Pinneo, who lives nearby and his five sons...ran to the rescue and succeeded after several minutes hard work in releasing Mr. Stanton, who was terribly burned and unconscious.... The escape from death is considered miraculous."[3]

The view from the high tower that carries power cables across the Niagara River to Buffalo. It is more than 200 feet down through the steel to the river below, on the right, flowing towards the top of the picture. Near the top centre a building is visible, and on the left is the intersection of Niagara Boulevard and Princess Street. The first time Sid Morris went up the high tower, he was hanging on so tight he felt he was squeezing the steel hard enough to make it thinner. Later he felt comfortable enough to take photos like this one. (Reproduction by permission of Sid Morris)

Clearly, there should never have been a live wire leading to the ground. It was the early days of the company and of electric power. Safety procedures were in their infancy, and the public was unaccustomed to the dangers. Over the years, the company developed safety procedures to protect both employees and the public. However, not all members of the public learned to protect themselves. No matter how hard you try to tell them, some people just do not understand how dangerous electricity can be. And some people, drunk or sober, are just careless or unthinking.

During his 45 years with Canadian Niagara, former line superintendent Harvey McAteer saw "maybe six or seven Canadian Niagara men get a jolt or thrown, but none of them got killed." He couldn't say that about others who climbed towers. Some were unscathed, some were badly injured, some died.

One night he was awakened by a telephone call. "We've had a trip on 6 line. We've got word that there's somebody up the tower. Will you go down and see what you can find out?" The high tower for the lines going across the river to Buffalo is close to Harvey's house. "I called in another guy, I think it was Bob Hanna. And so we went up. Sure enough, there's one [man] lying down and his shirt was on fire or had been on fire and he was burned pretty good." The man wasn't dead. "He was conscious. The deck planking was burning even."

When the firemen arrived at the base of the 20-storey-tall tower, "they came up with a stretcher and ropes to lower the guy down. He wouldn't go [on the stretcher]. He came down himself, his buddy behind him and one of the firemen in front of him so that he

couldn't fall, but he came right to the bottom and then he passed out. But all he could mumble, all the way down, was 'I'm on fire inside.'" That left Harvey and Bob to douse the smouldering deck that the current had set on fire. "We peed all over the goddamned platform to put the fire out. Because that goddamned tower's 230 feet high. You don't want to go down for water and come back up."

The night had been trying enough, but they had to go back up the next day. It wasn't a pretty sight. "We could see where there were pit marks on the railing where one hand had been and there were marks that you could see on the jumper conductor where he'd got the other hand. He took it right across."

Others who climbed the high tower were less fortunate. A few suffered even without climbing a tower. Sometimes an angler would cast up a fishing line where it became tangled in overhead lines and turned into a deadly conductor.

One of the most astounding features of the Great Ice Storm of 1976 is that no one was injured or killed from downed power lines. There were excellent warning systems in place with radio, newspaper, television, police, and fire services all cooperating with Canadian Niagara to warn people of the dangers of felled lines. The only known casualty was a dog.

Syd Freeman saw a "a line go down, land in a ditch. A man [who lived nearby] had power in his house, but his fridge would go on and off, his furnace too. He didn't have enough voltage to kick them in. [The power] was feeding through the water in a ditch. The man's dog went over and got a drink out of that ditch. It was the end of him." Although it is sad about the dog, it could have been a child who had stepped in the ditch.

The view from the top of the high tower was spectacular. Once he got comfortable with being on an open tower more than 20 storeys tall, Sid Morris liked to look around and take pictures. The world looked different from up there. The graceful lines of the Peace Bridge were clearer. The Niagara River Parkway was a ribbon of concrete that cut through the grass. (Reproduction by permission of Sid Morris)

The Dangerous River

Electricity is not the only source of danger lurking on the Niagara Frontier. The river and the falls pose enormous hazards to those who live and work by them. Everyone who has ever worked at the Rankine Generating Station is aware of the river's power and ever-present threat of death. Some have fished bodies from the forebay, some have watched in horror as a never-to-be-seen-again fellow worker slips into the tailrace. Others have dismantled a turbine to take out what is called "the remains." But one story of someone who was caught in the river has an improbably happy ending.

Sherry Vyverberg was a 20-year-old nursing student from Rochester who was visiting Niagara Falls with friends on May 30, 1983. She and her friends were standing on a concrete observation platform near the Toronto Power Company building, about 150 metres above the Horseshoe Falls, throwing stones into the water, when suddenly she lost her balance and fell into the river. She went under, but resurfaced. Although she couldn't swim, she knew enough to roll onto her back as the rushing water carried her along.

One of her distraught friends jumped into the water to try to save her but was forced back by the strong current. Another friend, who was hobbled by a leg cast, limped to the road and waved for help. Three workers in a passing Canadian Niagara Power Company truck — John Marsh, Peter Quinlan, and Joe Camisa — stopped. When they heard what had happened, John, a rigger, grabbed some rope from the back of the truck. He ran along the shore, trying to find a place to throw the rope to Sherry. But she had drifted too far away.

Suddenly, she hit the breakwater or weir and started drifting towards shore — and towards the intake to the forebay and powerhouse. Being sucked into the powerhouse would have been just as fatal as going over the falls. Seeing both his chance and the increased danger, John tied the rope around his waist and jumped in. Fortunately, he was an experienced scuba diver, a good swimmer, and a volunteer fireman who was trained to respond to emergencies. His co-workers anchored the rope to a railing. John swam out to Sherry, grabbed her by her long hair, and hauled her back to the shore where the other two men helped to pull her out. The two emerged from the river shaken, but uninjured. Sherry just had time to say "Thanks" to John before she was taken to the hospital, where she was treated for shock and cold. He spoke briefly to the police, and went straight back to work.[4]

The newspapers were filled with the story, and Sherry spent the next three weeks telling reporters over and over again about her experiences. John avoided much of the publicity, but at one point, the two were flown to New York to appear on "Good Morning, America." He also received countless awards, medals, and citations for bravery. The two stayed in touch for several months, and he spent that Thanksgiving with the Vyverberg family in Rochester.[5]

John downplayed his efforts and was embarrassed when people called him a hero. "She needed help," he said simply. "And I was best equipped to give it."[6] However, there was one additional hero who was never mentioned in the papers, never went to New York, and never received a medal. When the accident occurred, one of John Marsh's co-workers radioed the powerhouse from the truck to let the company know what was happening. The powerhouse operator realized the danger posed by the powerhouse intake, which sucked water in towards the turbines. The

operator immediately closed the plant down, thereby reducing the turbulence and the current of the waters near the intake. Without the operator's quick thinking, John Marsh might never have succeeded, and might even have been sucked into the intake himself.

Danger is a part of the job of linemen, riggers, and other power workers. Incidents such as these were all part of learning how to run an industry that served the public well. Today's systems are safer, but the need for constant public education and vigilance continue.

A painting and cleanup crew at Niagara Junction about 1946. *Left to right:* Sid Morris, Jack Lamb, Warren Kent (foreman), Jimmy Bird, Fred Gibbs, George Duncan (carpenter), Earl Snider, Ellis Holditch, Orville Heximer. Fred Gibbs used the scythe lying on the ground to clear grass and weeds from the right of way. Ellis Holditch never made it to retirement. A few years after this picture was taken, he was working at Rankine at the turbine-tailrace level. It is believed that he had gone to the bathroom and when he returned, before he could belt himself at his work position, he slipped into the tailrace and was swept through the tunnel to the Niagara River. His body was never recovered.
(Reproduction by permission of Sid Morris)

Chapter 15 Linemen Then and Now

The work of today's linemen is very different from that of the linemen who began work in the 1940s.[1] The tasks are still demanding and potentially dangerous, but the conditions, the equipment, the clothing, and the culture have all changed.[2]

Learning to Climb

One of the most noticeable changes is the modern use of hydraulic buckets or cherry pickers for many jobs. Linemen and riggers don't climb as much as they did when Sid Morris joined Canadian Niagara on May 1, 1939, as a meter reader. He soon joined the electrical gang, and stayed with the company for 44 years, retiring in April 1983.

Sid was familiar with construction work and had done some climbing, but he learned something new the first time he tried to climb a utility pole to change a streetlight. He strapped on his new spurs, nervously lifted one leg, and stepped down hard to dig his spur into the pole. He hung on tight and started climbing. Things seemed to be going fine, until he suddenly slid down the pole, back to the ground. Warren Kent, Sid's foreman, didn't seem surprised. He told Sid that he had slid down because he was frightened. Sid had stayed too close to the pole, hugging it, and this meant that his spurs were almost parallel to the side of the pole. The spurs were in the wrong position to get a good deep bite, so they had popped out and Sid had slid down.

They chatted for a little until Sid was more relaxed and then Warren said, "There's a light up there, Sid, that needs to be changed. You better just go up there and do it. Just lean out and back a little further." That was the beginning of many years of climbing. It was typical of how people learned to be linemen or riggers at Canadian Niagara — on the job. It was okay to make mistakes at first; there was always someone more experienced to offer advice and support.

Riggers and linemen both had to climb, but there was a difference. As former line superintendent Harvey McAteer explains, "Linemen work live stuff; riggers work dead." And if you couldn't climb, or didn't want to, you became a groundman, working on equipment and supporting those who did climb. There was a job for anyone who was willing to work hard and be part of a crew.

Soon Sid was climbing higher. Before he left to serve in the Second World War, he had helped put up new towers to replace some of the old original spider towers that ran from the transformer station above Rankine to Fort Erie. Sid soon became comfortable climbing poles, but he was still daunted by his first climb up the tallest thing in the Canadian Niagara system — the high tower at the edge of the river below the Bertie Street hill. It was about as tall as a 20-storey building. The first climb was intimidating, but eventually Sid got to like it up there. You couldn't beat the view. Sometimes he even brought along his box camera and took pictures of Fort Erie, Buffalo, and the Peace Bridge, as noted in Chapter 14.

Syd Freeman, another lineman, also liked the view from up there. "I thought it was gorgeous. I loved getting up there." His only regret about the tower was that he had to wait about eight or nine years after he was hired before he was allowed to go up it.

Rigger Harry Saylor Sr. first worked on the high tower, replacing some rotted-out rivets beneath the deck. To do that job, he had to climb to the top, then crawl under the deck on a little scaffold, where it was so cramped he could only work the ratchet one notch at a time. He had never climbed that high before, and he thinks his fingerprints may still be embedded in the steel where he was hanging on.

Not everyone could climb tall structures, and even some who had learned to climb met their match in a particular structure and realized they couldn't climb any more. When this happened, Harry says the company was "pretty good. They'd give you another job." For example, he remembers, one man had trouble on the water tank near the Rankine transformer station. "One fellow went up there and he froze. We had to go up and help him back down."

Canadian Niagara's antenna for its radio network was on the water tower. Tom Trasmundi normally didn't have to climb as part of his job in instrumentation and communications, but he had to go up it once. "We lost communications, so it's my job to go up there and find out why. Very reluctantly, I climbed up there and to my

In the late 1940s and 1950s, the demand for electricity kept growing. New homes and businesses meant more customers for Canadian Niagara, but it also meant a need for more trained linemen. Canadian Niagara employees took a variety of courses with Niagara Mohawk in Buffalo. Two Canadian Niagara employees, Syd Freeman and Walt "Babe" Douglas, attended Lineman's School "A" in November 1952. Everyone in the school came from some part of the Niagara Mohawk system. *Back row, left to right:* George Finkbeiner (Buffalo – instructor), Don Bull (Angola), Syd Freeman (Fort Erie), Walt Dryja (Niagara Falls, New York), Don Gritman (Angola). *Front row:* Walt "Babe" Douglas (Fort Erie), Jim Mirusso (Angola). (Reproduction by permission of Syd Freeman)

The ever-increasing demand for electricity meant more poles. More poles meant more hard work and muscle power. In this sequence of snapshots, the pole has already been unloaded by hand from the trailer, and a post hole dug, perhaps with some help from dynamite. In the first picture, the butt end of the pole is at the hole and the man on the left is holding the pole up with a pair of pike poles. (Reproduction by permission of Beverley and Ken Barnhart)

amazement, the coaxial cable had been chopped. I could see the chop marks at the connector there." It seemed a funny place for vandalism. Tom asked around and found that Stan Hall, the machine shop foreman, had seen somebody from the Niagara Parks Commission climb the tower. It turned out that the Niagara Parks Commission had sent a man up to disconnect a disused antenna owned by the Parks Commission. There was only one problem: "He mistook our cable for theirs." There was enough damage that "we had to take the antenna and everything down and get the cable repaired. It was a big job." When it was time to put the antenna back into position, the work was done by riggers. Tom was impressed. "To my amazement, they were walking along the edge of the tower like it was nothing, no fear of heights whatsoever." But Tom had one thing in common with some the riggers he admired. He loved the view and took some pictures.

Climbing was such an important skill that it could help you get a job. Syd Freeman started working for the Davey Tree Company in 1947. But after a few years he was "tired of travelling all over Ontario…I was looking for something permanent." Canadian Niagara often hired the Davey Company and in the fall of 1951 Syd was "trimming trees for the power company in Ridgeway. They'd had a storm and we were clearing some of the damaged limbs off the trees for the wires to be clear." He met the line superintendent Joe Murdock and asked about a job. "He asked me if I was afraid of heights. I said, no, I respected them, but I wasn't scared of them." Murdock brought him an application and a month later, November 19, 1951, he started work with Canadian Niagara.

Freeman had been vigilant but relaxed climbing trees. Now he had to make the transition from trees to poles and towers. As a tree climber, he'd got in the habit of wearing "a 9- or 10-inch-high boot. I always wore the high ones. It gave more ankle support for climbing. I've seen a few guys with the ones that were just barely at their ankles, but that has to give you no ankle support. To each his own, I guess."

Freeman took to the work naturally. "I've seen some guys climb poles so they look like a walking stick, they're so stiff. I always managed to be able to climb relaxed because Horace Heximer, one of the older linemen, showed me how to climb and I learned how to relax when I was climbing." Most days he didn't so much climb up a pole as walk up it. To watch a lineman walking up a pole is a wondrous sight.

"Lots of times I just hung on to the rope [safety belt] and just walked up the pole... without using my hands on the pole at all. Just use the belt. You just lean in a whisper and snap it up, just give it a flip up. And then step up. It's something that I guess was just natural to me." Syd Freeman is a big, strong man. Other linemen say he's the strongest man who has ever worked for the company, and Syd says he's "not going to argue with them over that."

Eventually, it was Syd teaching the next generation how to climb. He remembers helping them learn how to work without hanging on with one hand. "We had a few lads that had a tendency to want to hang on with one hand to the cross arm, or to their belt, or to a secondary bracket or whatever it was they were working on. There were not too many of them, but I showed a couple of lads how to relax and turn their belt...[that is] slide your belt around so you can work sideways and use both hands." Work was much safer when a lineman had both hands free.

However, each lineman had his own particular approach. Not every technique worked all the time. The oldtimers passed on what they knew; the newcomers experimented until they developed their own style. "If there was a trick or two that they [the oldtimers] knew that was easy for them, you would try it.... If it worked for you, great. If it didn't, you tried something else."

What impressed some new hires was the fact that many of the oldtimers were exactly that — old. As a physically strong 20 year old, Jim Nigh, who started in 1970, was astounded by the previous generation of linemen who "were crowding forty or something like that. They were the old guys to us.... Some of those guys didn't retire until they were 65." That generation was still fit and able to climb when they were in their sixties. The linemen had learned how to do it the right way from those who had gone before them. It was a physically demanding job, but they had learned how to do it in a way that didn't ruin their bodies.

When new linemen had learned to climb properly, the oldtimers had a little initiation ritual. Syd Freeman says, "They took a big staple and stapled my belt on the other side of the pole and said, 'You're now an official lineman.' I've done that to two or three of the other lads to tell them they were officially lineman. So I pass the tradition on." He adds, "I wouldn't drive them in full. That way they could get a hammer and knock them loose."

Above left:
In the next photo, the pole has been raised higher by pike poles and one man is guiding the butt end with his foot.

Above right:
In the third photo, longer pike poles push the pole closer to vertical as the men holding them walk forward. At the same time, the butt end slides into the hole. These men were rightly proud of their skills and coordinated teamwork.

(Reproductions by permission of Beverley and Ken Barnhart)

On Monday, October 17, 1949, Canadian Niagara inaugurated a new era in communications with two-way radios in two of their trucks. Before that, drivers went back to the office and used a telephone there, or asked a customer if they could use a telephone when they needed to make a call. The two-way radios were particularly handy during storms and emergencies. The event provided a photo-op for company management and local politicians. *Left to right:* General Manager Charles Clancy, Mayor Louis Ziff of Fort Erie, Joe Murdock (at the wheel), Reeve Ernest Gorham of Bertie, Reeve Norman L. Smith of Crystal Beach, CNP Manager Jack Babbit. (Reproduction by permission of FortisOntario)

Teamwork

Jim Nigh thinks the best way to develop a good lineman is to start with someone young "who is used to physical work, is clever and has some mechanical ability." A good lineman combines physical strength, intelligence, and horse sense. It's about a lot more than sheer brawn. Jim recalls one lineman who "was about as free from brains as a turtle is feathers. Anything that could be broke, he would break it.... He would try anything. Take chances. Boy there were some terrible things I learned from him. How *not* to do things." In a job where "you have to have eyes in the back of your head all the time," Nigh is still amazed that no one got hurt working around that lineman.

Syd Freeman wanted to work with linemen and foremen who had "common sense," who thought first and acted second. On one occasion, that meant doing the job slower than the foreman wanted, taking the time to chain down electric lines that were stretched tight by a split 55-foot pole before unbolting the lines. When they finally undid the bolts, the line that had been freed from the split pole jumped up. If the lines hadn't been chained when the lineman was working on it, "he'd have been gone." Looking back on that day Syd observed, "The foreman didn't think far enough ahead. Nice enough guy, but he just was in a big hurry just to get the job done, I guess." Syd emphasizes that there's no place for a lineman in a hurry. "You must stop and think what you're doing, because you're working with electricity. You can't see it, you don't hear it, but it does bite."

Syd recalls that "the guys I worked with were all excellent linemen. Everybody had good knowledge of what they were doing.... In the 37 years I was there, I enjoyed working every day. I had had good relations with everybody that I worked with. There was one that I was a little unhappy with, but he worked on another truck, so I didn't have to worry."

"Teamwork" today is a popular business buzzword, and anyone applying for a job claims to be a "team player." But the word has a different meaning in a job when the lack of teamwork can mean injury or worse. Linemen depend on each other in a very direct way. Walter "Babe" Douglas learned about teamwork from Syd Freeman.

In 1952 the two went to the Niagara Mohawk Lineman's School "A" in Buffalo for formal lineman training. The two-week sessions were grueling. Linemen had to show they could put up crossarms and buckarms. And as Babe recounts, "usually in those days two people put the buckarms on, one on each side of the pole." One had to hold it in position and the other had to bolt it on. In the outdoor session, Babe, 5'8" and about 127 pounds was assigned to work with Syd Freeman, 6'2" and about 190 pounds.

"We got up the 80-foot pole fine. And they sent up the 10-foot-long crossarms. The bolts are all in them and everything. I'm on the one side of the pole and Syd Freeman is on the other. They gave me that crossarm. The hole was a long ways up and boy my arms were just....I just wasn't going to make it. Syd just reached through the pole and put his hands under mine. The guy was so strong. If it hadn't been for him, I wouldn't have made it. He was tough enough to carry me. He's a guy that saved me from failing that course, because I couldn't have made it.... He helped me that time and I always respected him for doing that because he could have said, 'You're out,' just like that because I couldn't get that crossarm up."

A line crew posed in front of their Reo Speedwagon truck, probably in the 1930s. *Left to right:* Joe Murdock, Sam Kelley, Louis Rouse, Bill (last name unknown), Sam Hughes, and "Tug" (proper and last name unknown). Sam Kelley and Tug are wearing their climbing spurs. Louis has a brace and bit for drilling crossarms and Sam Hughes, smoking a pipe, seems to be cutting a wire with pliers.

(Reproduction by permission of Frank and Sandra Flake)

In 1949, this line crew took a break long enough to pose in front of their GMC line truck. *Left to right:* Horace Heximer, Bert Minor, Wallace Kelley, George Rose, Gabe Toth, Cec Hill, Sam Kelley ("The Boss"), Harry Pomeroy. Wallace, who is wearing his spurs, was Sam's son, and the father of Sandra Flake, whose husband Frank also worked for Canadian Niagara. The poles on top are pike poles. At the time, no truck was kept cleaner and no equipment in better shape, thanks to Harry Pomeroy and his legendary dedication to having everything shipshape.
(Reproduction by permission of Sandra and Frank Flake)

Fifty years later, Syd was astonished when he heard that Babe had remembered the story. "In all those years, Babe never ever told me he remembered that." Then he smiled, raised his massive arms, and held out the hands that after 75 years still radiate strength. "I just lifted the crossarm up and said, 'Put the goddamned bolts in.'"

Teamwork for a lineman is about stepping in and doing more than your job in those moments when the other person's job is a little too tough.

Friendships were often formed at the tops of poles. Tod Docker was a strong farmboy who had worked on electrical maintenance at Fleet Aircraft for three years before he got laid off. He'd heard that Canadian Niagara was looking for linemen, so he applied for and got a job there. That was in 1955. "I'd have to say our climbing days, pre-bucket days, brought out the best in everybody. You were dependent on your partner. Your life was in his hands and vice versa. And it's a rare occasion when two guys up a pole aren't able to talk about what they did last night or something friendly. There was never animosity when you're working together. You know, it was probably the greatest time of loyalty and friendship that you might experience in your whole life, except for maybe your family."

For line superintendent Harvey McAteer, there was no place for bad feelings on line crews. "If you haven't got good guys, guys that can get along, you've got nothing."

Tod Docker and other linemen knew that close-knit crews were an important safety device. When he was working with someone he knew well, Tod could "almost tell you what the guy's gonna do, except for blowing his nose.... I could tell you what everybody was gonna do if we'd worked together very long and I knew what to keep an eye out for."

If one of the crew members was new, or was one of the few who were careless, you had to watch them more carefully. Working with unfamiliar crew members added to the risk during bad storms, when men came in from other areas to work with Canadian Niagara linemen.

Rigger Harry Saylor Sr. felt privileged to work with a good gang. "I'll tell you how good they were. Earl Snider was the lead hand and he would take over for the boss when he went on vacation. [First thing in the morning] generally we used to meet in the machine shop up the hill, and then go from there down to the power plant. And so one day we went down to the power house. Nobody seemed to have any instructions on what to do, because Earl had gone on vacation. So the boss at that time called me up in the office and said, 'We're going to put you in change of the gang.' And I said, 'Oh, geez, I don't know about that, because I'm the junior man in the gang, and these guys have been here quite a few years ahead of me.' He said 'No, we want you to be the lead hand while Earl's away.' So I said, 'Well, I'll do it on one condition. The first time I hear somebody complain, then I don't want the job, because I've got to work with them for — what ? — maybe 50 years, and I don't want to cause any trouble.' So I went down to the guys and...told them what the score was. And I said, 'If anybody's not satisfied with what I'm doing, let me know now. I don't want to have any hard feelings with you guys, because I've got to work with you.' I was surprised. The one I thought maybe would complain, he was the first to come up and shake my hand. He said, 'You won't hear a word from me.' So that really gave me a lot of confidence."

From then on, Harry was lead hand whenever Earl went on vacation. Years later, he explained that as a lead hand, "you got a couple of cents an hour raise. It doesn't mean anything to have bad feelings for a couple of cents an hour."

Groundmen

Teamwork extended to the men who didn't climb. They, too, had an essential role to play. As Tod Docker puts it, "The guy on the ground some days was just as important as the highest trained journeyman lineman, because he had to help you."

Most of what the linemen needed on the pole had to be sent up by rope and pulleys. There was a right and a wrong way to send up things up. "There was a knack to being a good groundman, I'll tell you. That was to make it easy on the lineman when he got the equipment on the end of the rope in front of him." That equipment might be a 10-foot-long crossarm and the linemen depended on the groundman "to tie a knot that a guy could untie easily, not something you never saw before. You tie an old granny knot and your boss would be ready to jump." The ropes had to be placed on the crossarm so that when the knots were undone, the crossarm would swing the right way, away from the lineman, not towards him. Some groundmen were good at it and "some never did figure it out. They didn't work here that long."

Gabe Toth was one of the legendary groundmen. He arrived from the Ukraine in 1927, never got a driver's license, but learned everything about taking care of equipment. And the linemen showed their appreciation. When Gabe wanted to get a small house, Babe Douglas and two other linemen did a lot of work on it, as well as all of the wiring. When Gabe died, Babe and his wife Jean sold their larger house and bought Gabe's old house.

Harry Pomeroy was another member of the line crew who didn't climb. Babe remembers him as a perfectionist. "He used to come in on weekends and take the ropes off the trucks and recoil them if he didn't like the way they were coiled. And he'd dust the inside of the truck. We never had to worry about sitting in dirt, because this man cleaned that truck." Pomeroy also "was the last truck driver/groundman. They never had that classification in the book any more when he left. But he never wanted to climb…he never wanted to do anything but drive and care for a truck." All the time he worked there, linemen never had a problem with "equipment being dirty, or anything wrong with it." Gabe Toth and Harry Pomeroy made life better, safer and easier for linemen.

Bad Habits

When Babe Douglas was a rookie in the 1950s, his line foreman, Horace Heximer, was famous for a few things, including cigars. He "smoked a package a day," which probably meant at least five. However many it was, it seemed he always had one on the go in the rubber cigar holder that sometimes smouldered and smelled worse than any cigar.

The office kept track of transformers by location and serial number. "Horace would write everything down on his cigar box. We'd be driving down the highway and he always drove. He was the boss and he drove all the time. He'd light up another cigar on the way in and that'd be the last one. So out the window went the package. No matter where we were going, we had to drive all the way back, climb a pole — because we couldn't find the cigar package — and get the numbers to register the transformer…. The numbers were usually on the bottom [of the transformer], so you didn't have to go into the live area, you'd just go up and get them."

Horace wasn't in the habit of keeping the inside of the truck clean. One rainy day, Tod and Babe found themselves in a substation with Horace. There wasn't much linemen could do in the rain. Tod and Babe decided to see what colour the truck was inside. Tod said the "whole inside of the cab had turned yellow from the nicotine, and I mean *yellow*." But they didn't want to clean the whole inside, they just wanted to satisfy their curiosity. "We cleaned half the truck [with a dividing line] right dead centre, right down the dash." The clean half turned out to be white and that made the other half look even yellower. "Horace came out and I thought he would go right through the roof. Mad! He blew his top that day." Babe adds, "It stayed that way until it got dirty again."

Heximer certainly had a temper, and had been known to engage in fist fights on the job. It would start with disagreements over how to do a job, "and then they'd end up in an argument; they were probably drinking and they'd end up in a fight," says Jim Nigh.

Drinking? On the job? It was a different time and a different culture. What seems inconceivable now was nothing out of the ordinary then. As Jim Nigh puts it, when he started as a young lineman in the 1970s, "Back then it was almost a badge of honour for guys to be drinking and driving. That wasn't regarded as a bad thing." Harvey McAteer remembers that "liquid lunches were common. Especially in the summer. I've seen them [linemen] quit at noon, knock over 8 to 10 glasses of beer and then go back to work." Work included climbing. Harvey also remembers that after a liquid lunch, "lots of time they'd pee off the top of the pole."

Similar things happened in other companies. As Jim recalls, "It was accepted. It was common.... An awful lot of drinking went on and an awful lot of hard work. A tremendous amount of work. When I look back at it, the work that got done without the equipment they have today is incredible." The physical work was hard on the men, and drinking was partly an expression of macho culture and partly a way of letting off steam and dulling pain. "For lots of guys their knees were hurting, their shoulders were hurting, their backs were hurting. Fortunately none of the guys [linemen] today will see that kind of bull work; it was hard." Sometimes drinking was just a way to cope with boredom and inactivity. "If it was a rainy day...there'd be four guys stuffed in the cab of the truck and everybody smoking and drinking."

Despite the drinking, there were very few accidents. In trying to explain this paradox, Jim Nigh suggests, "I think that we were a lot more careful maybe because there wasn't as much safety [equipment] then, so you knew what could happen." Drinking wasn't an everyday occurrence, but it wasn't rare either. It was an accepted part of the dangers of the job. If one man had had too much to drink, no one expected him to do something that would be a danger to himself or his fellow workers. Excessive drinking, however, was recognized as a problem. Frank Flake, who worked at Rankine as a blacksmith and welder, remembers one employee whose drinking was getting a little too frequent. His fellow workers were worried. When he got blind drunk one day and told off the boss in no uncertain terms, they took him home. The company arranged for help. Eventually the man returned, and was a productive employee for many years after that.

In the 1970s, the crews did not always get on as well as their predecessors had done. When Jim Nigh looks back at those years, he says, "There was a lot more fighting and bickering. When I started, we were just getting over that. But cripes, they'd have a party and they'd end up in a fist fight just about every time." He remembers that most of the fights seemed to start over what was the best way to do certain kinds of work. It was a professional difference of opinion over how to work best, and perhaps who could do certain tasks the best.

As Jim sees it, in those days, "Every line foreman had a different way of doing things." When putting up a new line, two crews with two different foremen might start a mile or two apart and work towards each other. Then they'd discover that each crew was stringing the line at a different height "and they'd get in a fight over who was right." When Fortis assumed full ownership of the company, it brought in what he calls a "book of procedures for doing every type of job you can." It makes his job as line superintendent a lot easier. "It's real important now that we have a standards book and everybody's on the same page." It also makes things safer with the "multitude of voltages" in the Canadian Niagara system.

But when there was a real emergency, the men put aside their differences. As soon as the lights went out, "everything changed. The priority was for people to get 'em back on." There was no room for bickering, and no time to argue with the foremen.

The Long Road to a Culture of Safety

Pictures from the earliest days of the company show workmen in shirts and trousers, often with ordinary boots and soft hats, not the specialized work wear that is common today: hard hats, safety boots, and heavy-duty clothing. Over the course of a century, safety equipment has become an entire industry in itself, and decade by decade, the way linemen look and work has changed.

When Syd Freeman started work, it was up to him what he wore. He preferred cotton clothing, but always heavy enough to keep him covered and warm. Usually he wore "green pants, green shirt, and green jackets," all made of cotton. In the fall he put a liner in the jacket. In the winter, when it was cold and windy, he wore "long johns, pants, and overalls." When a Carhartt store opened in Niagara Falls, he says, "I swung over to Carhartt. I wore a lot of that: brown overalls and parka. That's basically all I ever wore." His clothes were not of the lightest materials, but "the wind was pretty raw up there on the poles" and he never worried about the weight. Syd never did understand anyone who would climb with a "short winter jacket" with "bare back" showing. He wasn't interested in the lightweight jackets made of nylon and other synthetics, even before he heard about "a couple of Niagara Mohawk fellows that had some of the polyester clothing on" while working on high voltage equipment. "The [polyester] stuff caught fire and it stuck right to their flesh." After that incident, there was a ban on polyester clothing at Canadian Niagara and throughout the Niagara Mohawk system.

When Jim Nigh started as a lineman, he and the younger guys often dressed differently from the older linemen. The younger men wore "blue jeans, and a short denim jacket over a hooded sweatshirt. Summer and winter were basically the same. That's how you looked." But now Jim is one of the older guys and the younger guys dress differently. "Now it has to be orange fire-retardant clothing. Long-sleeved shirts all the time if they're in the air." It is a far cry from when "it was a white cotton T-shirt in the summer and you'd climb with your jeans on. Clothes are important now. They have to be visible in traffic."

Today the ruling philosophy for lineman culture is, as Jim Nigh puts it, to eliminate unsafe situations as much as possible. "You try to rubber up everything if it's alive" and "you wear rubber gloves from the time you leave the ground to the time you hit the ground again." Things were very different earlier. The men who were nearing retirement in the 1970s were a "different generation altogether." Jim Nigh says, "Every culture takes ten years to change, from what I've been told. Hard hats came in in the 1960s — but they were still trying to coax us into wearing them in the 1970s."

These were the same linemen who often "didn't wear rubber gloves unless you were on primary. And if it was a dry day, you might try to get away with it even then." It was "just the way it was done then. If it was raining, your hands are soaking wet with leather gloves and you're still working lives" without rubber gloves.

Harvey McAteer started his lineman training after serving in the Merchant Marine during the Second World War. He remembers that for many years linemen worked live 2,400-volt lines with leather gloves. Rubber gloves came into more routine use when they moved up to 4,800 volts. During his 45-year career, he saw many changes in the use of rubber gloves and safety equipment, and he recognizes that "nobody wants to change. You have to overcome it." But he also wonders if safety precautions have gone "overboard the other way." Now linemen cannot work without "the whole shebang, the diving board [a temporary platform for standing on], the rubber sleeves, and the rubber gloves...all the hoods and toques and everything...before you could get ready to work, half the day was gone." And sometimes the equipment was so safe it was unusable. As rubber gloves became heavier, they got so thick "they were like car tires. Your hand would get cramped trying to use them." For some jobs, the Canadian Niagara linemen went back to using thinner gloves.

For as long as any lineman can remember, rubber gloves have always been tested in two ways. Linemen are supposed to test their own gloves every time they use them. Each lineman grips the cuff tightly, twists it and squeezes the glove to see if any air leaks out. If air leaks out, electricity can surge in. Gloves are also periodically shipped to the Lineman's Laboratories for a dielectric test, which tests their insulating properties by putting high-voltage electric current through them.[3] But simply having safe gloves doesn't mean that they always get used.

Back in the 1970s, when linemen were working on a live secondary line where the voltage was lower, Jim recalls, "If you called for your rubber gloves, they'd be calling you every kind of sissy you can imagine." In that kind of macho culture, linemen didn't put on rubber gloves until it was absolutely necessary. But, "you'd be really careful" and when you were "getting poked really good, then you'd finally call for your rubber gloves. Oh yeah, it was different."

Sometimes technology got ahead of the available safety precautions. Wally Farrell, general manager and a professional engineer, had the idea of using 34,000 volts (34 kV) for more of the system in order to increase overall system capacity. Jim Nigh remembers, "We didn't get any training. They just said, 'Here's your gloves,' and it's 34, so we treated it like 5,000 volts (5 kV). You can go up on 5,000-volt delta, when it's pouring down rain and you can work away with it. You get a little buzzing; it's not a big deal. But with 34, if it's overcast, calling for rain, you're going to get poked all over the place. Incredible. It could be a little damp and start to snow. You go out and touch it and it would hit you like you wouldn't believe.... So we found out our voltage was running really around 38 but they called it 34. Our gloves were only good for 40. We were right at the limit. We stopped working with rubber gloves [on 34 kV], and then we went to hot sticks."

Hot sticks were for working live lines. These insulated sticks are the "size of a broom handle, six feet long with tools at the end." Hot sticks came in a wide variety of designs, depending on the job needed, such as opening and closing a switch, removing cotter pins, or changing transformer fuses. Some linemen were masters of the hot stick. Everyone could do the easy operations like pulling a cotter pin, but Jim Nigh points out that only the masters could use hot sticks with "mirrors that look behind stuff" and work where they couldn't even see, rather like large-scale outdoor dentistry.

When hot sticks became everyday equipment, they were used to change fuses in pole-top transformers on live circuits. Under the right circumstances, the work was safe and uneventful. The fuses were in a box. As soon as one hot stick — called a switch stick — opened the cut-out box door, the circuit was deactivated. The linemen removed the fuses in their U-shaped holders with a hot stick, put in new fuses, and replaced the holders with a hot stick. The circuit was made live only when they took their switch stick and closed the door.

Two views of Canadian Niagara Power's Reo Speedwagon line truck.

Left: From the back on the outside, ladders on one side and pike poles on the other. Inside, neatly coiled ropes hang from the side and a wooden nail keg sits beside coiled cable. The bracket to hold the spare tire is underneath.

Below: Here, the truck is at the Queen Street loading dock. Unlike later line trucks, this one has no second seat. However, the canvas top provided some protection from rain and snow. Men would sometimes sit on top of the racks and canvas as the truck went down the road.

(Reproductions by permission of Sandra and Frank Flake)

It all seemed simple and safe. But there could be a problem when it was raining hard. Even though the pole-top equipment was insulated, the electricity sometimes "tracked" — that is, stray currents ran along the surface wherever there was a conducting path. Sometimes that conducting path was the body of a lineman. Syd Freeman has experienced the feeling of current running down with the water. "I've had it where it trickled down my arm, and down the rain suit and off the spur and down the pole." Wherever the external insulating path was incomplete, the lineman's body became an alternate conducting path. How did Syd react to this? "Grit your teeth and keep working.... I've had a couple of them, not too many. Once or twice a fellow doesn't mind one of them, but it's not something you look forward to."

Today's linemen work with a wide range of insulating devices from hot sticks and cover-ups for live wires to insulated booms on bucket trucks. All of these are tested routinely; it is a radical change in how Canadian Niagara and the industry ensures greater safety. Blankets and other cover-ups are tested with a dielectric test to see if any current gets through. The booms on bucket trucks and cranes are also tested periodically. Jim Nigh explains, "If they're working a bucket truck with high voltage, they have to have a test before you even start. They shove the boom into the wire, and they have a meter on the bottom of the truck so see if it's leaking [any current]." This safety culture even extends to jumper wires on the trucks. "They even have to test it out to make sure it can take the amperage for which it's designed. This kind of stuff was unheard-of when I started. You never tested jumpers and stuff like that." Moreover, safety equipment is now much lighter and easier to handle.

Sounder procedures complement the improvements in equipment. Clearly written and carefully thought-out work procedures have added immensely to safety for linemen. These leave less room for inadvertent error due to personal judgment on what might be the best way to, say, make a certain type of connection. But the greatest safety innovation for linemen is doing far more work on dead lines. Not only is it safer to work on dead circuits, it is also much faster. Jim Nigh compares "doing it dead versus doing it live. It would take you four times as long to do it with hot sticks. It's slow, very slow."

The macho culture has given way to a culture of safety. Linemen still do hard physical work, and face difficult situations in emergencies, but many dangers have receded.

CHAPTER 16 A Rich Social Life

The late Sharon Bews worked in both the billing and the service departments. Colleagues remembered her as a perfectionist who would stay late or arrive early, if that's what it took to get the job done right. But her dedication went beyond the workday. Maria McKinnon remembers that she "was the one who baked for us. She made the special birthday cakes, from scratch. That's definitely missed. Even after she retired, she'd have birthday parties for us. She was very thoughtful that way."[1] Sharon oiled the social network that made Canadian Niagara such a special place. She wasn't alone.

Mary Genco-Brant entered the 25-year service club in 2004. She remembers that when she arrived, she worked with the soon-to-be-retired Dorrie Dunn. She was "a sort of a gruff kind of lady, right to the point. I remember her just sitting there chain-smoking at her desk, constantly. She was chain-smoking, that's when you could smoke right at your desk." Dorrie often made meals and snacks for the Board of Directors when they met on the second floor of the Bertie Street Service Centre; after all, she had given cooking lessons in the company's demonstration kitchen. But she made her biggest contribution to the social life of Canadian Niagara's employees at her home on Buffalo Road. Dorrie's house was the place for great parties. When she died, the house was sold. But a few years later, a young woman who worked for Canadian Niagara bought the house, not knowing it had been Dorrie's or even who Dorrie was. At Canadian Niagara it is still referred to as "Dorrie's house" and the new owner is expected to carry on Dorrie's inherited traditions.

Suits, white shirts, and ties for men and long skirts, white blouses and hats for women indicate that Canadian Niagara's "First Annual Outing" in July 1913 was a big affair. The large number of CNP Co. pennants also speak of company pride. The location of the outing, Ontario Beach, New York, reflected the strong American influence in the company, as well as the desire to make the day extra special. (Reproduction by permission of FortisOntario)

Traditions are funny things. Without any, we feel lost. With too many, we feel constrained. As Canadian Niagara entered the 21st century, many employees wondered about the place of tradition and social life in a rapidly evolving company. Would it be lost in the rapid pace of modern life?

On the fourth Monday of every month, a group of retirees, sometimes as many as twenty, get together for coffee and toast at a restaurant in Stevensville. They've been meeting for ten years, starting in Hector Smith's garage, which the group has since outgrown. There is no agenda, no formal organization, no dues. They get together because Canadian Niagara is still an important part of their lives and who they are. For them, work and social life have never been completely distinct. They exchange news of their lives and health, vacations, operations, and grandchildren, and they reminisce.

Somewhere between coffee and settling up at the cash register after the March 2005 meeting, Stan Hall told Syd Freeman he planned to go bowling that night. These two had been on the first Canadian bowling team to win the Niagara Mohawk System Bowling Tournament. Stan was the principal organizer, and each year he would put together a pickup team from whoever bowled and was available to go to New York State for the Niagara Mohawk bowling tournament, which attracted about a hundred American teams.

Canadian Niagara bowling teams found their way into other leagues, such as the Little Anthony [Restaurant] Bowling League in Niagara Falls, and tournaments such as that put on by the Buffalo Chapter of the Utility Employees Association. But the Niagara Mohawk System Bowling Tournament, which dated back to 1952, was the biggest. In 1987, Ron Katz, Miles Sabine, Tony Cerminara, Harry Saylor Jr., and Stan Hall drove to Syracuse to represent Rankine Station under the name Canadian Marvels. They didn't win, but the goal remained.

Above left:
The Niagara Mohawk Power Corporation started its system-wide bowling championship in 1952. In 1994, Stan Hall, who had retired in 1993, put together a team he called the CNP Oldtimers. They beat out 88 other teams to become the first Canadian team to win the championship. Shown here with their trophy, *left to right*, are: Stan Hall, Syd Freeman, Tony Cerminara, Harry Saylor Jr., and Tim Hayes. (Reproduction by permission of Tony Cerminara and Stan Hall)

Above right:
After the Niagara Mohawk Power Corporation bowling championship banquet in 1994, Syd Freeman seatbelted the trophy into his Oldsmobile 98 and brought it home. It was the trophy's first trip to Canada. (Reproduction by permission of Syd Freeman)

Then in 1994 Stan Hall put together a pickup team under the name Canadian Niagara Power Oldtimers, paid the $145 entry fee, and made the trip to the Sheridan Lanes in Amherst, New York. This time Syd Freeman drove back from the banquet at the Buffalo Airport Sheraton Inn with a huge trophy seatbelted into the back seat of his Oldsmobile 98. Thanks to Stan Hall, Syd Freeman, Tim Hayes, Harry Saylor Jr., and Tony Cerminara, the trophy spent its first year in Canada.

Picnics, Parades, and Annual Parties

Shirley Athoe retired in 1995 after 30 years with the company. Shirley fondly remembers the picnics at Rock Glen Park (Niagara Glen) that punctuated life at Canadian Niagara, and she even organized some after she retired. These events were an integral part of work life. Shirley was allowed time to do all the food shopping and the power company truck was there to haul around whatever needed hauling. "The guys were great; they hooked up these huge metal containers as big as a picnic table and they'd load them with ice and then we could put the salads in them in bins and keep them nice and cold. Somebody even brought a little freezer down and we kept the ice-cream in it for the kids.... We had this huge big company barbecue where we did everything from ribs to hot dogs, hamburgs, corn on the cob. Everybody had their job to do." Somebody even knew where to get "a little kiddy train that would run around with the kids and we'd have a jumping thing for them and races. Every child got a toy of some kind. Yeah, we had some great times."

For Wendy Spry, the early years were about spontaneity, togetherness, and the Canadian Niagara family. "We'd go to a union meeting and make a do out of it. You know, everybody would go out for dinner before the union meeting and then we'd all go out afterwards for a few drinks. We always did everything together. If you were moving, you called the boys up, and they'd run by with a truck. Everybody helped everybody; we were a big family."

The Friday night Christmas parties held upstairs in the Chippawa Fire Hall were legendary. It seemed a good compromise location, about halfway between Niagara Falls and Fort Erie. When you walked in at about 6 o'clock, Shirley Athoe remembers, "there was always a huge tub of Manhattans. They were very strong, so you only wanted one of those." Upstairs it was "crowded, you're elbow to elbow, but it was wonderful" with delicious meals made by the volunteer firemen and their wives. "They had a live band and you danced all night. Sometimes you'd be having such a good time the guys would all chip in and keep the band for another hour or two. And then we'd all go out for Chinese food.... Some of the guys always knew how to square dance and they would teach you, and we'd have someone who would call it." The Christmas party would go on until "maybe two or three in the morning, depending on how long the band stayed."

No one seems to know how it happened, but one day the venue of the party changed from the crowded informality of the Fire Hall to the greater formality of hotels or restaurants, alternating between Niagara Falls and Fort Erie. These parties were nice, but they weren't quite the same. It seemed less like the employees' turf and more like the company's turf.

Sandra Flake also has fond memories of children's Christmas parties. Sandra's father, Wallace Kelley, worked for the company for 37 years and his father, Samuel Nelson Kelley, also worked there. The children's Christmas party had the best Santa Claus imaginable, played by one of two brothers in Chippawa who had genuine full white beards and little glasses. Every child got a silver dollar and a gift. One year it was Etch-a-Sketch, when that was the hottest toy around.

In her early years at Canadian Niagara, Wendy Spry remembers making a float for the annual Christmas parade. It was more of an employee effort than a company-sponsored matter. "In those days, the company did not pay for anything. So we'd gather bits of lumber from somebody's house, and the girls would go out and buy fabric for a dollar a yard and we'd all chip in and make costumes and somebody else would throw some money in, and we go and buy candy to throw off the float. We did that for years. And we spent nights working out in the garage doing all these things together.... Back then, we never thought of asking the company to pay for it. You know, if we took scrap lumber from down at Rankine, or got something done in the machine shop, boy, we were just tickled pink that somebody would give us that. We never assumed they would pay for these things. We felt we were well paid, and it was just something we enjoyed doing." Then things changed, and the company paid for everything. Wendy thinks that took something away from the experience. "Nowadays they tend to look at it as an obligation."

Company funding may have reduced the element of adventure in building floats, but it allowed larger and more elaborate ones. Nothing topped the float built for the Canadian Niagara Centennial in 1992. With a little company money and countless volunteer hours, a show-stopping float emerged from the shops at Rankine featuring a seven-foot-diameter layered birthday cake, the Falls, the rainbow over the Falls, and a hidden fog machine to create the mist rising from the Falls. After all, if there's no mist, it's not Niagara Falls. It awed crowds on both sides of the River — startled customs agents took just one look and waved it on. The float took on different lives over the years. The cake turned into a pot of gold with Irish dancers for St. Patrick's day. Later everything got remade for Christmas and a Santa Claus parade. One year, it was even made to resemble the machine shop. It also appeared at Fort

In 1992, the centennial of the founding of Canadian Niagara brought out the biggest and most elaborate float ever. It was made by CNP employees and featured the Falls, complete with fog machine to create the mist and a rainbow, along with early electrical equipment. The float even crossed the Rainbow Bridge to Niagara Falls, New York. Customs officials seemed too dazed to do anything other than wave it through. (Reproduction by permission of Sandra and Frank Flake)

For years, Shirley Athoe kept a record of social events such as Christmas parades or decorating the Christmas tree on Queen Street. She glued photographs onto sheets of paper, wrote captions, and stapled them together. Sadly, these mementoes have disappeared, with the exception of CNP Christmas 1974, 1975, 1976, which show Christmas parades. A pickup truck provided a good base and employees created costumes with whatever they could find. The event relied on people being good sports, willing to contribute. There are no photos for 1977, only the caption, "No good sports available to continue. Budget nil."
(Reproduction by permission of Shirley Athoe)

Having lunch together was an important part of social life. On an April day in 1944 in the Rankine Station offices over the inner forebay, Marg Tolan, Caroline Krass, and Stella Learn enjoyed tea and sandwiches. The picture of the anti-aircraft gun on the Canadian General Electric wall calendar was a reminder of the urgent need for every kilowatt-hour of power Canadian Niagara could pump out. In addition to her regular work, Marg helped Bill Rickard get out the *CNP Tattler*, which kept employees in touch with their enlisted colleagues. (Reproduction by permission of FortisOntario)

Erie's July 1 Friendship Festival that symbolically stretched hands across the border between Buffalo and Fort Erie to celebrate American–Canadian Friendship.

People today have less time for socializing, but employees like Maria McKinnon, Jeff Hoover, and Courtney Bonito are breathing new life into the Social Club. At work there are five-dollar "all-you-can-eat lunches" cooked up by fellow employees. Jeff spearheaded the drive to get more off-the-job events than the annual Christmas party. Easter-egg hunts and summer picnics bring children and families together. Bus trips to everything from football games to shopping destinations bring employees and spouses together. These events let people get to know each other and see each other as more than a job description. Canadian Niagara seems to be recapturing and creating new 21st-century versions of the events that made the company such a great workplace for most of the 20th century.

Sandra Flake uses the word "supportive" to describe the company, "incredibly supportive of the workers themselves, and of the extended family too." This includes retirees. Her husband Frank remembers that "the pensioners would come into the machine shop, or the blacksmith's shop for a little job, wanted to get their knives sharpened or wanted to get a little bracket made. You would just drop what you were doing and look after that guy. He's put in his time. Pensioners come first."

Marg Tolan, who started with the company in 1940 and retired in 1985, seems to have held a special spot in many people's hearts. Frank remembers that every spring, "it was automatic" that "the line crew or brush crew from Niagara Falls went to Marg Tolan's to trim her trees." They would take down her storm windows too. As Frank explains, "We just looked after each other, it wasn't a big deal, but you couldn't do it today."

A More Leisurely Era

One key to understanding the changing social life at Canadian Niagara and the current quest for a richer social life is the simple fact that life is now much faster paced, both on and off the job.

When Annette Brinda started with accounting in 1974, the department was in the attic of Rankine Station, with windows looking out over the Falls. When accounting moved to Fort Erie, that area became a boardroom. Annette liked working in the powerhouse and had the time to enjoy it. She worked 8:30 to 5:00 with an hour and a half for lunch. Most people went home for lunch, and the extra half hour was to compensate for the time it took the get through the tourist traffic in the park during good weather or to scrape the ice off the car in bad weather. This meant she had a 35-hour work week.

When accounting moved to Fort Erie, the lunch period dropped to an hour, but the work hours were the same. This meant she had a 37 1/2-hour work week. Then, in order to get home earlier in the day, everyone at the Queen Street office agreed to "a half-hour lunch, so we could get off at 4:30." In 1992, the work day got even longer, when Pay Equity came in. Pay went up for the women in the office, but so did the hours. "Our hours are now 8:00 to 4:30 with a half hour for lunch. You work 40 hours."

The pace of work increased at the same time. "I think when I was hired, there were 110 people on staff. And each year that was dropped and dropped," while the company increased its customers and corporate income. Computers made some of the work easier, and enabled each person to do more. But paradoxically that combination made the job harder, because expectations for productivity rose. "You don't feel like you ever get finished."

Hockey and Baseball

For decades, hockey and baseball were an important part of life at Canadian Niagara. Harvey McAteer's father Percy came to Niagara Falls from Cornwall, Ontario, in 1922 or 1923. He worked with Canadian Niagara until he retired in 1957, but it was hockey that brought him to Niagara Falls. John Lawson, general manager at Canadian Niagara (1936-1946) and later president (1946-48), was also part-owner of the Niagara Falls Cataracts hockey team. He brought in other players such as Pug Herrington, Obs Heximer, and Bill Hunter. Hockey didn't pay well, so McAteer, Herrington, Heximer, and Hunter got jobs at Canadian Niagara as well.

Jack Irish and his brother Duncan, who ran the Esso gas station by the Peace Bridge before opening a hardware store where Canadian Niagara employees cashed their weekly pay cheques, remember the glory days of Niagara Falls and Fort Erie hockey. Jack keeps a newspaper clipping from the time when the Niagara Falls Cataracts won the O.H.A. 1932-33 Senior Championship. The cup they won was probably the one that Harvey McAteer remembers being around Canadian Niagara. John Lawson was secretary-treasurer of the team and Bill Hunter played left wing. Hunter, who came originally from Sault Ste. Marie, was general manager from 1969 to 1979, one of Canadian Niagara's most respected managers.

When an on-ice injury ended Percy McAteer's hockey career, he continued working at Canadian Niagara, starting in the electrical gang in maintenance and ending up as electrical foreman. When his son Harvey was about two years old, the family moved into one of the two company houses. "You got a reduced rent [$19 a month], but you were on call all the time." Young Harvey practically "lived in the machine shop." One machinist made him a little hammer; another a little boat. When the company needed strong workhorses to haul something heavy, two Percherons, Crawford and Walters, were hitched up. "I used to ride with them. My old man would say 'Oh, come on. Get on and go with them.'"

Then there was the old Pierce–Arrow chain-drive truck that Louis Frommert drove. Everyone called him Bunny. It "couldn't go any more than about 20 miles an hour, but oh, it was just like a tank. And they'd take me, 'You want to go? Come on, get in.'" The Pierce–Arrow was a great big thing with "racked sides and a canvas cover that you could move back. And in the corner they had a roll of toilet paper. So I figured that was where you went, so I went while they were driving, used the paper. And afterwards, Louis says, 'What the heck did you do?' I said, 'I used the washroom.'" Harvey did that "just the once."

Hector Smith Sr. and Jr. are second- and third-generation Canadian Niagara employees. Hector Sr.'s father worked as a labourer at the Rankine Station. He died at age 54. Hector Sr. started in 1931 as a janitor, mopping the floor of Rankine, "all 583 feet of it. That's a lot of mopping." He was a top-notch baseball pitcher and the 1930s were the heyday of company leagues. The Canadian Niagara team played teams from other companies, such as Carborundum, Norton, and Dominion Chain.

Canadian Niagara was the only Canadian team in the Niagara Mohawk baseball league. The parent company even offered Hector Sr. a job in New York State when he was at the peak of his pitching career. He turned it down, and when he pitched in Buffalo at the Offerman Stadium before a crowd of 12,000, it was for Canadian Niagara. His best year was probably 1932. One of his clippings reads: "Hall of Fame now has another occupant since Tuesday night [October 16, 1932] when the CNP hurler Smith pitches no-hit, no-run game." He also played hockey and considered a pro hockey career.

Rick Heximer joined Canadian Niagara in 1965 after graduating from Providence College with an accounting degree. He had been there on a hockey scholarship. His father had played hockey in the late 1920s and early 1930s with the Bruins and the Rangers. When he retired from hockey, he came to the powerhouse as a rigger.

For Percy McAteer, Bill Hunter, Obs Heximer, Pug Herrington, and Hector Smith Sr., excelling at sports was an important part of their early lives at Canadian Niagara. They weren't hired just to play, but it made sense to hire good athletes. Many of the jobs at Canadian Niagara were very physical, and good hockey and baseball players were strong. They were also good company ambassadors at a time when local amateur and semi-pro sports were an important part of community life.

But not all of the athletes were in the limelight. Barry Bracken remembers that his father Don, who had been a talented football player, coached football at Stamford Collegiate, something many people didn't know at the time. Some of the games were just for fun. Frank Flake remembers "we played Slow-Pitch with the Niagara Parks Commission and the Niagara Parks Police, and we had a little round robin and the company was very good as far as sponsoring us with shirts."

Gradually the emphasis on organized sports leagues disappeared. Television brought professional sports into the home. Cheering the pros became a substitute for playing yourself. Regular leagues and games disappeared and those who longed for older times had to make do with one or two pickup games a year — Niagara Falls employees versus Fort Erie employees or Canadian Niagara versus some other part of the Niagara Mohawk empire.

The Power Pioneers

The Power Pioneers was a creation of Niagara Mohawk. Some years, the dinners were held in Buffalo and sometimes in Niagara Falls, Ontario. But the format was always the same. Canadian Niagara guests were picked up and returned home by charter bus after an evening of dining, awards, a dinner speaker, and socializing. Family and spouses were not invited. There were no gifts other than pins to mark the years of service. The pins were not expensive, but they were precious, because you couldn't buy them. You had to earn them.

When Canadian Niagara distanced itself from Niagara Mohawk and then severed ties completely, the Power Pioneers dinners fell by the wayside. Now there are no long-service pins. There is no employees-only dinner. Nor is there a charter bus. Family is invited and employees arrange their own transportation to the Canadian Niagara Social Club Anniversary Picnics. They can choose gifts from a catalogue divided into categories by years of service. Meanwhile, the Niagara Mohawk Power Pioneers service pins have become heirlooms. Mary Lou Smith had her husband's made into a ring to pass on to the next generation.

The Machine Shop as Social Centre

The machine shop, which has since been torn down, was once a major gathering point for Rankine employees. What is now the Rankine lunchroom used to be the garage, where Barry Bracken remembers going in on weekends with his dad to get the company car washed. The older lunch room, complete with a proper kitchen, was up on the bluff at the machine shop building, on the second floor. And a lot more than lunch went on there.

The welding shop, garage, and paint shop were nearby. The guys in the machine shop and welding shop made a barbecue that could hold about 50 burgers at a time. The Niagara Parks Commission had a 9-hole, par-3 golf course nearby and once or twice a month Stanley Hall would organize something after work on Friday. Stan usually managed to negotiate a special rate at the golf course and so for six dollars each, as many as 18 employees would get a game of golf. "Most of us were just duffers, but we had a great time," Bob Douglas remembers. They would return to the

Retirement parties were important social events, whether they were held at the machine shop lunch room and kitchen at Rankine, the Queen Street Service Centre, the Bertie Street Service Centre, or a restaurant or hotel. Louis Rouse retired on March 31, 1967, during Canada's Centennial year, when a group of employees decided to grow beards. *Left to right, back row:* Jack Barton, Harry Saylor Sr., Louis Rouse, Tod Docker, Joe Parisi, Ray Miller, Syd Freeman. *Front row:* Garry Bowman, Wally Kelley, Earl Snider, Dave Durward, Ed Staedt. (Reproduction by permission of Syd Freeman)

One of the famous machine shop parties. After women employees complained about being excluded from these events, they were invited to join in. (Reproduction by permission of FortisOntario)

garage to play euchre, have a few beers, and enjoy barbecued burgers made by the guys in the machine shop. Sometimes they'd have a corn roast. Frank Flake also liked the horseshoe pitch up there. As Stan Hall observed, these kinds of get-togethers helped work go more smoothly and made the company a better place. "It's the softer side of working, you know."

For years, retirement parties were held in the "gang room" above the machine shop. John Fogg remembers that "Some of the guys were pretty good cooks" who could make a full meal in the kitchen — "sometimes fish, sometimes roast beef with potatoes, veggies and stuff." At first these events were for men only, but when the women complained about being left out, they were invited too. When the Bertie Street Service Centre expanded, these parties eventually migrated there. Now they are held in restaurants or hotels.

"Game nights" weren't about bingo or poker. They were about men cooking up food that you didn't buy at a store. Quite a few of the guys liked to hunt and fish and some were legendary cooks. Ed Staedt, Second World War veteran and painter, was reputed to have been offered a job as a chef at the Royal York in Toronto. Harvey McAteer, who served two years in the Merchant Marine, loved to cook and even became a partner in a restaurant at one time. Depending on the season, the menu ranged from local fish to beaver, wild turkey, rabbit, venison, or moose. Dave Durward says the feasts included "beaver, rattlesnake, God knows what-all." Whatever it was, it tasted great and the parties sometimes went on long into the night.

Parties for the Bigwigs

When the Queen Street office opened in Fort Erie in 1928, the United States was in the middle of Prohibition, an unsuccessful social experiment that made it hard to get a decent drink. The Niagara Mohawk directors and senior managers from the United States were thirsty and needed somewhere to hold meetings. So they came to Fort Erie.

But Ontario had its own laws covering the consumption of alcoholic beverages. One was that you needed a license or permit to serve alcohol in any place other than a residence, even if you were giving it away. That explains a curious architectural feature on the second floor of the Queen Street building — a bedroom where no one ever slept. It was, however, rented annually to a board member just to keep everything legal.

Harvey McAteer and his wife Shirley both worked at the Queen Street office and remember parties for the top brass. "Christmastime, it was wild. I stocked the bar for quite a while. I can tell you, there was big money involved. The best. Crown Royal, all that stuff." Shirley added, "If they ran out during the evening, they'd call him to go someplace to borrow it." Usually he found what he needed at the Cherry Hill Golf Club. One time, he had to get crème de menthe and called the bartender at Cherry Hill, who said, "'Yeah, come on up, I'll get you some. How much do you want?' I said 'Three, four bottles.' So I went out and got it, brought it back. I took it up [to the Canadian Niagara bartender] and he told me, 'I didn't want that, I wanted green.' It was white. I had to go back and get two greens."

Harvey recalls that Niagara Mohawk people "were heavily into politics" and entertained everybody from "police chiefs" to "Monsignors" at Queen Street. That was a time when Niagara Mohawk was said to have three Lear Jets. But there was only one way to get to Fort Erie — chauffeur-driven limousine. Jack Irish, who lived nearby, remembered, "All up and down the street and around the corner and down Goderich Street there would be cars with chauffeurs. The chauffeurs brought them over and waited out in the car. Sometime after midnight, the party would start to break up and they they'd be singing away, coming down, out to their cars."

Few people knew more about how to make things smooth for Niagara Mohawk directors than David Gordon, Toronto lawyer and later member of the Board of Directors. When he started as a junior lawyer with McCarthy and McCarthy (now McCarthy Tétrault), Beverley Matthews, the senior partner, was chairman of the board of Canadian Niagara. Matthews was one of those legendary figures "who could sit at the head of a round table." He terrified junior lawyers. He told David Gordon, "I'd like to get you involved on Canadian Niagara Power matters. It's one of my favourite clients. It's not big, but it's important... do a good job." David Gordon's relationship with the company lasted over a quarter of a century.

At first, he was less there for legal advice than for logistics. "When they [the directors] would come to Toronto for a board meeting, the Niagara Mohawk plane would come into the Island Airport. I would arrange for two or three limousines to be out there to pick them up, to bring them back." One of the senior board members from Niagara Mohawk, Walt Machold, was in a wheelchair. When Machold came to Toronto, David Gordon knew what to do. On the east side of the Royal York Hotel, there was an oversized elevator for trucks that went to the laundry room in the basement. When Machold arrived, David made sure the way was cleared for Machold's limo to enter the elevator and take him to a convenient level. Canadian Niagara even installed an elevator in the Queen Street Service Centre to allow Machold access to the second floor.

Meetings in Fort Erie were memorable. David Gordon remembers "the big board room. A gorgeous big table and at the end a picture filled the whole end of the wall, picture of Hiawatha in a canoe or something like that." (Accounts of that now-missing painting vary. One person says it was of Father Hennepin, the first European to see the Falls, another just the Falls, another an Indian by the Falls, and someone even remembers a "guy with a lightning bolt in his hands.") At the end of the meeting, somebody would push a button and the picture slid upwards. The company had rigged up an electric motor similar to that on a garage door. The painting disappeared and behind it was a fully stocked bar.

The painting had not always moved up electrically. When Frank Flake first saw it, "somebody would have to go around behind it and pull it like an old garage door on a rope. Chick Clancy said, 'This is ridiculous. This is an electrical company, and I want it electrified.' So he had the machine shop in Niagara Falls devise a mechanical opening device, so all you had to do was turn the key." It wasn't an expensive mechanism — bicycle gears and a chain, along with a lead counterweight that worked well enough, "until one day the chain came off the sprocket and the lead ball came through the wall."

There was a bartender and wait staff to serve food, often prepared by Dorrie Dunn. Shirley McAteer remembers, "They'd call her up and say "'Well, Dorrie, we're going to have a dinner. We want lobster or we want steak or whatever.'" And Dorrie had to round up everybody and everything: food, servers, whatever was needed. "I'd always go up in the afternoon, about 3:00 o'clock, when we'd have our coffee break, and they would have made up the hors d'oeuvres by this time. Dorrie would always make us up a little tray to bring downstairs." Curiously, the privileged treatment of the board members and Niagara Mohawk people didn't seem to alienate Canadian Niagara employees. When Rankine no longer needed a senior hydraulic operator, Bob Douglas went to Bertie Street on the four-to-midnight shift. There he saw that the people with real class and style have time for everybody. "There was one quite elderly gentleman, one of the top dogs over the river. He always brought me down a drink or two and my supper." Canadian Niagara was a good place for everybody; different people just enjoyed it differently. Moreover, people were proud of the fact that these events took place on home turf.

The Herschel Fishing Club

The Herschel Fishing Club on Herschel Island in Baptiste Lake, 12 miles from Bancroft, Ontario, is a shadowy part of the history of Canadian Niagara and Niagara Mohawk. The main 15-acre island was joined to another island by something called the Kissing Bridge. Board member Donald Smith remembers that locals thought it was a Mafia hideout. Men arrived in black limousines, got into a private launch, and headed off to a private island. "The Canadians didn't really participate in it that much. It was more for the guys from [Niagara Mohawk Headquarters in] Syracuse," who used it as a base to fish and to party.

The origin of the Club is briefly mentioned in the minutes of a 1934 Board of Directors meeting, when the board approved $8,500 for "providing health and recreational facilities for the officers and employees of the company."

Sid Morris was the man who could wire anything, and he first went to Herschel in 1946 to look after the generators, which were two cantankerous, 32-volt DC affairs. Both generators were kept in working condition, so that if one failed, the other would be available. This sometimes meant that Sid had to work at 2 o'clock in the morning to keep them running. He went back every summer until 1954, when Hydro came to the island and the company bought a 60-cycle army surplus generator for back-up. This was popularly known as "Generator No. 12" (in addition to the eleven generators at Rankine).

The rest of the staff for the island consisted of a year-round caretaker and summer help that included three guides, two waitresses, a cook, and a cook's helper, who also did the laundry. Visits to the island were not about roughing it. Cynthia Durward, who processed some of the bills for the Club, notes, "They ate very well." There was the main lodge, a guest house, a boat house, and a bunk house where the staff slept. The main lodge had a big lounge, a dining room, a large kitchen, pantry, and four bedrooms. Sid remembers people relaxing in wicker chairs in the lounge.

For Canadian Niagara employees, the pride in the Herschel Fishing Club came from the things they made for the place. The passenger scow and the work scow were both made in the carpenter's shop at Rankine. People also like to tell the story of Jack McLeod, the Fort Erie meter reader, who in 1953 brought a 1-ton truck full of building materials to the Club to repair the damage caused by Hurricane Hazel. He returned with a load of lumber – many red pines had been felled by the storm, and he took them to a local sawmill.
(Reproduction by permission of Sid Morris)

The Herschel Fishing Club needed ice to keep food and drinks cool during the summer months. In February 1949, Sid Morris and Chick Clancy drove to Lake Baptiste, where local residents Allen, Hugh and Stan Peever cut thick ice from the river by hand and moved it into position where a horse pulling a rope could draw it up skids and into the insulated ice house. There it was packed in sawdust to await the summer. (Reproduction by permission of Sid Morris)

Canadian Niagara employees were proud of having built the flat-bottomed boats for Herschel, as well as a system of narrow-gauge tracks and carts that were used to winch supplies and luggage from the dock up to the lodge. However, some employees were a little disapproving, since the place had a reputation for loud drinking parties. A few were not happy about having to work there, and one even quit on the spot. As far as Cynthia Durward could tell, Herschel was where "Canadian Niagara spent a lot of time and money making things comfy for the Americans. Not many Canadians went there." Those who did sometimes felt uncomfortable. Herschel was something of a symbol of Canadian-American tensions in the company that rarely surfaced but were always present.

Getting Herschel ready for the season was like opening a cottage, only on a much larger scale. Every spring, everyone who could be spared from Niagara Falls would go there to paint and repair the place and make it welcoming to visitors. But the Club was not supposed to interfere with other priorities, and there was an unwritten rule that if there were any problems at Rankine, the employees had to drop everything and go back to Niagara Falls. But once, when Don Bracken ordered the men back to Rankine to fix a problem, the visitors from Niagara Mohawk refused to let them go. Bracken complained to top management, and management reprimanded the people involved. But the next time he went to Herschel, three or four Niagara Mohawk visitors threw in him the lake in retaliation.

Times change, and what seemed unremarkable in the mid-20th century became unacceptable a few decades later. Herschel was sold off in the 1970s.

Will We Still Care for Each Other?

Canadian Niagara tended to feel like a family because it was a small organization. It also had a local base. When customers called in about a problem, quite often the person they spoke to was an acquaintance or a neighbour, and the employee usually knew exactly the location of the problem.

Today, the people in the Bertie Street Service Centre may be fielding calls from Welland, Port Colborne, or as far away as Cornwall. They cannot respond in the same way to calls from areas they do not know, and they worry about being out of touch with customers. But within the company, the family feeling remains in many ways.

Shirley Athoe has many memories of that feeling. Some years ago, only a few months after her daughter died, her son was admitted to Buffalo General, where he remained for about seven months. It may have seemed a small thing, but it meant a lot to her that during that time, the company let her leave work early enough to avoid the traffic jam on the Peace Bridge, so she could be there to feed him and spend time with him. Meanwhile, fellow workers pitched in, making casseroles and cakes to feed her other two children.

She also remembers happier times: "No matter whose birthday it was... there was always a cake and some silly gift. And we played jokes on the fellows too — we'd switch their boots or we'd stuff their gloves full of paper. It was all good fun." It was also the social cement that holds a company together.

Other employees remember how the Canadian Niagara family was there to help when times were tough. They want a FortisOntario family that will care for and support employees as the old family did, even though the old family was rooted in the Niagara Frontier and the new, extended family is spreading across Ontario.

Stan Hall worked hard as a machinist at the Rankine Generating Station. He also found time to organize many events that enriched life for Canadian Niagara employees. When he retired in June 1993, the "boys in the shop" showed a local cartoonist a picture of Stan and told a few tales about him. Today the framed caricature looks down over Stan's home workbench. (Reproduction by permission of Stan Hall)

In 1957, the Rankine Generating Station was shut down, and the water flow reduced almost to nothing, to allow for inspection and maintenance in the discharge tunnel that took water to the Niagara River gorge. After more than 50 years of use, the tunnel was in excellent condition and needed very little repair work. Tours of the tunnel were very popular. *Left to right:* Unknown, Stan Hall, Ray Nicholson, Don Fraser, John Fogg, Don Bracken, Jim Bird, George Wilkinson.
(Reproduction by permission of FortisOntario)

Maintaining and Updating an Aging Power Plant

CHAPTER 17

When the Rankine Generating Station opened in 1905, it was the most advanced hydroelectric generating station in the world. Engineers and electrical specialists travelled to Niagara Falls just to see it. Other tourists paid admission to view the generators (Canadian Niagara donated the proceeds to the local hospital). A fortunate few had season passes. It was thrilling just to stand on the balcony overlooking the generator floor, to hear the loud hum of electric power being generated before it surged out to homes and factories in Canada and the United States. The brass gleamed, and the control panels, which had not yet been moved out of sight to the second floor, looked like visions of a mechanized electrical future.

There were eleven generators in all. Unit No. 1 started supplying power to customers on July 27, 1905, and two more came on stream later in the same year, No. 2 on September 11 and No. 3 on September 27. Nos. 4 and 5 came on stream on April 21 and May 17, 1906. The building had to be extended before the next ones could be installed. No. 6 started generating power on December 5, 1910, and No. 7 on July 28, 1913. Nos. 8 and 9 came into service on September 7 and October 22, 1916, in time to supplement wartime production. On January 13, 1917, No. 10 came on line. By this time, the station had the capacity to produce 5,000 electrical horsepower over the limit of 100,000. The addition of No. 11 in 1924 provided reserve capacity to compensate for times when another unit was out of service.

Running the station was the work of a small army of employees — from hydraulic and electric operators, through riggers and machinists, carpenters and blacksmiths, to painters and janitors. Rankine employees did not have formal apprenticeship programs, but their training was rigorous and systematic. Most men started at the bottom, as janitors, and worked their way up, learning more than one job on the way. Women usually started on the telephone switchboard. This meant that the employees were flexible and able to handle a variety of tasks. As former machinist Doug Jacobi put it, "When you worked at Canadian Niagara, you did what had to be done. It didn't matter what your trade was. If you didn't know anything about carpentry, you were going to work with somebody that did and you'd learn. That's why I think everybody enjoyed it."[1]

Learning the Ropes

Bob Douglas started as a janitor in 1947. One of his first assignments was to polish the "great, big brass door" at Rankine. That's where he started to learn that no matter how simple a job might look, there was a right and a wrong way to do it. At the end of the week it "wasn't right and I did it again. You didn't ask or say no; they said that's not right, do it again. So I kept doing it and got to know how to polish brass."

He used something called Monton, "a little bit heavier, creamier than Brasso" and learned to pay attention to the seasons and weather. "It was easier to do in the fall and spring, because it wasn't too cold or too hot. In the wintertime if you put on too much and let it sit there it would really get hard. So, you had to kind of put it on, take it off, put it on, take it off. In the summertime it would get wet and damp, and the door sometimes had a green tinge to it. It was just one of those things that you got to know." On the midnight shift he did brass rails, "a certain section, like one long rail and 2 or 3 short ones" each shift.

In those days, "the place was spotless. They had a day man. His job was to keep the floor clean, and he would go back and forth, back and forth, just keep that spotless. People came in from outside, they would look at this in utter amazement."

After Bob Douglas had been a regular janitor for a year or so, he became "electrical janitor" which was a "step up, not in pay, but in learning." There was still dusting, cleaning, and floor washing, but you "would learn how to operate the oil switches and things like that." At that time, "there were seven men on a shift and the same seven guys were on the shift all the time. So you got to know them pretty well."

Learning the ropes on the job from experienced people was essential, because as the plant aged, the crucial knowledge about machinery dating from before the 1920s became scarcer. Parts could not be bought; they had to be made — another specialized skill.

As a machinist, Doug Jacobi appreciated "excellent welders." He recalls one difficult job. "A welder we had at that time was Frank Flake. He got made a foreman, but he was a hell of a good welder." Frank was assigned to work on an elderly part made of cast iron. "I don't know if you've ever had any experience with [cast iron], but being old, it was porous as hell and it was full of oil. He would try to preheat it to weld it, and oil kept bubbling out of it. It was a hard thing to do. It took a lot of perseverance, but he managed it. Then, of course, we [the machinists] would… put it back up again."

Whatever you did in the machine shop, you did using "very old machinery." Jacobi remembers that the boring machine was a "rotten old machine." The "damned thing wouldn't bore straight," but he was proud of learning to work around its idiosyncrasies.

Hard, Dirty Work

Some of the work at the plant not only required specialized knowledge of old machinery, but simply hard physical labour. One tough job was repairing or replacing babbitt thrust bearings. These were the life-support system for the generators. They carried the weight of the massive rotating shaft joining turbine below and generator above. A bearing should not be the same metal as the one it rubs against. The softer babbitt bearings rubbed against a harder steel surface. This meant that if anything went wrong, the babbitt would be damaged and have to be fixed or replaced. But compared to steel or cast iron, the metal of the bearing has a lower melting point, is easier to pour, as well as being softer and therefore easier to shape.

If anything went wrong, the riggers had a big job ahead of them. As rigger Harry Saylor Sr. explains, the thrust bearing carried a tremendous load, almost all the weight of the rotating parts of the generator, down the entire length of the shaft to

The water-driven air compressor in the pit at Rankine Generating Station provided air for many uses from shutting down the turbine-generator units to blowing out lines. Here, oiler Ken MacIntyre is topping up oil reservoirs while George Wilkinson, Operator-in-Charge, looks on. George had worked his way up to a suit position but he still had to wear safety shoes.

It is difficult to appreciate the size and complexity of the many parts below the generator floor. Sometimes they had to be brought up from below by crane for repairs and maintenance. As with so much in the Rankine Generating Station, they are bigger than the men who maintain them.

Stan Hall and Jim Mooney are hard at work in the machine shop. The equipment was sometimes politely described as idiosyncratic, but the machinists knew how to get the most out of it.

As the equipment in the Rankine Generating Station aged and parts became harder and harder to find, the machine shop became even more important. Employees throughout the company were proud of the machine shop: it was a good place to go for an industrial-strength barbeque grill for a company picnic, or to get knives sharpened, or have parts made for restoring antique cars. It was also the site of some legendary cookouts. But serious company work was done here. In this photograph, John Fogg poses with two of the important parts of his trade: micrometer and lathe.

(Reproductions by permission of FortisOntario)

Generators seem to go on forever, but not the inner workings of the turbines. The runners and turbine blades inside the scroll cases usually chew up ice quite handily, but sometimes meet their match when tree trunks, oil drums, or truck tires mounted on rims shoot down the penstocks. It is a tough job to get the debris out. If the damage is bad enough, the overhead cranes have to haul the runners up to the station floor for repairs. The runners shown here with Harry Saylor Sr. and Ivan Bradden will be repaired and balanced before being returned to service.

This is an extreme case of damaged blades in a turbine runner that had been removed from the Rankine Generating Station.

(Reproductions by permission of FortisOntario)

the turbines, a weight of about 198,000 pounds. The wooden, lignum vitae bearings at the bottom carried a much smaller load. Getting a thrust bearing out was "a lot of work." Just loosening the bolts took a wrench about three feet long. "One guy would handle the wrench and there'd be three or four guys pulling on the wrench" with a rope. If the bearing had got too hot, "it would melt, then you'd have to scrape a lot of the metal off. But generally if everything went well, it would be a good day and a half, two days maybe, that's working overtime." The sense of urgency varied over time. When Harry started with Canadian Niagara in 1946, everything "was going 24 hours a day. If they had to shut a machine down, then lots of times you had to work right through [without a break] to get it back in operation, otherwise it would cost a lot of time and money."

Babbitt bearings helped teach Frank Flake the value of learning from older colleagues. He started after blacksmith Johnny Kossip had a heart attack. When Kossip returned, he started to teach Frank the tricks you need to successfully pour "six hundred pounds of liquid Babbitt that you bring out of the fire." The "pour is done in 15 minutes" but the "preparation could take three or four days." Doug Jacobi recalls that pouring a thrust bearing "took about four of us" including those on the two-man ladle. John Fogg remembers that when Frank was ready for the pour, three or four men from the machine shop would come up, put on gloves and face shields and get to work. "We all knew what we had to do."

The bearings sit in an oil bath, but the oil tends to get squeezed out by the weight when the bearings and shaft are not rotating. Because of the size, and the need to fit the bearing around a shaft without removing the shaft, the bearing is cast in wedge-shaped pieces. However, the babbitt surfaces did not go directly from pouring to installation. Each piece was cast about an inch thick and then had to be machined in the "rotten old" boring machine to make it dead flat.

There was a lot of scraping, testing, and more scraping. "You had a big surface plate and you would put bluing on it, move it around. If there was too much blue in here and none on the inside you had to scrape it out until" it was perfect. "It was a long project." It might take about three days for scraping. "It was a tiring job. You had to sit on a stool and work it. By the end of the day your neck and shoulders were sore."

You hoped that after you'd done all that work, the thrust bearing would last a while. There was no set life expectancy. Doug Jacobi has "seen them replace two of those bloody bearings in the same week because one didn't do the job it was supposed to." That was very unusual.

Another tough job involved clearing out penstocks and turbines. After a major storm tore through the area and toppled many trees, the forebay was almost choked with trunks, branches, and debris. The trash racks are the final line of defence to keep debris out of the equipment, but for some reason, the trash racks were not in place. "We had every man in the place working, clearing those machines out. It took days." Trees had gone right down the penstock and into the turbines. "You know there's lots of room for them damned trees to come down in there. But there ain't no room to get the damned things out at the bottom. There's this little manhole you crawl into to get in the turbine. You had to handsaw everything. They wouldn't let you use chainsaws because of the exhaust. Oh, it was a hell of a job. And it was a wet job; it never quits dripping and it was dirty, and it was hard work."

Passing On Knowledge

Rankine was a unique workplace that demanded unique skills. Although most people were willing to share their skills, some people felt that teaching others would undermine their own position. When Frank Flake started in 1975, he wanted to learn things from older employees, particularly the operators and machinists. But it took a while before they stopped feeling threatened. One machinist, for example, was very skilful at keeping his shoulder between Frank and the work. When they were at a critical point in the work, he would send Frank off to get something. When Frank got back, the work was finished and Frank hadn't learned what he wanted to learn.

As a hydraulic operator, Bob Douglas worked closely with the electrical operator to bring generators on and off power. For a year and a half Bob worked with a senior hydraulic operator who "would never let me know the sequence of operations" and how this was related to what the electrical operator was doing. It was actually quite simple, but when he was assigned to work with a different gang and was told to take over, he had to admit he didn't know what to do. Given the person he had worked with, the other senior hydraulic operator understood and set about teaching Bob about more than just the immediate operation he was doing. Bob at last was learning about the system, not just a small part. When he went back to his old shift, he went ahead and used his new knowledge. The rather secretive senior hydraulic operator "kind of looked at me sheepishly. But that was one of the problems."

Sometimes the problem was simply that employees could not articulate their knowledge. They would feel or sense problems from a sound or a vibration or a smell. Frank Flake talks about walking by the case for the thrust bearing with a certain "awareness": "You look at the gauge, but then you just put your hand on the thrust case." If "there's no hit, no thump," he knew it was working well, but if he felt something out of the ordinary, he would check it out.

Some people said Hector Smith Sr. just needed to smell the air carefully to tell you what had gone wrong, or might go wrong, at Rankine. But employees couldn't learn these tricks overnight. There was also a whole range of knowledge about the

Some of the men who kept Rankine running gathered for a photo in the machine shop. *Left to right:* Charlie Vasbinder, Leonard Hunt, Bob Ramsnee, Unknown, Willard Benson, Bill Macrathy, George Vasbinder, Henry Witt. The date of the photo is unknown, but the men's attire suggests the 1940s or 1950s.

Good men who knew what they were doing, worked hard, and watched out for each other were the lifeblood of the Rankine Generating Station and the entire Canadian Niagara system. *Left to right:* Orville Heximer, Percy McAteer, Ed Staedt, Fred Gibbs, Earl Walmsley, Fred Binder, George Duncan, Bill Rickard.

(Reproductions by permission of FortisOntario)

differences between apparently identical pieces of machinery. For example, over time, the workers learned that what you did with Governor No. 5 to control the speed of the turbine, and with it the generator when starting up or shutting down, was very different from what you did with Governor No. 7. And as they worked through a succession of jobs — governor man, thrust deck, turbine man, hydraulic operator — they understood at first hand how each part interacted with others.

Bob Douglas appreciated the value of learning by doing, and picking up on the experience of older workers, but he also felt there was a need for some written procedures, particularly for emergencies.

The huge shaft that ran up from the turbine to the generator had to be lubricated properly or things would grind to a halt. Oil was pumped out of the tanks at the turbine deck level and up to the bearings. The filters that kept the oil clean and free of water had to be changed periodically. As senior hydraulic operator, Bob wanted to know exactly how that was done. He asked a rigger take him through the job. Bob recalls, "There's all kinds of valves above the floor and below the floor" and there was no room for error. "If you closed one valve manually, something was going to happen that you wanted. But the one that was just down from it, you didn't touch." He asked how the riggers knew which valve to turn. The answer: "'I just know.' 'How?' 'Somebody told me.' "

Bob knew that not everyone wanted to put information on paper, but he thought there was less chance of costly or even dangerous accidents if he started making things clearer. After all, this was no trivial matter. Some of the lines had high-pressure oil pumped through them to open and close valves remotely. Bob likes things to be organized and clear. "So I drew up a schematic of the whole thing, in colour, and outlined how to do every operation. Each valve had a tag on it. Open Red A. Close Blue A. That type of thing." He did that for every deck, or level, in the generating station. When he was finished, "Everybody knew, if they looked at those things, where the valves were and where they came from and where they went." Perhaps his career pattern was one of the reasons Bob could see the need. He had worked in the line department and then in the electrical department before becoming a hydraulic operator. He knew first-hand that the level of formal training varied across the company. "In the electrical end of it, the training was very rigid," but "there was a less formal training program for the hydraulic operators." It was an important change and also a useful starting point for automation.

The Beginnings of Automation

When Harry Saylor Sr. joined the company as a rigger in 1946, the goal was to keep every unit running 24 hours a day, as it had been during the war. Counting the one man who worked in the transformer station, Rankine had an eight-man shift. Several decades later this would be reduced to two.

In the interim, Rankine went through a period of uncertainty, particularly in the 1960s and early 1970s. The Generating Station had always turned Canadian Niagara's water rights into electricity and therefore into money. But would it continue that way? People at the company wondered if it would make more sense for Canadian Niagara to get money directly from selling its water rights to Ontario Hydro and stop generating its own power. While this idea hung in the air, there was

At the portal or discharge end of the tunnel, the brick walls and floor are replaced by granite, and the floor drops sharply. In this photo taken in 1957, Don Bracken and George Wilkinson stand at the transition point where water from the tunnel empties into the Niagara River. Those who have stood at this point say the view and the sounds are awe-inspiring.

With the Rankine Generating Station shut down, men could work in at the bottom of the pit, an area that was normally a raging, deadly torrent of water that would have swept them through the tunnel and into the Niagara River.

(Reproductions by permission of FortisOntario)

little investment in the station, and some people, assuming that the plant would eventually be closed, moved away to other jobs.

Canadian Niagara was also becoming uncomfortably conscious of its reliance on Ontario Hydro for its supply of 60-cycle power. Getting power from Ontario Hydro had not provided the savings the company had expected. Moreover, letting Hydro use some of its water allotment was not saving Canadian Niagara money. Even though generating equipment was not running as much, it still cost a lot to have the equipment standing by ready to generate power. The company was also facing rising projected costs on the new Service Centre. And although Canadian Niagara was producing less power, personnel and maintenance costs were not falling. In fact, they were rising. Canadian Niagara asked the Niagara Mohawk engineering department to see if monitoring equipment could reduce the number of people needed to run Rankine.

The first priority was monitoring in the station. When Hector Smith Sr. smelled the air and knew which equipment to look at carefully, or when Frank Flake touched a piece of equipment to see if it was working properly, they were gathering information. Canadian Niagara needed more modern ways to gather information. Although sometimes mistakenly called automation, these earliest changes should really be called monitoring or remote sensing. The company installed sensors to take the temperature of bearings, or measure oil flow and water flow past certain points. The information went to the system operator electronically. However, this was not full automation, because the system operator could not send a signal back to correct the situation. This was information at a distance, not control at a distance.

By October 2, 1972, all of the monitoring equipment had been installed, except for one pump. That connection needed a 24-hour shutdown and Canadian Niagara couldn't shut down until Ontario Hydro had settled a bitter strike, then its 100th day. With the monitoring system in place, Rankine required 13 fewer employees, but no one lost their jobs. Instead they were reassigned to other work without loss of pay. Some went to work for Niagara Mohawk.

Entering the Computer Age

In 1979, Canadian Niagara and Niagara Mohawk started a joint study of computer needs and capacity. The first area to computerize was accounting and billing. In 1981, Wally Farrell, Vice President and General Manager, reported to the Board of Directors that the company was considering buying a Honeywell Level VI computer at an estimated cost of $150,000 and that with this he would consolidate accounting operations at Queen Street. The computer was on its way by early January 1982 and the cost was less than expected, $130,000 for hardware, software, and consulting services. Meanwhile, at Rankine, after years of partial operation, all 11 units were being operated at full capacity with no increase in staff. Things seemed to be going full steam ahead.

The first project ran into difficulties, and Canadian Niagara fired the software consulting company, but things were back on track by 1983. The company acquired a large mainframe computer. The next step was personal computers, but at the time, many people — including Wally Farrell — couldn't see their potential in the workplace.

In 1975, when Blaine Desrosiers started with Canadian Niagara as a draftsman in the planning department at the new Bertie Street Service Centre, drafting meant drawing with ink on linen. The most advanced technology available consisted of rotary "electric erasers that worked almost like a drill." His colleague, Dave Durward, had joined Canadian Niagara as a draftsman in 1958. They worked in a cramped office on the second floor of the Queen Street Service Centre. Dave's drafting table was under the sloping roof and he had to bend down to get at it. With the blueprint machines in a crowded space "it did get smelly." The "one small air conditioner really just sagged out a lot. It was hot in there in the summertime."

Dave remembers, "We had a small office, we had a lot of work ahead of us, and I could see a computer might be the thing." At the time, computers were used only in billing and accounting. Dave decided he needed to learn about computers, so he "went out and bought a computer from Radio Shack" with his own money. He brought it into the office, and even though it wasn't the best computer available, Dave's investment paid off, because it helped convince Wally Farrell that personal computers had a place in the company.

Gino Visentin had studied broadcasting at Niagara College. He too became a draftsman at Canadian Niagara. He remembers when the company bought some Leading Edge brand PCs with green screens, 5 1/4-inch floppy discs, and 20-megabyte hard drives. They were "really low end," but they opened the door to a new era.

In the mid-1980s, no one was really in charge of computers, but employees were finding new uses for them. For example, a program called Symphony made by Lotus was adapted to do work orders for linemen. It reduced paperwork, and eliminated the slow two-stage copier that spat out brown copies on waxy paper. More important, it made inventory control easier.

When staff moved from the Queen Street Service Centre to the Bertie Street Service Centre, most of the computing power and needs were housed in one building. In accounting and billing, several versions of Honeywell DPS6 computers had replaced the original computer. The pace of change picked up even more in 1992 with the arrival of a Dell server running UNIX software. Gino Visentin recalls "That's when everything took off and we really got into more computers throughout the company." It was also when the company began to realize the implications of being computer-dependent.

Although it wasn't part of his job description, Gino had developed "a great interest in computers." He learned to use various kinds of software, read about computers, and took some courses. Pretty soon, he found he was the "planner/computer guy" — the one that other employees in the planning and engineering department looked to for everything from sorting out day-to-day printer and computer problems to "scoping and purchase of computers."

Meanwhile, Al Stouffer had emerged as the computer guy in billing and accounting. Al and Gino first learned just how much the other was interested in computers when they saw each other at a computer course each was taking on his own time. Computers were important for customer service, but Al was the only one in his department with high-level computer skills. If he went on vacation and something went wrong, employees would call "the fellow who wrote the program," who lived in North Carolina. That was Ralph Bard, who had lived in the Kitchener-Waterloo area when the company first turned to him for programming. Even though Ralph was now far away, "he could dial in and fix the problem, re-queue a print run or a billing cycle or whatever." But he wasn't always available and Canadian Niagara was concerned about outside costs. Most of all, Canadian Niagara was worried about its vulnerability when only one person knew how to run critical software.

To remedy the situation, the company appointed Gino "Assistant to the Customer Service Supervisor/Computer Application Specialist." He stayed there from 1992 to 1994, when he returned to planning and took the computer work with him. He describes himself as Niagara's "one-man computer show right up until 1998." But other people were beginning to get interested in computers too. The company provided training for new software and systems and encouraged employees to take courses to learn more.

Automating the Substations

Tom Trasmundi joined Canadian Niagara on March 3, 1980. He had studied electronics and computer technology at Niagara College and worked for Transport Canada. Tom started working for John Parisi in the Meter Department, which also looked after substations and relaying. Tom discovered that Canadian Niagara thrives on flexibility. "They wanted me to be multitasking. You know, able to do everything. But my main job was to start automating their substations."

Before Tom arrived, John Parisi and Harry Beer had already automated Station 15. The project had worked out, but it had also convinced them they needed to hire someone with more training in electronics. Tom was assigned to automate Station 11. Power for Fort Erie came into Station 11 from Ontario Hydro at 115 kilovolts (115,000 volts) and Canadian Niagara's transformers reduced it to their transmission voltage of 34.5 kilovolts (34,500 volts).

Automated substations were run by a SCADA system, which is variously interpreted as Supervisory Communications And Data Acquisition, or sometimes, Systems Control And Data Analysis. As Tom explains, SCADA "allows substations, generators, or whatever to be remotely monitored and controlled from a remote site control room." Control at a distance was the new element that took this project beyond mere monitoring of equipment.

For SCADA to work properly, the control room at the Rankine Generating Station needed equipment that the operator could use to open and close breakers or switches and to get readings on voltages and currents. The system also had to detect anomalies, such as oil that had leaked from a transformer or an unusually high temperature reading, and automatically sound an alarm in the control room at Rankine.

The custom-designed equipment came from RFL in the United States, but Tom had to make it work. John Parisi, who had worked on the first automation job, was now in management and Harry Beer, whom Tom had hoped would be able to help bring him up to speed, retired earlier than anticipated. As a new employee, Tom had a lot to learn about Canadian Niagara's circuits and systems. Jim Whitehead and Norm Damato came to the rescue. Tom still remembers that "they were excellent" at sharing their knowledge and helping him learn. Tom quickly came to rely on them and says with their help, "we were able to do it."

It was not all smooth sailing. In some cases, he had to go to Niagara Mohawk to borrow or use equipment. For example, "We had to bring a relay over for calibrating, because we didn't have the test equipment at the time." But he found that their employees "worked right to the book as far as coffee time, lunch time, whatever. They would not go out of their way to do anything extra. It was very frustrating for us. ... We wanted to get the job done, and we'd look around — coffee time, guy's gone. By the time they'd get back, it'd be a half hour wasted." And then when he found out how much Niagara Mohawk had charged Canadian Niagara for using equipment for a week and occasionally seeing a technician, he realized, "we could have bought our own test equipment for less money" if the purse strings hadn't been closed so tightly.

Tom wasn't the only employee who could see that sometimes they were treated as if they were just occupying an outpost that wasn't as important as the centre across the river. But he persevered and went on to other substations to perform the same task.

With the SCADA system installed in substations linked to the control room at Rankine, automation had arrived. But that was not the same as automating Rankine itself.

Automating Rankine

Tom Trasmundi sees the decision to automate Rankine as one of the most important events in Canadian Niagara's history. It wasn't because of the technology, advanced and exciting as it was. He remembers, "When they finally told us they were going to automate Rankine, we felt pretty good, because we felt the place is going to be around still. They're not going to shut it down. They're automating it for a reason."

It was a high-risk, leading-edge project. Canadian Niagara worked with North American Hydro, a Wisconsin-based company that had done similar projects to retrofit smaller hydroelectric generating plants. Although no one had ever attempted a plant as large as the 85-year-old Rankine station, Wally Farrell had faith in North American Hydro and in Bill Harris, its brilliant engineer and co-founder. Harris had worked for Wisconsin Hydro when it was getting rid of small-scale hydro sites that it regarded as too uneconomical to operate as manned stations. Harris saw an opportunity and formed a company to acquire such sites and automate them. It was not a simple task. In addition to the electronics end of it, North American Hydro had to redesign critical mechanical components such as the ring-valve gate controller for the water turbine.

Sadly, Wally Farrell, who handled the negotiations with North American Hydro, died before work really got under way, but he could see that the two companies were a perfect match. Canadian Niagara needed someone to take them into uncharted territory, to make Rankine a world leader again, just as it had been in 1905. North American Hydro needed a showcase project that would allow it to grow.

The automation project gave a number of people what they had been looking for. The first thing was pride in the company. They were going to prove that Rankine still had a lot of potential. The other was the sense that the company cared enough about its employees to give them the equipment they needed. The culture of "if it ain't broke, don't fix it" had given way to a culture of investment. Bob Watson, who succeeded Wally Farrell, was prepared to invest in equipment. "You still had to justify that you needed it, but if you needed it, he budgeted for it, and you got it." With better equipment, Tom found that his productivity rose because he could troubleshoot faster and better. His earlier experience with Niagara Mohawk illustrates why Canadian Niagara needed to create its own culture of investment and become less reliant on its parent company.

North American Hydro and Canadian Niagara each brought expertise to the job. At Canadian Niagara, Watson put together a team of people to work on the project. Rankine already had good mechanical and electrical gangs who were doing a great job of taking care of things, but they needed support. Tom Trasmundi was one of those sent to Niagara Falls to help.

Bill Harris of North American Hydro concentrated on the Programmable Logic Controllers (PLCs), the "brains" of the automation that were linked to the controls, relays, transducers, and sensors that made up the monitoring equipment. Canadian Niagara had lots of work to do before the more delicate equipment and components could come in. Tom, Dennis Levesque, and others "had to do all the infrastructure. Cables had to be pulled through, ducts had to be drilled and installed. Racks had to be hung."

The best way to understand the goals of automation is to consider several examples. The generator is the heart of Rankine. As Tom Trasmundi explains, "There are quite a few transducers or sensors that monitor the voltage of the generator, the speed of the generator in r.p.m. [revolutions per minute]. They are all set up so if the generator exceeds these set limits, it shuts down to protect itself.... If the generator went overspeed and hit, say, 500 r.p.m., it could self-destruct."

However, shutting a generator down is not a simple matter. The brakes cannot be applied too early. The automatic controls have to reduce the water flow quickly. When the generator is working, "the headgate is open and the [turbine] ring valve is open so water is flowing through the penstock to the valve, turning the turbine and the generator. If a fault occurs, the ring valve shuts itself down automatically within seconds. That stops the flow to the turbine. The headgate in the meantime has also started to shut down, so that will shut off the flow into the penstock. So by that time, hopefully, the unit is starting to slow down and then it will keep on spinning until it hits another limit." This next limit is the point at which it is safe to start putting the brakes on. The brakes grip the shaft joining turbine and generator, but the sensors must not activate them until the shaft is going at 100 r.p.m. or less. Moreover, the brakes have to be pumped, that is, alternately applied and released, so they won't burn. All of this can be done without human intervention, or even any knowledge that it is happening.

To make all of this work required attention to an extraordinary number of details. Mechanical and hydraulic parts and system paths had to be replaced by electric motors and electronic pathways. The guiding design philosophy was to use existing equipment as much as possible and to add automation to it rather than replacing it before automating. No one had ever automated on this scale before, and the 11 units were not identical. The team started with Unit No. 1, perhaps not the best choice, as it had long-standing vibration problems. Much of the design work had to be done on the job — a heavier motor here, changing wiring there, and always refining the "programming of the PLCs to make sure everything worked."

Walter Lord Charest recalls one time at 2:30 a.m., when he just wanted to be at home sleeping, Bill Harris wanted to try one more thing. That one more thing led to a runaway generator that created such a windstorm that a distracted Harris was running between his computer and the papers flying through the air while Walter kept yelling that the turbine ring gate and head gate weren't closing, no matter what the computer said. Walter was yelling that the generator was going faster and faster, 275, 300, 350 r.p.m., and climbing; it was going to blow up. Bill kept hunting for papers and pecking at his computer. Walter shut the unit off manually and went home to sleep.

One early incident tested everyone's faith. When management wanted a demonstration of progress on Unit No. 1, the team started it up. All went well. To get a better look, some of the visitors went into the chamber underneath the generator, where they could look up and watch it spinning. They shouldn't have been there. "Shortly after they walked out, the commutator ring exploded" sending deadly bits of shrapnel-like jagged metal into the chamber. If it had happened a few minutes earlier, the visitors could have been killed. The hard-working team was "pretty demoralized, but management said they were still committed to automation." It turned out the accident had nothing to do with automation; it was metal fatigue on a rotating part more than 85 years old.

With each unit, the team gained experience and confidence and had fewer troubles. The team members also learned an important lesson about the limits of automation, and their admiration grew for the men who had run the station manually since 1905.

Operators had told the automation team about the difficulty as well as the skill and experience needed to get a generator running at the right speed and frequency, as well as properly in phase before bringing it on line. The synchroscope made the job easier, but it was still not easy. One of the project goals was to make it possible to synchronize generators automatically, with no human intervention or oversight. Bill Harris had done all of his programming and everything was in place. It was time to test the equipment.

The test started, and Tom recalls what happened next. "There was a pretty big bang that shook the place up pretty good." In fact, the tremendous forces released by the mistaken timing "almost blew the generator off the shaft."

In a generating station, the operators are equivalent to pilots in airplanes. They

Bill Hunter Jr. started with Canadian Niagara on August 8, 1954 as a janitor at the Rankine Generating Station. The next month the photographer caught him washing windows. It was a good idea to be buckled up. Below, swiftly moving water was drawn from the outer forebay to the inner foreaby and down to the turbines. The window ledges were narrow but the small moveable platform provided better footing. Bill Hunter Jr. retired in 1981. (Reproduction by permission of FortisOntario)

must be cool, unflappable, never panicking. The operator that day, Walter Lord Charest, ran down the stairs from the control room, and said quietly, "Please. Don't do that again," or something to that effect. A badly shaken automation team never did. After the unpleasant surprise, and Walter's calm request, the team members investigated. They conducted test after test. There was nothing wrong with the program, sensors, or anything else. Finally, they realized there was something an experienced operator could do that they could not automate.

On each generator circuit, the old oil-switch circuit breaker is the interface between the generator and the outside world. If there is a fault outside, from say a lightning strike or a downed line, it interrupts the circuit from the generator to the outside world of feeder lines, substations, and customers. Similarly, when the generator is off line, the circuit breaker stands ready to link the generator to the outside world, but the power going out and the power it is joining must be at the same point in the ever-varying AC wave form which cycles through from maximum negative voltage to zero to maximum positive voltage and back again, endlessly. The timing is all-important. When the automated system told the oil-switch circuit breaker to close and it closed to compete the circuit and joined the generator to the already energized outside world, the timing was off and the generator tried to self-correct. Self-correcting can be dramatic. "You've got about 100 tons of rotating metal there trying quickly to adjust, the stresses will damage it."

The problem the automation team faced was that the old oil-switch circuit breakers were not uniform. Automated synchronization is based on predictability: every time you ask the switch to close it must take the same predictable amount of time to close. They already knew that each generator had its own idiosyncrasies, but they were surprised to find that each circuit breaker had them too. "Even when we did our testing on individual breakers, the cycle times [the time to close] varied." Moreover, they found that "electricians could do maintenance on the breaker one day, test it, then test it the next day and [the cycle time] has changed already." From an automation point of view, "there were too many variables" to understand, predict, or control how quickly or slowly the oil-switch circuit breaker worked on a given day. But the operators knew how to make the system work. "They have a feeling for it, because they have their hand on the control and their eyes on the synchroscope and they know." It might have cost around $1 million to replace the oil-switch circuit breakers. That step was not automated and everyone gained an even deeper appreciation for Rankine's operators.

We are Ready to be a Centre, Not an Outpost

The learning, the experience, and above all the confidence that came with Rankine's automation opened the way for further developments in computers and information technology. The company eventually became so adept at remote monitoring and control that in the 1990s, a wind farm in Alberta was being controlled from the Rankine station. The story of how this came to be is told in the next chapter.

But the company was still looking for a sense of direction. The employees were ready to make Canadian Niagara a centre, not an outpost. But a centre of what? Answering that question was a job for management and directors.

William Birch Rankine was a key figure in creating the Canadian Niagara Power Company in 1892 and shepherding it through the tumultuous early years. His memory is justly served by the generating station named after him. However, his premature death in 1905 thrust much of the responsibility for the future of Canadian Niagara onto the shoulders of the young lawyer A. Monro Grier (*above*). For decades Grier guided the company and served as its most public face at everything from court cases to town council meetings. He was also the one who provided decades of experience and continuity at meetings of the Board of Directors.

By the 1990s Canadian Niagara was deeply mired in another tumultuous era. But Grier was not there; the man who had joined the company in 1901 as Secretary and Solicitor, became Vice-President in 1901, President in 1919 and retired as Chairman of the Board in 1947, had died in 1954 at the age of 94.

Fortunately there were others to guide the company through its darkest times. The task of helping the company regain its soul and chart new directions for a new era fell to Mardon Erbland and Bill Daley, both worthy successors to William Birch Rankine and Alexander Monro Grier. (Reproduction by permission of FortisOntario)

What's to Become of Us? Searching for the Future of Canadian Niagara

CHAPTER 18

Over 100 Years of Searching for the Future

In some ways, Canadian Niagara has always been searching for the future. The Rankine Generating Station is a powerhouse in a park that was born out of a growing conservation movement and a reaction against the despoiling of the natural beauty of one of the great natural wonders of the Western world. It was also born during a bitter fight between Edison on one hand and Westinghouse and Tesla on the other over the future of electrical systems. Edison, the brilliant inventor, had done much to open up the electrical future with direct current (DC). Westinghouse was on the side of alternating current (AC), based largely on the work of Tesla, which allowed large-scale development by making it possible to step voltage up for distant transmission and step it down again near where it would be used. Power generation and power use could be separated by great distances.[1]

At the Columbian Exposition in Chicago in 1893, Westinghouse, Tesla, and alternating current clearly triumphed over the DC system that Edison had pioneered and invested in so heavily. With the prospect of large-scale electrical development at Niagara Falls, the Battle of the Currents was on.[2]

Nevertheless, Albert D. Shaw, the determined promoter and founding president of Canadian Niagara, had great difficulty finding backers for his vision of an electrical future on the Canadian side of the falls. No one knew if markets would emerge or if the technology would be reliable. Even when these questions seemed to have been answered, Canadian Niagara didn't meet critical construction and generating schedules, not only because of financial difficulties but because it was waiting for the future, waiting for the next generation of bigger, more efficient, better-designed equipment. Even on January 2, 1905, the traditionally accepted opening date of Rankine, the company did not generate electricity for use by customers. That had to wait until July 27, 1905.

With the emergence of what became known as Ontario Hydro, the vision of a largely privately owned power industry in Ontario vanished. There was still a place for Canadian Niagara, but that place kept getting smaller. Ontario Hydro grew rapidly, while Canadian Niagara was limited to generating 100,000 horsepower or 74.6 megawatts of electricity. In many ways, the Depression and the Second World War were the most straightforward periods in the company's history. The company prospered, contributed to the community, provided power at a better rate than its rivals, and offered employees good jobs and the prospect of employment for life with one of the area's most respected employers.

The return of peace in 1945 was supposed to mean the return of security and economic prosperity, but it seemed to threaten Canadian Niagara's future. The demands for electric power were so great that efficient use of water became a public issue. Ontario Hydro could use the water of the Niagara River more efficiently than Canadian Niagara could. Before the 1940s were out, Hydro had announced — without any consultation — that under a new plan to meet Ontario's need for electricity, Canadian Niagara would be closed and its water allotment used more efficiently. The plan was quickly retracted and ridiculed in the press. But once again the seeds of uncertainty had been sown.

In 1950, a new international treaty allowed more water to be taken from the river by generating stations. Not only did Canadian Niagara not get any of the water, but it was taken out upstream from Rankine, thereby reducing the water level at Rankine. After 1950, Canadian Niagara sometimes could not draw enough water from the river: the water level was simply too low. Then came the province-wide conversion to 60-cycle power. Canadian Niagara continued to generate at 25 cycles for a dwindling market and depended on Ontario Hydro for the 60-cycle power it needed for its customers. It was getting harder and harder to see the future.

During the 1950s, the company explored ways to maximize its apparent value, because the future seemed to hold a takeover by Ontario Hydro, which was using Canadian Niagara's water more efficiently and providing power in return. In the 1960s and early 1970s the company prospered, but the Rankine Generating Station fell into a period of neglect and very limited use. In 1973 there was again talk of an Ontario Hydro takeover.

But by this time Rankine was undergoing a period of renewal. New monitoring equipment allowed for a staff reduction of 13 in 1972, although true to its traditions, Canadian Niagara didn't lay off employees. Either retiring employees were not replaced or employees were reassigned to other jobs. The three who left the company did so voluntarily. The opening of the new Service Centre on Bertie Street also pointed to a bright future.

By the mid-1970s, Canadian Niagara's future looked bright. Rankine had been revived and made more cost-effective. It looked as if the company would retain the role it had played for most of its history — a "money machine" for its wealthy American parents. Canadian Niagara made a tidy profit exporting power to its parents and selling the rest to Canadian customers more cheaply than any neighbouring utilities. The reason was simple. It had a single, paid-for generating station and was not trying to build more. Ontario Hydro had a constantly growing empire and faced rising construction and transmission costs.

But Canadian Niagara's rosy future darkened as the company entered its most strife-ridden, tumultuous, and controversial years. While those years were not easy for employees to live through, they turned out to be the company's most creative era, and set Canadian Niagara on the path to the future it faces today.

During this period, five individuals stand out as creators of the new Canadian Niagara: Wally Farrell, Harry Macdonell, H. Stanley (Stan) Marshall, Mardon Erbland, and Bill Daley. Others were influential, but these five made the company what it became in the opening years of the 21st century.

Wally Farrell: Visionary Engineer

Wally Farrell and Harry Macdonell were both devoted to Canadian Niagara, but in utterly different ways. Wally, the electrical engineer, wanted to revive the company by doing more on the river, by making Rankine greater within its small world. Harry, Toronto lawyer, business manager, and frequent board member, admired Wally's tenacity and engineering skills, but wanted to make Canadian Niagara part of the larger world. Each thought his own vision was the way to a better future for Canadian Niagara.

No one knows if Canadian Niagara deliberately hired Wally Farrell as General Manager to shake things up. But Farrell certainly put a "different spin on things." He brought a different management style to the job from his predecessors. William J. Hunter (1969-79) and before that Charles A. Clancy (1948-69) had been what board member Donald Smith calls "caretaker managers." This is no reflection on their abilities; they represented what an essentially static company needed. With perhaps a hint of exaggeration, Donald Smith summarizes the company history: "From when it came on line in '05 and through the wars ... you're well into the 1970s before anybody is paying any attention to CNP. It's just there, it's grinding away. It pays money to the Parks Commission. It makes money and it gives it to Niagara Mohawk. And nobody says boo about it." It was a "traditional money machine." Smith remembers Farrell as "an electrical guy. He started lifting the rocks and seeing what was underneath them and trying to turn the company around. I think he did a superb job of controlling costs."

Jimmy Fretz also remembers Wally Farrell fondly. Like so many other employees, Jimmy was born into Canadian Niagara. The Fretz family came to the Fort Erie area in the early 1800s. His father joined Canadian Niagara in 1937 and stayed there for 44 years. His uncle George Wilkinson worked there from 1917 to 1964, the year Jimmy started with the company. Jimmy was a little surprised when Canadian Niagara hired an outsider as general manager. Wally Farrell was a Maritimer who had graduated in electrical engineering from St. Francis Xavier University. In 1978 he left his position as chief engineer of the Hydro Electric Commission of Thunder Bay to become general manager of Canadian Niagara. Over time, as Fretz got to know Farrell, suspicion turned to admiration. An unusual on-the-job friendship developed between the two. Unusual, because Fretz was very junior. "Between '64 and '73 I probably did nothing but sweep the floors and learn a little bit about the plant." In 1978 he became second operator. Farrell was very senior, a professional engineer, and an outsider. He came from away and hadn't worked his way up through the ranks.

Farrell didn't know the company lore as the insiders did. He also didn't know what was impossible, what was unthinkable. He was an intensely bright, quiet individual who didn't excel at small talk or stand out at a party. But he was responsible for a number of important innovations, such as the early automation in the powerhouse and the extension of the 34-kilovolt system designed to increase system capacity economically. However, in terms of the larger corporate history, his most important contributions came from his fresh questions.

Wally Farrell started out slowly, "trying to get a good understanding of all the water power contracts and agreements and the operations. Then he started cutting costs" with automation and reducing the number of men per shift. That was a significant move, because it meant that Rankine had a future. Farrell's greatest contribution to the company was a renewed sense of confidence in the future of CNP and Rankine.

Farrell arrived at a company that had been fundamentally unchanged for many years and that lived in the shadow of Niagara Mohawk. He could see that those days had to end. Canadian Niagara and Niagara Mohawk were moving apart, not out of animosity but out of necessity. For Canadian Niagara to thrive, it needed to look at the world more positively and shake off some old fears. Jimmy Fretz recalls, "There was a fear of the Niagara Parks Commission first of all, and Ontario Hydro second — like, don't do anything to tick off Niagara Parks because they could revoke our water licence, and don't do anything to Ontario Hydro, because you know we're working for them. That was the mentality at Canadian Niagara until Wally came on. Wally dug in his heels and said, 'Now wait a minute, we have agreements and we're entitled to this water and we're entitled to generate power. You're taking our water, so you have to replace it. We belong on this river.'"

In the very best sense of the word, Wally Farrell brought a visionary spirit back to Canadian Niagara. He didn't win all of his causes, but he helped shape the future. Above all, he helped instill confidence in the future; something the company sorely needed.

Farrell was convinced that Canadian Niagara was being treated unfairly. In 1957-58, Canadian Niagara had begun supplying its customers with 60-cycle power, but it continued to generate at 25 cycles to supply customers such as Carborundum in Niagara Falls. The rest of its power went to Ontario Hydro, which supplied the 60-cycle power. However, Hydro levied a "wheeling charge" for moving power around. The wheeling charge, which kept going up, was the same no matter how far the power moved, from one street to the next, or from one end of the province to the other. It seemed unfair to Wally. Ontario Hydro had been built on the premise and promise of "power at cost." But by Farrell's time "power at cost" had given way to what economists call "policy-induced distortions in spatial cost surfaces." This means that power costs and rewards were determined by government policy, rather than by something as simple as proximity to power sources and the actual cost of delivering power.[3]

Wheeling charges weren't the only thing that grated on Farrell's ideas of logic and natural justice. Canadian Niagara had the right to take enough water to generate 100,000 electrical horsepower or 74.6 megawatts of electricity. However, low water caused by a combination of water taken out to supply the Beck II station, natural fluctuations in lake levels, and intermittent ice problems during the winter, meant that Canadian Niagara couldn't always take the water it needed out of the river. But if Rankine couldn't use its water allotment, it wasn't wasted. It went to Hydro, and in this arrangement, Farrell thought Canadian Niagara was being shortchanged. Hydro could take the unused water and send it downriver, where there was a higher head and it generated more power — that is, it was used more efficiently. What did Canadian Niagara get in return? Only the power it would have generated if it had used the water itself. Hydro kept the excess power. Farrell thought there should have been some sort of bonus or profit-sharing plan for Canadian Niagara. He also disliked the fact that if Hydro couldn't use Canadian Niagara's water allotment, it sold it to the United States but did not share the profits with his company.

It seemed to Farrell that Ontario Hydro had Canadian Niagara's cake and was eating it too.

After a great deal of consultation and preparation, the matter went to arbitration. Canadian Niagara lost. It was hard to win against Ontario Hydro. The loss seemed to confirm that the commitment to never ruffling feathers was the right way for Canadian Niagara after all.

The loss was an intense disappointment for Farrell, who had wanted prosperity, continuing employment, and a fair deal. He looked at the antiquated plant with its turbines and generators that had once been the largest in the world. Now they seemed like miniatures. He also looked at the franchise agreement; the way he read it, the company could apply for another 20 years of generating past 2009. But it would need a plant capable of functioning well past 2009. Wally had an idea: replace the 11 generators and turbines with two large 60-cycle units using water drawn from further upriver to give it a higher head. It seemed a wildly optimistic, wildly expensive plan. However, the company was searching for a new future, and the Board of Directors asked Acres International to conduct a feasibility study.

But in the end, the Board of Directors turned down Farrell's proposal. Jimmy Fretz believes it was "one of the biggest disappointments of his life." Nevertheless, this master at containing costs had already done much to ensure the future of the company at a critical time in its life. With the austerity program he launched in 1987, he was able to contract out tasks such as tree trimming, meter reading and cleaning, to staff the plant without replacing those who retired, and to reduce the number of company cars. He also instituted greater use of computers and information technology.

Finally, Wally's premature death even robbed him of his retirement plans. Jimmy Fretz calls him one of the many "downeasters…who had to leave for employment, but as soon as they get a chance, they head right back there." Each year, early in July, Wally and his wife would drive to Cape Breton. He'd leave the car and fly home, and then in August fly back to the cottage on Lake Bras d'Or that was to be their retirement home. But eight months from retirement, in the summer of 1990, he died there of a heart attack.

Wally was a quiet man who was more interested in achievement than in publicity. No one at Canadian Niagara seemed to know of his $1 million legacy to the university he had graduated from in 1947 — the J. Wallace Farrell Memorial Fund, which supports scholarships in engineering. His wife, Margaret Martell Farrell, who died in 2001, also bequeathed $1.5 million to St. Francis Xavier University to fund scholarships, bursaries, and library acquisitions.

Wally cared about life and people in the Maritimes. He probably would have got along with the people from Fortis, the people from Newfoundland who eventually came to own Canadian Niagara — but that is getting ahead of the story.

Wild West Adventures

Wally Farrell and Harry Macdonell both wanted a bright future for Canadian Niagara. Wally saw it in a modernized Rankine Generating Station, perhaps also in a gas-fired generating plant at Fort Erie. Harry Macdonell, Toronto lawyer, board member, and eventually chairman of the board, saw the need to diversify to reduce risk.

Members of the Board of Directors, Canadian Niagara management and others happily gathered in front of a generator and governor at the Rankine Generating Station. On other occasions they posed in front of the now demolished Queen Street Service Centre or in the Board rooms in the Rankine Generating Station or Niagara Mohawk headquarters in Syracuse. Many of the directors such as Bev Matthews, David Gordon, Don Smith and Harry Macdonnell were members of the Toronto legal firm McCarthy and McCarthy, later McCarthy–Tétrault. But wherever they came from, board members worked hard and served the companies well. They also had some very memorable parties for which no photos are available. *Left to right:* Pat Vernon, Unknown, Bob Dixon, Don Schott, Rick Heximer, Bill Donlon, John Powers, Bev Matthews, John Haehl, Dick Torrey, David Gordon, Don Smith, J. Wallace (Wally) Farrell, Vice-President and General Manager 1978-90, and Harry Macdonell.
(Reproduction by permission of FortisOntario)

Over the years, Toronto lawyers have served Canadian Niagara well. In the early years, William H. Beatty served as president from 1901 to 1912, followed by the Honourable Wallace H. Nesbitt, who served until 1919. A. Monro Grier, who was president until 1933, was also a Toronto lawyer. In later years, in addition to legal advice and services, partners from McCarthy–Tétrault provided Canadian Niagara with chairmen of the board, most notably Beverley Matthews, Harry Macdonell, and Donald Smith. Members of the firm also acted as unofficial management consultants and advisers. When asked what made the firm take such an active, almost proprietary, interest in Canadian Niagara, Harry Macdonell suggested, "I think law can be a pretty dull practice…the idea that you can possibly be involved in a completely different organization that's actually doing something is very appealing to lawyers."

The chain of events that would lead to so much turbulence and eventually to new owners started at Niagara Mohawk and its regulator, the New York State Power Authority. In the 1960s and 1970s, the disclosure requirements of the regulatory agencies were not onerous. Regulated utilities in the United States had a built-in incentive to keep costs high, since their rates were determined as a percentage of costs. The higher your costs, the more you could charge your customers, and the higher your profits.[4]

By the late 1970s, regulatory commissions, including the New York State Power Authority, were becoming much more inquisitive. Niagara Mohawk, under newly appointed CEO John Hale, needed to scrutinize its books more carefully. For example, the Herschel Fishing Club, which was wholly owned by Niagara Mohawk through its ownership of Canadian Niagara, came under scrutiny. As Donald Smith sees it, the Club could not be satisfactorily explained to the regulators. "It was just a boondoggle." The Club was quietly sold off.

Given the more active regulatory regime, Niagara Mohawk also wanted to deflect attention away from its profitable Canadian subsidiary. Early in the 1980s, Niagara Mohawk decided to leave the Canadian Niagara profits in Canada. It was the biggest change in direction in the history of Canadian Niagara. The company would never be the same.

Suddenly Canadian Niagara became a very busy company. In Wally Farrell it had had a general manager hell-bent for change on the home front. However, it needed to find investments for all of the money, and that meant that for the first time in its history, the company would not be focused solely on the Niagara Frontier.

Canadian Niagara was completely unprepared for this move. Throughout its history, it had done nothing but generate money and ship it to the United States as generous dividends for Niagara Mohawk. It had never had to think about investments and returns, never had to look at the big financial picture or consider other possible areas of business activity. No one at Canadian Niagara had been groomed for such responsibility.

Rick Heximer, who started in accounting at Canadian Niagara in 1965 and worked his way up to assistant treasurer, knew first-hand how little financial analysis took place in Canada. Billing, for all but three Niagara Falls companies (Norton, Carborundum, and Shredded Wheat), was done by the Fort Erie office. The accounting department, however, was in the Rankine Generating Station. Before it moved to the Bertie Street Service Centre in the 1980s, accounting and financial reporting were done with a pencil and a Burroughs adding machine. As Rick remembers, "We did our financial reports in pencil and sent them to Niagara Mohawk who typed them up from our figures. Everything was done manually.... It was kind of quiet. Things tended not to change for a long, long time." The Canadian Niagara accountants didn't even need to talk to the people at Niagara Mohawk unless they had a problem with the figures. Canadian Niagara bore no responsibility for significant investment and business decisions.

Suddenly, the company was able to keep its profits. Not only did it have to increase its financial expertise, but it needed to find investment opportunities that would allow it to build on past experience and expertise. But hydroelectric investments in Ontario were still largely off-limits. The company would have to turn elsewhere. It was the beginning of an unexpected adventure.

Before 1995, when a private member's bill freed Canadian Niagara to engage in a wide range of businesses, its activities were restricted. The company could get around this restriction only by forming and owning other companies. The first and most important was the holding company Opinac, wholly owned by Niagara Mohawk. Some people assume this entity was named after an obscure First Nations group, but the name simply comes from **CA**nadian **NI**agara **PO**wer, written backwards. Opinac held Canadian Niagara and all Canadian Niagara dividends flowed into it. These dividends provided the funds to start new ventures through other newly formed companies wholly owned by Opinac. In terms of Canadian Niagara's history, the most important of these new companies was Calgary-based Opinac Energy. Other companies included Opinac Investments Ltd., Opinac Exploration, Opinac Acquisitions, and Caraugh.

Canadian Niagara has always carefully chosen its board members. When it started down the road to diversification and created Opinac, Don Bean, president of Wood Gundy, was on the board. At first the company had a balanced portfolio. The advice from Wood Gundy was excellent, and so were the returns. There was more money to invest and at that point, the company decided to "get into oil and gas, because at least it was energy-related, which is sort of in the bailiwick, if not in the same neck of the woods," as lawyer David Gordon puts it.

This move took the company into uncharted territory. It needed someone on the ground in Calgary. Donald Smith, who was on the board at the time, remembers how they found their man. "There was a guy with quite a reputation; I don't mean that in a negative sense. He'd been a very active oil and gas guy. And, he'd got into a fight with his then employer. Anyway, he was singled out as the guy [to get] if you want expertise in oil and gas." His name was Bob Dixon. "We took up an equity position in a number of small oil and gas companies. Bob Dixon had a group of guys with him at his former employer, and when they left, they all went in different directions, all in Calgary. But they all formed little companies. Dixon had the greatest trust in these guys. So he had us buy shares in a company called Renaissance Energy, which became one of the sweethearts of the oil patch. There were two others and they both did very well.... The problem, if there was a problem, was that Niagara Mohawk knew absolutely nothing about oil and gas.... [Bob Dixon was in Calgary] doing things with our money, and no one back east really knows what the hell's going on."

But the money was rolling in, so Opinac invested more money. And in what in theatrical terms is called a "willing suspension of disbelief," everyone was supposed to believe that the utility mindset, which needed steady, predictable returns, could be happily married to the cowboy mindset of the oil patch, which could handle rollercoaster returns. In the meantime, investors were expected to keep quiet and not get pushy with things like interference and oversight.

For a while, all was well. Opinac Energy acquired excellent gas reserves and the decision was made to take the company public with a stock offering. That's when things started to unravel.

About the same time, Beverley Matthews stepped down as chairman of the board of Canadian Niagara and recommended as his successor another McCarthy lawyer, Harry Macdonell. It was a good choice. Macdonell had wide corporate experience and knew how to run a company.

When Canadian Niagara needed a new president and CEO, Macdonell agreed to let Bob Dixon become president, but he withheld the CEO position. He wasn't quite sure of Dixon; nor were the senior people at Niagara Mohawk. With the president of Canadian Niagara in Calgary, busy running Opinac Energy, it must have seemed that Canadian Niagara was subservient to Opinac Energy and was being run from the West. Maybe Bob Dixon could be forgiven for thinking that too.

One of Dixon's first decisions was to ask headhunters to find a successor to Wally Farrell, who was then about a year and a half away from retirement. The board members wanted to give Farrell time to train his replacement. They also wanted time to get a closer look at the new hire to make sure they had made the right decision. Dixon chose another Westerner, Bob Watson from Kootenay, B.C. No one knew very much about him. In 1990 when Wally left to go on his last vacation, Jimmy Fretz recalls that Wally "was down a little bit." Wally was worried that people no longer cared enough about Canadian Niagara. And he had been wondering if Watson was right for the job. It was too late for second thoughts.

After Wally's death, Bob Dixon, who was in Calgary, seemed to lose interest in Canadian Niagara, except as a source of investment money. This is not surprising: his heart was in the oil and gas industry. Dixon couldn't have been more different from Harry Macdonell. David Gordon calls it "a fundamental chemistry problem." Macdonell says that he and Dixon "didn't see eye-to-eye," particularly over who was responsible for what and who reported to whom. Normally, the president reports to the chairman of the board, but not all presidents see it that way. Dixon didn't. He didn't even want Macdonell to talk to Niagara Mohawk people in Syracuse. It all came to a head when Dixon issued the classic business ultimatum to Niagara Mohawk: Harry goes or I go. Harry stayed. Dixon left the company in 1992, and Bob Watson became president.

But even before that showdown, Dixon had been in difficulties. He had indicated that there were promising gas reserves, and on this basis the board decided to take Opinac Energy public. But first, Opinac needed independent confirmation of the reserves. Dixon stalled. Finally he admitted that the reserves weren't there. Donald Smith remembers that "the biggest gas find that we had — and at first blush it seemed to be a huge find — turned out to be one of those anomalies in the oil patch where just through sheer luck" the initial drilling hits something. "So, using the conventional estimates of reserves, you come up with a very big number." Subsequent drilling told a different story. There was no fraud; the first tests had indeed hit gas, but to everyone's dismay, there was very little gas around it.

A Turbulent Period

With Bob Dixon gone, the willing suspension of disbelief ended. It had been a good try. The company had made money, and had explored other investments, such as a co-generation plant at Whitecourt, Alberta. Canadian Niagara even ended up with the Cowley Ridge wind farm at Pincher Creek, Alberta. But Macdonell wanted to reassess the company's long-term future. He hoped that promoting Watson to president would keep things quiet on the home front.

Unfortunately, on the home front, it was a time of change, rather than quiet. Line superintendent Jim Nigh says that Bob Watson was "pretty controversial," but he also feels that part of the problem was a long-standing tension between the people who worked in the Rankine plant and those in the Fort Erie Service Centre. "For the operations and trades people, he made the biggest change at Canadian Niagara since I've been here." Watson "treated the linemen like they were an important part of the company." He was willing to spend on equipment: "Our line trucks, all new. Our safety equipment, all new. Anything we needed to make the job easier, all new." Watson believed that linemen need good equipment and with equipment they could be proud of, they would take care of it better.

This was a huge change from the days when Canadian Niagara had trucks in which you could watch the road going past by looking through the holes in the floor. Canadian Niagara was also infamous for bucket trucks that leaked hydraulic oil so badly that linemen up in the air had to keep raising their arms to work as the bucket slowly fell. When Watson arrived, that changed.

Watson also shepherded through the expansion of the Bertie Street Service Centre, which at the time some regarded as extreme extravagance in terms of design and size. They called it the Fort Erie Taj Mahal, but it has served the company well and today no one regards it as larger than necessary.

Donald Smith describes Watson as an "experienced guy" who "had had a responsible position with a private electric company" in British Columbia. People like that were hard to find. Once on the job he got things done, although he "wasn't nearly the engineer that Wally Farrell was." Because he had performed well as General Manager, his promotion to president seemed a good decision. The board even noticed that he was trying to put some of the polish back in the Rankine station by getting the paint stripped off the beautiful copper doors and trimwork inside the Rankine offices. They also liked his work in making better use of information technology.

But others took a different view. A company that had always managed to weather economic downturns and troubles without laying people off suddenly began to lay off long-service employees. For those who had expected to spend their lives working in the same place, this was a painful period. It was even painful for those who kept their jobs, because they experienced a form of "survivor guilt" that one associates with wartime.

What bothered Frank Flake most was the declining image of the company in the eyes of the townsfolk, customers, neighbours, and the members of the Chamber of Commerce or the Lions Club. For example, it was an honour to be part of the Supervisors' Club, and members wore a suit and tie to the dinner meetings. A number of people disapproved of the head of the company showing up in a plaid shirt and blue jeans. What Watson may have seen as an attempt to be informal and friendly was seen by others as less professional.

Morale sank to the point at which some employees sent anonymous letters or made phone calls to board members. The biggest problem seemed to be extreme polarization among employees. As Donald Smith recalls, employees "were either on the ins or the outs." Clearly it was time for a change.

The board had to take a stand. The first order of business was to find a new president and general manager for a badly wounded company with demoralized employees. Second, and most important for the long term, was to figure out where the company was going and what it stood for.

Difficult Decisions

Donald Smith remembers agonizing with Harry Macdonell over what to do after Watson left. "You have a vacuum. You've got to put someone in.... You can plug in a lawyer, God forbid, who knows absolutely nothing, but would exercise judgment." But they had both been around Canadian Niagara long enough to know how employees felt about outsiders. They needed an insider. And that meant "you take the most senior guy who appears to be responsible." Jimmy Fretz, a vice-president who had worked his way up from the bottom, seemed to fit the bill. They knew Fretz had no financial training or overall business experience, but he had what the employees needed: he was trusted. The board moved cautiously, at first appointing Fretz acting general manager. Fretz himself was surprised, and even more surprised when the appointment became a permanent position.

Fretz faced a daunting task. Some employees had become used to certain perks and a fairly free hand with expenses. That had to be scaled back. And while he was busy trying to recreate a workplace like the Canadian Niagara of the past, Macdonell was trying to find a workable future for the company.

Of all the long-term board members, Harry Macdonell brought the widest business experience to the table. He sat down and "tried to put together some sort of business plan" for the future of Canadian Niagara that he could present to Niagara Mohawk. There were just three options.

1. Carry on as is.
2. Sell it.
3. Merge it.

Meanwhile Niagara Mohawk, 100 per cent shareholder of Canadian Niagara, was not what it had been. Once the envied giant of the American electrical world, Niagara Mohawk now needed money. David Gordon remembers that many people probably thought that "for a company like Niagara Mohawk, a tiny little dividend of a million-or-whatever a year from Canadian Niagara surely would get lost in the rounding. Not so. In some years that was the difference between a profit or a loss for Niagara Mohawk."

Legislation and new regulations had not been kind to Niagara Mohawk. In a well-intentioned but misguided attempt to encourage small, independent power producers, the New York State Power Authority had forced New York State utilities to buy power at a fixed price. Unfortunately, that fixed price was higher than the price at which the utilities could sell it. They lost money on every sale. At the same time, Niagara Mohawk was having difficulties with the Nine Mile Point nuclear plant. As Donald Smith puts it, "They just went from one crisis to another; lawsuits up the yin-yang. They were distracted. Canadian Niagara was a blip on the screen as far as they were concerned." Whatever Macdonell recommended, it had to increase cash flow for Niagara Mohawk. And by now it was clear there were no pots of gold in the West.

There was also another important factor to consider. Although it was commonly believed that Canadian Niagara's water lease would expire in 2009, that was not quite accurate. In 2009, the lease would be up for a 20-year renewal, but, unlike earlier renewals, it would not be an automatic renewal. No one knew how that issue would be resolved. Investors dislike uncertainty, and there was plenty of uncertainty surrounding the future of Canadian Niagara's right to Niagara River water.

Given Niagara Mohawk's need for money and the uncertainty about the future of Canadian Niagara's water rights after 2009, Harry Macdonell rejected Option 1, carrying on as usual. Harry likened the situation to owning a mine, "a wasting asset, where the company's value is declining with the passage of time." A better strategy would be to increase the value of the company.

Option 2, sell the company, would have been an easy route, and Canadian Niagara would have been an ideal purchase for a large utility that wanted a presence in the Ontario market. But there wouldn't be much of a future for Canadian Niagara employees, as a new owner would undoubtedly bring in its own people. As Harry put it, both Canadian Niagara and Niagara Mohawk were "very protective of their employees, and the employee interests were very important to them."

We Merge Only If It Is Fair to Canadian Niagara Employees

Merger was the only fair path to pursue, fair to employees and fair to Niagara Mohawk shareholders. But how to pick the right partner? Harry Macdonell had a few ideas. The company had to respect Canadian Niagara employees, but at the same time it had to change direction. A merger partner had to be a company that could bring in the kind of expertise Canadian Niagara needed to survive, but Harry and other board members wanted to handle the situation carefully. They wanted nothing to do with management by hatchet men.

To get the necessary management experience, the merger had to be with a utility. This limited the field. Board members talked to many companies that used the word *merger* when they meant *takeover* — buy the company, strip off the assets, and dump the employees. But one company was different: Fortis. "It was small, it was well run, and very successful in building up its operation from the old Newfoundland Power." Moreover, Harry knew Dr. Angus Bruneau, one of the founders and chairman of the board of Fortis.

They had met many years earlier on the Churchill Falls project and had a strong relationship built on mutual trust and respect. Also, they were both on the board of SNC Lavalin. Dr. Bruneau was a former University of Waterloo engineering professor who had moved to Memorial University of Newfoundland in St. John's as its first Dean of Engineering. Dr. Ray Gosine, the current dean, has known Bruneau for years and regards him as a "man of vision" who "sees beyond technology" and "sees as part of a bigger plan what others see as independent opportunities." Through Fortis, Bruneau became a legendary figure in Newfoundland business. Harry simply asked if Fortis was interested. Bruneau saw more than a small Ontario power company. Not long afterwards, Harry, Jimmy Fretz, Bruneau, and Fortis president H. Stanley (Stan) Marshall met in Montreal to begin discussions. Things went well. Fortis had "a real desire to create an independent operation in Ontario." Other possible merger partners "had a real desire to be operating in Ontario but not to create an organization."

John Powers, chief financial officer for Niagara Mohawk, who had become president of Canadian Niagara after Watson left, "really ran the negotiations." Macdonell, as a longtime friend of Bruneau, also wanted to avoid any appearance of conflict of interest. Therefore, though he and Powers talked regularly about the matter, Harry remained in the background.

Fortis considered a straightforward acquisition of the company, but Macdonell wanted to structure the relationship as a merger. If things worked out well, then Fortis could acquire the remaining shares or Niagara Mohawk could exchange its interest in Canadian Niagara for an interest in Fortis. But for the moment, he felt both Fortis and Canadian Niagara needed a link to Niagara Mohawk's energy marketing expertise. So in 1996, Niagara Mohawk sold 150,000 common shares to Fortis Inc., giving it a 50 percent interest in Canadian Niagara. As part of the deal, Fortis had a veto over the appointment of Canadian Niagara's CEO, while Niagara Mohawk had a veto over the appointment of the chairman. While neither company had the exclusive right of appointment, it worked out that Fortis appointed the CEO of Canadian Niagara and Niagara Mohawk appointed the chairman of the board. Stan Marshall, who as president of Fortis had done much to bring Fortis and Canadian Niagara together, was for a time president of Canadian Niagara. He was also instrumental in Fortis buying out the remaining 50 percent of Canadian Niagara shares in 2002.

Meanwhile, Canadian Niagara continued to improve operations under Jimmy Fretz as general manager. At first, employees were glad to see one of their own as general manager, but their enthusiasm waned when he began to cut costs. Fretz traces the change to the point at which when he started to reduce the number of company vehicles. He remembers "driving up the Queen E[lizabeth Way], and every third vehicle that passed me had a Canadian Niagara emblem on it. I said to myself, 'Geez, we have to cut back.' This made some people unhappy, when you start taking things away or saying, 'No more.'" Trucks had become a particularly contentious issue in the company. It seemed that whether or not an employee had a company vehicle depended less on need than on connections. But trucks weren't the only issue.

At the same time, Niagara Mohawk was fighting off bankruptcy and the oil and gas venture out West hadn't worked out as planned. Fretz tried to dispel rumours that Canadian Niagara was about to close. Although the Cowley Ridge wind farm was making money, and keeping jobs at Rankine — the wind farm was monitored remotely from Rankine, and the machine shop employees made parts — the anxieties persisted. Fretz encouraged people to be positive about the future, but employees were suspicious of management, and he was now part of management. Even the pensioners felt something had been lost. Niagara Mohawk had treated them well until after Wally Farrell died, but now they felt neglected.

The problems went beyond the visible day-to-day difficulties. Canadian Niagara had grown since the days of managers such as Lawson, Bracken, Hunter, and Clancy. Management issues were far more complex and demanding. Other companies, such as Fortis, had switched from the lone manager model to a management team approach. Canadian Niagara now needed an organized management team led by someone with high-level systematic management training and experience. Instead, it chose Fretz, who had little management experience, and put him into a position based on the lone manager model. Even the board members who had selected Fretz had been concerned about his lack of formal training for the many requirements of the job. It didn't look like a long-term solution.

Fretz was stressed and felt isolated. He had few, if any, people in Fort Erie or Niagara Falls with whom he could discuss management problems. He often called board member Donald Smith for advice and the two talked on the phone for hours at a time. But there were simply too many problems. In 1997 an employee formally complained about an alleged threat by Fretz. Jimmy Fretz took an early retirement package to clear the air and prepare the way for a different kind of manager and management structure. After he left, board member Donald Smith remained a personal friend, and Fretz's daughter still works for the company — as does Bob Watson's son. Canadian Niagara is that kind of place.

A New Start

Canadian Niagara was a valuable company with two 50-percent owners. Niagara Mohawk — busy with problems of its own — represented the past, and Fortis, the new co-owner, represented the future.

Fortis president Stan Marshall had to protect a valuable asset and help Canadian Niagara become another growth centre. Canadian Niagara and its employees needed, and deserved, more than just another manager. It was time to end the turbulence and restore employee morale. Marshall turned to Mardon Erbland, an electrical engineer from Ohio who had been with Fortis for so many years that Newfoundland had become his home.

In St. John's, Mardon had been in charge of Human Resources where, as Bill Daley put it, "he had a reputation for creating team environments and building consensus." With retirement only five years away, Mardon would finish his career as a CEO and Fortis would get a proven executive to take on the daunting task of putting the "soul" back into a deeply troubled but valuable company.

Mardon and his Newfoundland-born wife Doreen hadn't planned to move to Fort Erie five years before Mardon's retirement, but a lot of what had happened in the years before he arrived at Canadian Niagara hadn't been planned either. The Board appointed Erbland president and CEO of Canadian Niagara as of December 31, 1997. Shortly thereafter, Bill Daley of Niagara Mohawk became vice-president, human resources and corporate services. A management team was emerging.

Wendy Spry, a dedicated, long-serving employee, remembers a company that had been dragged down. "By the time Mardon came, things were probably at an all-time low.... Our reputation in town was not the best." The most painful memories from the period after the death of Wally Farrell are of the large number of people who were let go. This was particularly painful in a company where employees traditionally worked until retirement age.

A few years after Mardon started work at Canadian Niagara, he presented Wendy with her 25-year service award. By that time, the rift between management and employees was starting to heal. Mardon asked her, "'What's it like to have been here 25 years?'" Tactfully, she replied, "Kind of a rollercoaster. There were good times and there were very bad times." She was trying to be fair but not hurtful, since he was finally restoring what had been stolen from many Canadian Niagara employees — senior managers they could respect. Many things have changed in the past few decades, people lament that no one polishes the brass railings at Rankine or the wonderful bronze front door that is now green, but at least the reputation of Canadian Niagara management is no longer tarnished.

A Return to Basic Values

By time Mardon Erbland joined Canadian Niagara, the company had followed many trails, most of which had gone cold. In the late 1980s and the early 1990s, when Canadian profits were suddenly available for investment, the company had lost focus. It was not surprising: there was a lot of money to invest, and regulatory and legislative barriers to investing in what Canadian Niagara knew best — medium-sized hydroelectric utilities in Canada. The quest for oil and gas in the West had made local employees feel that Rankine and electricity were no longer the heart of the company. The company got out of oil and gas completely in 1993, and when Mardon arrived, its only Western venture was the wind farm at Cowley Ridge, which the company later sold. But it had investigated dozens of other opportunities, from glove and meter testing (the glove testing went ahead, the meter testing prospect was abandoned), to buying generating stations on the Ottawa River and in Bombay, India.

During this period, the company seemed to have lost sight of fundamentals. In response to limited opportunities in the business it knew best, it tried to diversify so widely that it did not build on its strengths. In the process it had lost one of its historic strengths — mutual respect between employees and management.

When Mardon arrived, "restructuring of the industry was on the horizon, but it wasn't a big driver... I really had to focus on the core business stuff.... I wanted to make my mark in terms of leaving a company that had some strong core values. When things got uncertain, we'd go back to those." Responding to restructuring would be Bill Daley's job.

Mardon hadn't met former employees or executives and wasn't interested in trying to dig up all the details of who had done what to whom. What mattered to him was that employees believed they were no longer respected. "I had to change that." And change it he did. Wendy Spry used the word "firm" to describe his way of doing things. Mardon has a reputation as a consensus builder and sees himself as "generally a big believer in collaboration," but you have to know where to draw the line. "I think you have to avoid paralysis of decision making, because you can collaborate too much.... [When] I said I wanted to do something with core values, there were

Employees, board members, Canadian Niagara management and visitors all fell under the spell of Rankine Generating Station. And the deeper one descends into the pit, the more one travels back in time to another era, the more one feels humbled by the boldness and ingenuity of an earlier age that still serves the present. Board members often talked about how refreshing it was to escape from their offices and travel to Rankine in Niagara Falls or the Service Centre in Fort Erie. *Left to right:* Jack Delaney, Dick Torrey, Bev Matthews and Bill Hunter.
(Reproduction by permission of FortisOntario)

some people who said, 'Let's get a team together right away, and we'll talk about it.' I said, 'No. Here's what they are and I expect people to follow them. If they don't, well, we'll have to deal with it.'" The values were straightforward:

Respect for people
An unrelenting commitment to safety
Quality customer service
Financial success criteria
Productivity
Commitment to community

This was not a public relations exercise. Mardon didn't hold a press conference. He didn't plaster the points all over the walls. "People said, 'Let's get a frame to put on it.'" He countered with "'No, no, no. that's counterproductive... you've just got to live it.'"

He knew it would take time. After years of unstable leadership, employees weren't immediately going to jump on the next bandwagon. But over time, people did, indeed, "live with it." Mardon found this out about a week before his retirement and return to Newfoundland, "I had people come up and say, 'Well, you know, I guess maybe you were for real. I hate to see you go.' "

He had helped people get over a very rough time, but they had also worked hard themselves. Wendy Spry thinks it took "all of Mardon's five years to bury the rage. It was anger at losing fellow employees because in those years, we weren't just fellow employees. We chummed together. We were close friends. And so it wasn't just dragging the company down; it was a personal shot at us. When Mardon came in, he made it clear that everybody got a second chance. We would start again. There was to be no animosity."

Animosity never ends if you try to get rid of everybody who had become unpopular during a difficult period. Wendy and others had to "learn to deal with it. We would put it back together. And that was fine. You could do that during the day at the office, but you had to go home at night and face those who had been left behind. That was tough. Because it's a 24-hour-a-day thing. Fight here at the office to get along with people. And go home and deal with the guilt that you were a survivor."

Mardon helped set the stage for employees to rediscover for themselves just how strong they were and how deeply they cared for the company and each other. But he also had to get the business back on track, refocused and ready for the future.

When Mardon looks back at reorienting the company, he says, "If there was one thing that really helped the company, it was the people. You know, if you don't have the right people who share the vision, you can't do it." To get there, he had to choose his "own team." Eyebrows were raised when about four months into the job he had to let one of his senior people go. But he insisted on making sure that people were right for the jobs they did. And of those he chose, one would have to be groomed as his successor. This was extremely important. The company hadn't always been good at preparing for leadership changes.

Restructuring and Expansion

Bill Daley started work at Niagara Mohawk after graduating from the State University of New York at Buffalo. He worked his way from productivity analyst to human resources at the natural gas division. Although he enjoyed the work, he was thinking about leaving to study labour law. The regional manager in Albany suggested another plan. If Bill would come to Albany to head up the regional service organization — 140 people, 11 supervisors, $12 million budget — he would sponsor Bill in the two-year executive MBA program at Rensselaer Polytechnic Institution. Bill accepted and made the dean's honour list at Rensselaer, while running the regional organization. But he yearned for something more entrepreneurial, something he could build rather than just run. Niagara Mohawk had created a new company, Plum Street Energy Marketing, later Niagara Mohawk Energy Marketing. It offered a challenge. "I basically had to go out and build a business. In about two years we went from zero to 140 accounts."

Achievements like this attract attention. One day, the president of Niagara Mohawk Energy told Bill "some guy named Mardon Erbland wants to meet you to talk about an opportunity at Canadian Niagara Power, a vice-president's job." When Bill met with Mardon they ended up talking about "values. We talked about people. We talked about our philosophy of management. We really hit it off." Mardon arranged for Bill to take a temporary 11-month leave from Niagara Mohawk to become vice-president of human resources and corporate services at Canadian Niagara. When he started on April 14, 1998, it was a leap of faith: he was married with three kids, and there was no job guarantee on either side of the border when the 11 months was up.

For the first few months Bill felt like a coroner: he had to deal with the bodies left over from earlier. And his responsibilities seemed rather diverse: human resources, customer service, information technology, and strategy. Things changed again with the passage of Bill 35 in Ontario, the *Electrical Energy Competition Act*. Fortis wanted to take advantage of the opportunities for restructuring and consolidation of municipal utilities that the new act was supposed to create. Mardon asked Bill to take on this responsibility at the start of 2000 as vice-president of corporate development. It was both a great compliment and a tremendous task. Mardon said, "There's no pressure, but we're putting you in charge of the future of the company." It was no joke. The new legislation had torn things apart. Generation and transmission and distribution were now separate spheres that had to be run by distinct companies. Fortunately, Daley would be working with a management team to create the future, not as a lone manager; those days were gone.

Daley had to bring the company back to basics. "If you have the foundation, it's really easy to put the walls and the ceiling on. But if you don't have the foundation, the walls are going to tumble. So my number one goal was to look after the foundation."

Canadian Niagara, under FortisOntario, had to become two companies. Canadian Niagara Power Inc. is in the distribution business. It makes its money by owning and operating the system required to distribute electricity — the poles, wires and transformers. The Ontario Energy Board sets the profit the company is allowed to make, or what is called "a regulated return." The company buys electricity but must sell it at cost — no profit, no markup, pure flowthrough. That is what it does in Fort Erie and Port Colborne. The other company, Canadian Niagara Power Limited, inherited Rankine. It is in the generating business, the business that the legislation says can generate and sell electric power in the so-called open market.

In Ontario's newly created competitive electric market, municipal utilities could now merge or be swallowed up by other utilities. Smaller utilities could see that the new regulations would make it difficult for them to economically provide the required level of service and safety. Daley saw an opportunity. FortisOntario (that is, Canadian Niagara) knew how to run small utilities safely, reliably, efficiently, and economically. It was ready to buy or provide services to other utilities. There was one catch: if municipal utilities were sold to the private sector, the seller paid a 33 percent transfer tax. This tax did not apply to sales to Ontario Hydro (now Hydro One) or municipal utilities.

The transfer tax was a formidable barrier. But Canadian Niagara was prepared for change. It had excellent employees, a flexible administrative structure, and excellent computer and IT skills and capability, whether for office procedures, SCADA or remote communications and interconnections. Canadian Niagara also knew how to deliver electricity cost-effectively to customers and how to repair, automate, maintain, and operate hydroelectric generating stations from an earlier era. Moreover, it knew how to work in smaller markets.

The transfer tax created an unlevel playing field if Canadian Niagara hoped to buy any municipal utilities. So the company asked the provincial Minister of Finance if it could buy 10 percent, if the municipality retained the other 90 percent. After all, the *Municipal Act* said the municipality must own 90 percent of the utility, but put no restrictions on the remaining 10 percent. A favourable ruling led to two important business deals being signed in October 2000.

On October 31, in the smaller of the two deals, FortisOntario acquired 10 percent ownership of Rideau St. Lawrence Holdings Inc., an electricity distribution company serving about 6,000 customers in eastern Ontario.[5] The larger deal was signed a few days earlier with Westario Power Holdings Inc. FortisOntario acquired 10 percent of this electricity distribution company with about 21,000 customers in southwestern Ontario.[6]

The 10 percent ownership in two companies serving 27,000 customers was more important than it might seem. It represented six former municipal utilities that had opted for a public/private partnership. These former municipal utilities gained the commercial discipline, management experience, and risk management they wanted. FortisOntario demonstrated that it was serious about growing in Ontario and finding a way around the 33 percent transfer tax, and that it had more than a little administrative and financial ingenuity. The deals also gave the company a seat on the boards of both companies, and Westario gave FortisOntario an IT services contract.

The two deals represented a good start, but only a start. Quite naturally the company wanted a larger percentage of the ownership. However, patience is an important part of the Fortis way of doing things. For example, today Fortis Inc. supplies electricity to Grand Cayman in the Cayman Islands through Caribbean Utilities. It happened 10 years after Stan Marshall first started to pursue that market.

The municipal electric utility in Port Colborne had about 10,000 customers. The city wanted to concentrate on its core services, but did not want the quality of electric service to decline. Mayor Vance Badaway thought a local or regional company would provide the best service. Mardon Erbland had established a good relationship with the mayor, but an outright sale was out of the question because of the 33 percent transfer tax. In any case, the mayor preferred a revenue stream from a sale spread out over years, rather than one large payment. All that remained was to find a way to structure the deal so as not to run afoul of all the regulations. On April 15, 2002, CNPI (Canadian Niagara Power Inc., the distribution company) entered into the Port Colborne Lease. On November 22, 2004, the Canadian Council for Public-Private Partnerships honoured that 10-year lease with its Chuck Willis Award for Innovation and Excellence in Public-Private Partnerships for the tax-efficient way it put together a complex deal. In brief, Canadian Niagara and the City of Port Colborne set up a lease-to-purchase arrangement whereby Canadian Niagara leased the assets of Port Colborne Hydro for a certain percentage of the value of the business and then gradually took over the costs in return for the revenues from the business. At the end of the 10-year period, Canadian Niagara will make a final lease payment and become full owners of the Port Colborne distribution system. It is currently working to improve the system and service.

Events in the United States have always influenced the history of Canadian Niagara. Fortis had bought a 50 percent stake in Canadian Niagara from financially pressed Niagara Mohawk. Subsequently, National Grid acquired Niagara Mohawk and with it the growing embarrassment of 50 percent ownership of Canadian Niagara. National Grid was a distribution company and it was under such criticism from the New York State Power Authority for owning a Canadian generating company that it briefly considered giving Canadian Niagara profits to charity. Instead, it agreed to Stan Marshall's suggestion that Fortis acquire the remaining 50 percent.

It was more than a financial transaction. It was an important symbolic event. It broke Canadian Niagara's last tie with the original American investors who had pioneered large-scale hydroelectric power development on both sides of the river at Niagara Falls. The pioneering Adams Generating Station in Niagara Falls, New York, had been demolished. Its creator, the Niagara Falls Power Company, had disappeared into Niagara Mohawk, which had been acquired by National Grid. On the Canadian side, the Rankine Generating Station was still active and the Canadian Niagara Power Company was now wholly owned by Fortis, a Canadian company.

It was also an important moment for Bill Daley. When entrusted with the future of Canadian Niagara, he chose the name FortisOntario, because he wanted it to be "a mini Fortis, a growth company." Now it was fully a Fortis company on a good growth path. Although he felt the legislative and regulatory environment left him with "two hands tied behind his back with duct tape," there was still wiggle room.

Canadian Niagara had long had ties with Cornwall, Ontario. Percy McAteer had come to Canadian Niagara from Cornwall in the 1920s, Canadian Niagara had sent crews there to help recover from storms, and Sid Morris used to go to Cornwall to help do inventory.

Cornwall also gave FortisOntario its first clear ownership deal. On October 17, 2002, FortisOntario acquired Cornwall Street Railway, Light and Power Company Limited — also known as Cornwall Electric — from Enbridge, which had bought it three years earlier. This Ontario-based transmission and distribution utility, which had not been municipally owned, supplied about 23,000 customers in the City of Cornwall, South Glengarry, South Stormont and the Ontario portion of the Mohawk Territory of Akwesasne. As part of this acquisition, FortisOntario also acquired a

As Canadian Niagara searched for a new place, and eventually new owners, in the rapidly changing world of public utilities, work life continued. People did their jobs, people were hired and some such as John Crotinger (*at left*), retired. John joined Canadian Niagara in 1946 as janitor and worked his way up to Operator-in-Charge.

July 27, 2005 marked the 100th anniversary of the day of commercial electric power generation at the Rankine Generating Station. Canadian Niagara threw a party at the generating station, the weather co-operated and the Niagara Parks Commisision catered the event. People are sill talking about the food. Many retirees returned to Rankine to share memories, tell stories and admire wonderful equipment in a stunning building in a park near the falls.

John Crotinger and his wife Joy were there. The luckiest visitors were those who heard the retirees telling stories of what it was like when the storms blew in, when the wind howled and the tiles rained down from the roof. Or maybe they heard stories of trying to break the ice with dynamite and a tug boat. Or maybe they came to understand why so many people are in love with the generating station in a park at Niagara Falls. (Reproduction by permission of FortisOntario)

five-megawatt natural gas fired co-generation plant which heats a hot-water district heating loop and provides five megawatts of electricity to the Cornwall Electric grid.

Another acquisition was Granite Power, which had been founded by the Campbell family. It was a third-generation family-owned company with no interested fourth generation. Mardon Erbland and later Bill Daley had tried to purchase the company, but the two parties could not agree on price. A subsequent effort also looked like a lost cause until some eleventh-hour negotiations and concessions led to a deal. On April 1, 2003, FortisOntario acquired Granite Power, which consisted primarily of Eastern Ontario Power (formerly Granite Power Distribution) and Granite Power Generation. Eastern Ontario Power distributes electricity to about 4,000 customers in Gananoque. The five hydroelectric generating stations, which have a combined capacity of 5.6 megawatts, are now owned and operated by FortisOntario Generation Corporation. The acquisition made economic sense because of Canadian Niagara's advanced skills in computers, information technology, instrumentation, and remote operation of unmanned generating stations.

In 2004, FortisOntario's newly named Eastern Ontario Power Company won the Business/Industry of the Year Award from the Thousand Islands Gananoque Chamber of Commerce for improvements to the Service Centre and involvement in the community. Achievements such as this and the creation of Energy Park in Fort Erie demonstrate the drive and commitment behind the phrases in Daley's everyday

Long after the turbines and generators in Rankine have been stilled, the Niagara River will flow past just as it did before the Canadian Niagara Power Company was created in 1892 and Rankine opened in 1905. Some of the water will flow over the weir opened in 1937 to help divert water through the inlets and into the forebay. The electric-powered ice drag has not moved in years but its track bed over the inlets is still one of the best viewing points for the water that jumps from foaming turbulence to glassy smoothness as it races past the rapids to the falls. Twentieth Century-Fox film crews set up their cameras there and captured the most dramatic shots in the film that helped catapult a young actress to fame. Her name was Marilyn Monroe, the film, *Niagara*. In its present condition the site is too dangerous to allow the public to enter. (Reproduction by permission of FortisOntario)

speech: "We're going to grow. We're going to be good operators. Wherever we live and work, we're going to make a difference."

In between acquiring Eastern Ontario Power and getting the Business/Industry of the Year Award in Gananoque, FortisOntario Generation Company acquired a two-megawatt hydroelectric generating station at Rideau Falls in the City of Ottawa. It is in a beautiful park overlooking the Falls and the Ottawa River. Canadian Niagara had started in a park at Canada's most famous waterfall. With Eastern Ontario Power, it has a generating site at Parks Canada's Jones Falls Dam on the Rideau Canal. The Jones Falls Dam is one of Canada's most important nineteenth-century engineering achievements.

Early in 2005, Bill Daley reflected on the company's achievements. Earnings had grown from $9 million in 1999 to $16 million in 2004, while controllable expenses had been reduced by $2.3 million. Assets grew from $68 million in 1999 to more than $367 million. In terms of system reliability, measured by the frequency and duration of power outages, FortisOntario had taken Canadian Niagara from the bottom quartile of the Canadian industry to the top quartile. Employees were working with state-of-the-art systems and a modern data centre. The company's safety record was outstanding. These things were possible because Daley brought Canadian Niagara back to what it does best — operating as an electric utility company. In spreading across the province, the business had in another sense come home to its real business. Daley and Canadian Niagara had learned the same lesson that Welland newspaperman and merchant George Sawle had learned at the time Rankine was under construction: "If you want to get anywhere, you have to stay in one place."[7]

But Rankine was the biggest challenge. It had a curious spot in the generating business. Its legal water allotment often was not available, because of ice problems or the low level of the water, and Ontario Hydro used the water instead. Then there was the question of the efficient use of a scarce resource, that is, the amount of Niagara water designated for generating hydroelectric power. The Ontario Hydro plants further downriver could generate more power from a given amount of water than Rankine, because plants downriver benefited from the greater head of water, or distance the water fell before it entered the turbines. Although rarely described in such terms, the question of how to use the water allotment was one of environmental ethics.

It made sense to reduce the amount of water used for generating at Rankine and increase that diverted for more efficient use. To understand the changes, it is helpful to remember the original 1892 charter. Under the charter, the company received the right to use water from the Niagara River to generate power and in return took on the obligation to build a generating station. The province wanted power produced from the waters of Niagara, the investors wanted to make a profit, and the Queen Victoria Niagara Falls Park Commission needed the revenue from water rights, but also wanted to preserve the beauty of the area and enhance the quality of visits to Niagara Falls. Canadian Niagara's greatest asset, aside from its employees, is still the contract or water rights.

Over the years, Canadian Niagara's water contract has been viewed differently, but the original aims were honoured and solidified in the Niagara Exchange Agreement of February 2005. The company has always maintained that it has the right to request and receive a final 20-year extension of its water rights from 2009 to 2029. Others disagreed. No one benefited from the uncertainty, and the Parks Commissioners and the building itself stood to be the greatest losers. Whenever Canadian Niagara's water lease expired, the Niagara Parks Commission would become owners of the Rankine Generating Station. The earlier the Parks Commission could take ownership of Rankine, the better the condition of the building and the equipment would be. An additional 20 years after 2009 of underuse or inactivity at the powerhouse would not be good.

The Niagara Exchange Agreement was a sound, amicable way to end Canadian Niagara's history of generating power on the Niagara River. Under the agreement, FortisOntario Inc. transferred its water entitlement to Ontario Power Generation Inc. (OPGI). FortisOntario receives in return a firm commitment for 75 megawatts of wholesale power supply until April 30, 2009, in addition to a one-time payment of $10 million to settle various contractual matters. OPGI and FortisOntario both get what they need — certainty and the ability to plan ahead. Moreover, the Parks Commission has a limited right to start preparing the building for its new life as a Niagara Parks Commission property.

The Niagara Exchange Agreement came about after many, many meetings and much negotiation. In terms of historical importance, it ranks with the company's 1892 charter. The company is still in the electricity business, but under the changed circumstances of the 21st century. Employees, not only at Rankine and in Fort Erie, but in all of the FortisOntario family now have a clearer picture of the future.

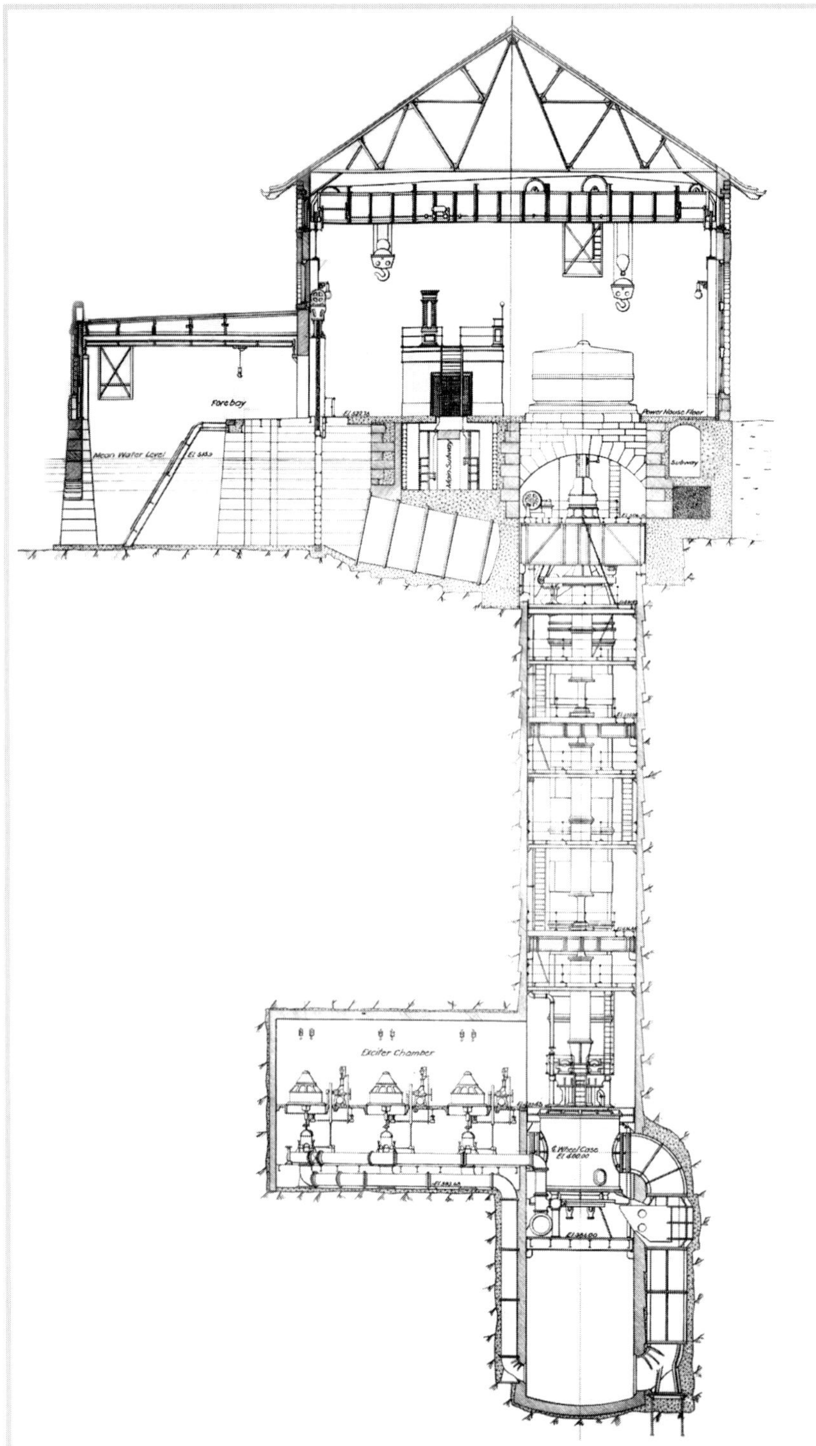

A 1913 sectional view of the Rankine Generating Station power house, pit and tailrace tunnel. People with vision, knowledge and skill, along with a large measure of courage and persistence, turned engineering drawings into the Rankine Generating Station. Then, for over a century other people with knowledge, dedication and ingenuity kept it running to provide electric power to commercial, industrial and residential customers along the Niagara Frontier in Canada and the United States. It will take further ingenuity, imagination and courage to usher the Rankine Generating Station into a new life after it becomes the property of the Niagara Parks Commission and is no longer generating electric power. (Reproduction by permission of FortisOntario)

Will the Brass Be Polished Again?

The Adams Station in Niagara Falls, New York was the model and testing ground for much that was built into the Rankine Generating Station. Craig Woodward, who at the time of writing is in his eighties, spent his entire career as an electrical engineer with Niagara Mohawk. His father had worked for Niagara Mohawk, and Craig grew up hearing stories of what a wonderful place Adams was. They were the same kind of stories you hear about earlier days at Rankine — gleaming brass railings, sparkling windows, spotless equipment, a shining symbol of the early years of electricity in Niagara Falls. But the Adams station he actually saw was well past its prime and the stories seemed exaggerated. Eventually, the Adams station was demolished and the company seal and doorway re-erected by volunteer labour in a park on Goat Island near a monument to Nikola Tesla. But many years ago, Craig visited Rankine when it was still in its prime and for the first time, he could imagine the Adams station as it had once been.

What will happen to Rankine? Today the powerhouse is fundamentally sound, but its looks already betray its dwindling importance as an actual generating site and a Niagara Frontier pioneer. And it is no secret that unused industrial buildings can go downhill quickly. Nevertheless, the Niagara Exchange Agreement, which some see as Canadian Niagara surrendering its place on the river, could be the saving of the powerhouse.

Before this decade ends, the Rankine Generating Station will become the property of the Niagara Parks Commission. It remains to be seen what will happen. Perhaps one day the brass will shine again.

Notes

Chapter 1

Niagara Falls, Where All Is Not as It Seems

1 Dow, Charles Mason, *Anthology and Bibliography of Niagara Falls*, Albany: State of New York, 1921, Vol. 1, pp. 25, 28.

2 Berton, Pierre, *Niagara: A History of the Falls*. Toronto: McClelland & Stewart Inc., 1992, p. 67.

3 Dow, *Anthology*, vol. 1, p. 230.

4 *The Niagara Falls Electrical Handbook: Being a Guide for Visitors From Abroad Attending the International Electrical Congress, St. Louis, Mo., September 1904*, Niagara Falls: American Institute of Electrical Engineers, p. 32. For further information on historical and current rates of recession, see John Jackson, John Burtniak, and Gregory Stein, *The Mighty Niagara: One River, Two Frontiers*, Amherst, New York: Prometheus Books, 2003, p. 426. As more water was taken from the river for power generation, the rate of recession decreased.

5 Seibel, George A., *The Niagara Portage Road. A History of the Portage on the West Bank of the Niagara River*. Niagara Falls, Canada: The City of Niagara Falls, Canada, 1990, p. 4.

6 In 1929 the Historic Sites and Monuments Board of Canada erected a bronze plaque in Stamford to memorialize this significant road. The plaque reads: "Niagara Portage Road. East of this Memorial passes The Portage Road Queenston-Chippawa opened by United Empire Loyalists in 1788. Government Portage from 1791. Principal route of travel and trade to Upper Lakes Region until opening of Welland Canal, 1829 and Railways 1854. Important strategic position and line of communication During War of 1812-14 and Rebellion of 1837-8."

7 Jackson et al., *The Mighty Niagara*, pp. 130-131.

8 Once home to 1.2 million people, Buffalo now has only 300,000 residents. However, in terms of architectural legacy, it is regarded internationally as one of the great American cities. Frederick Law Olmsted's park system still speaks of beauty and grandeur in a city whose plan won an honourable mention at the Paris Exposition in 1878.

9 Ronald L. Way, *Ontario's Niagara Parks: A History*, Niagara Parks Commission, 1946, p. 5.

10 J.B. Harrison as quoted in Greenhill, Ralph, and Thomas D. Mahoney, *Niagara*, Toronto: University of Toronto Press, 1969, p. 117.

11 Greenhill and Mahoney, *Niagara*, pp. 81-82, quoting a report by William Lyon Mackenzie.

12 Greenhill and Mahoney, *Niagara*, pp. 114-115.

13 Irwin, William, *The New Niagara. Tourism, Technology, and the Landscape of Niagara Falls 1776-1917*. University Park, Pennsylvania: The Pennsylvania State University Press, 1996, p. 8.

14 Irwin, *The New Niagara*, p. 25.

15 Adams, Edward Dean, *Niagara Power: History of The Niagara Falls Power Company 1886-1918: Evolution of its Central Power Station and Alternating Current System*. Niagara Falls, N.Y.: privately printed for the Niagara Falls Power Company, 1927, on the Fiftieth Anniversary of its Foundation, in two volumes, Vol. 1, p. 23.

16 Berton, *Niagara*, p. 174.

17 Spieler, Cliff, and Tom Hewitt, *Niagara Power from Joncaire to Moses*. Lewiston, N.Y.: Niagara Power Publishers, 1960, revised edition, p. 4.

18 Hulse, A.E., *The Story of Niagara Falls Power Company & William B. Rankine*. Southampton, N.Y.: The Yankee Peddler Book Company, printed at the Steamboat Press, 1965, p. 7.

19 *The Niagara Falls Electrical Handbook*, pp. 37-38.

20 Greenhill and Mahoney, *Niagara*, p. 154.

21 Hulse, *The Story of Niagara Falls Power Company & William B. Rankine*, p. [8].

22 Jacob F. Schoellkopf, father of Arthur Schoellkopf, remained as president of the company until September 15, 1899, the day he died.

23 Greenhill and Mahoney, *Niagara*, p. 154.

24 *The Niagara Falls Electrical Handbook*, p. 39.

25 Greenhill and Mahoney, *Niagara*, p. 155.

26 Hulse, *The Story of Niagara Falls Power Company*, p. [8] lists the original investors as Jacob Schoellkopf, George Matthews, W.D. Olmsted, Benjamin Rhodes, and Arthur Schoellkopf.

Chapter 2

The Birth of the Canadian Niagara Power Company

1 Queen Victoria Niagara Falls Park, Annual Report for 1889, p. 4.

2 Ball, Norman, and John Vardalas, *Ferranti-Packard: Pioneers in Canadian Electrical Manufacturing*, Montreal and Kingston: McGill-Queen's University Press, 1994, p. 10.

3 George A. Seibel, *Ontario's Niagara Parks: A History*, Niagara Parks Commission, 1985, pp. 33-34. Seibel as drawn from John M. Mills, *History of the Niagara Falls Park and River Railway Company*, Toronto, 1967, p. 93.

4 Queen Victoria Niagara Falls Park, Annual Report for 1892, p. 4.

5 The full text of the agreement is found in Queen Victoria Niagara Falls Park, Annual Report for 1892, pp. 4-8.

6 Schedule A to 55 Victoria Chapter 8 (Ontario). "An Act to confirm a certain agreement made between the Commissioners of the Queen Victoria Niagara Falls Park and the Canadian Niagara Power Company, and to enable the said company to carry the agreement into practical effect." [Assented to, April 14, 1892.]

7 Mavor, James, *Niagara in Politics: A Critical Account of the Ontario Hydro-Electric Commission*. New York: E.P. Dutton & Company, 1925, p. 24.

8 Albert D. Shaw received 1,900 shares, valued at $100 each for total value of $190,000 and 30 debenture bonds valued at $1,000 each with interest of 6 percent, value of $30,000; Francis Lynde Stetson received 3,050 shares, valued at $100 each, total value of $305,000 and 45 debenture bonds valued at $1,000 each with interest of 6 percent, value of $45,000; and William B. Rankine received 50 shares, valued at $100 each, total value of $5,000.

9 Minutes of First Meeting of Incorporation and Directors, Clifton House, Niagara Falls, Ontario, July 16, 1892. See bound collection of minutes covering July 16, 1892, to June 2, 1905.

10 "New Power Company," *Niagara Falls Gazette*, Niagara Falls, New York, July 16, 1892, p. 1, as quoted in Dumych, *The Canadian Niagara Power Company: One Hundred Years*, p. 5.

11 "To Utilize The Falls," *Toronto Mail*, July 19, 1892, p. 3, column 1.

12 "Power From The Falls," *The Globe*, Toronto, July 18, 1892, p. 7.

13 "News of the Day," *The Globe*, Toronto, April 18, 1892, p. 1.

14 Minutes, Directors' Meeting, April 6, 1895, and Minutes, Executive Committee Meeting, August 17, 1895. The latter meeting was attended by Shaw and Rankine.

15 "Buffalo and Power," *Niagara Falls Gazette*, Niagara Falls, New York, July 19, 1895, p. 1, quoted in Dumych, *The Canadian Niagara Power Company*, p. 7. The railway was represented by Director H.C. Hammond of Toronto and general manager Ross Mackenzie.

16 Dumych, *The Canadian Niagara Power Company*, p. 7, states that an agreement was signed on September 7, 1895, but provides no source. Seibel, *Ontario's Niagara Parks*, p. 387, gives a date of November 27, 1897. The Park Commissioners' Annual Report for 1896, p. 28, mentions that an offer had been made to the railway company by Canadian Niagara, but nothing was signed until the following year. The agreement also stated that if at the end of the three years Canadian Niagara were in a position to do so, it could take over the power contracts from the railway.

17 Minutes, Executive Committee Meeting, Niagara Falls, N.Y., August 17, 1895.
18 Seibel, *Ontario's Niagara Parks*, p. 153.
19 "The Canadian Plant," *The Electrical World*, p. 48. Undated clipping in scrapbook, National Archives.
20 In January 1897, the Board of Directors met to talk about getting an extension of the obligation to begin construction "on or before the first of May, 1897."
21 Queen Victoria Niagara Falls Park, Annual Report for 1896, p. 28.
22 Queen Victoria Niagara Falls Park, Annual Report for 1897, pp. 10-11.
23 Queen Victoria Niagara Falls Park, Annual Report for 1897, pp. 11-15, plus map.
24 Queen Victoria Niagara Falls Park, Annual Report for 1898, p. 3.
25 Queen Victoria Niagara Falls Park, Annual Report for 1898, pp. 3-4.
26 62 Victoria 2, chap II, par. 35.
27 Queen Victoria Niagara Falls Park, Annual Report for 1899, pp. 3-4.
28 The full text of the agreement is in Queen Victoria Niagara Falls Park, Annual Report for 1899, Appendix A, pp. 11-18.
29 The new rates effective May 1, 1899, consisted of: (1) Basic annual rent, which included the right to generate up to 10,000 horsepower: $15,000. (2) Additional annual rent of $1 a year "for each electrical horse power generated and used and sold or disposed of over ten thousand electrical horse power up to twenty thousand horse power." (3) Additional annual rent of 75 cents per horse power over 20,000 and up to 30,000 electrical horse power. (4) Additional annual rent of 50 cents for each horse power over 30,000 electrical horse power. Under the old scheme, rent for the first 10 years was fixed at $35,000 a year. Thereafter it dropped to $26,000 a year for the eleventh year and rose $1,000 a year until it reached $35,000 in the twentieth year, where it would stay.
30 As originally outlined, the water would be generated at two sites, one above the Falls with a head of 40 feet. The same water would flow to a second generating station in the gorge below the Falls, where it would use the full hydraulic head of the Horseshoe Falls. The proposal was later changed to eliminate the first generating system and draw water from the Niagara River.
31 "Big fire at the Falls" *The Thorold Post*, Friday, September 8, 1899, p, 1.
32 "Powerhouse Burned at the Falls," *The People's Press*, Welland, Ontario, September 5, 1899, p.1.
33 Untitled article, *The People's Press*, September 26, 1899, p. 3.
34 "Niagara Falls," *The Thorold Post*, September 29, 1899, p. [8].
35 Untitled article, *The People's Press*, Welland, October 3, 1899, p. 3.
36 "Rebuilding the Powerhouse," *The People's Press*, November 14, 1899, p. 3.
37 Seibel, *Ontario's Niagara Parks*, p. 387.
38 "Niagara Falls Power," *Welland Tribune*, Friday, May 24, 1901, p. 2. Underneath the headline is the line: "*Canadian Electrical News*, May, 1901."

Chapter 3
Building the Powerhouse

1 Daniel M. Dumych, *William Birch Rankine: One in Ten Thousand*, Drumdow Press, 1991, p. 6, calls it the construction arm. It is called a holding company at
http://ublib.buffalo.edu/libraries/projects/cases/niagara.htm.
A third source credits Stetson as the inventor of the holding company, but states that the first-ever holding company was the Metropolitan Street Railway Company, formed in 1893. ("The Age of Big Business, A Chronicle of the Captains of Industry,"
http://www.globusz.com/ebooks/BigBusi/00000015.htm)
2 *Niagara Falls Gazette* editorial, September 30, 1905, collected in *Memorabelia* [sic] *of William Birch Rankine*, compiled by deLancey Rankine, Niagara Falls, New York: Power City Press, 1926. The comments from Tesla and Ely are also from this source.
3 Regrettably there are no known copies of the Niagara Falls, Ontario, newspaper for the period covered by this chapter. For this chapter, I have relied heavily on a number of key papers on the technology and construction of the station. These include the papers listed in the bibliography by: A.H. Van Cleve (1909); Cecil B. Smith (1903 and 1905); *Scientific American* (1905, 1906); Imlay (n.d.); *Canadian Electrical News* (June, 1906); and H.W. Buck (1906). These articles are complemented by corporate correspondence in the Canadian Niagara Power Company papers in the National Library and Archives of Canada and reports of the Commissioners of the Queen Victoria Niagara Falls Park.
4 "Power," *Welland Tribune*, Friday, October 25, 1901, p. 5.
5 Grier to Rankine, August 2, 1903, Box 3 at National Archives.
6 Grier to Rankine, January 25, 1904, Box 3 at National Archives.
7 Based on Dumych, *The Canadian Niagara Power Company*, pp. 40-42, and "Fiendish Attempt To Blow Up The Canadian Power Tunnel," *Niagara Falls Gazette*, Niagara Falls, New York, April 26, 1902, p. 1; "No Clue To Person Who Attempted Wreck of Tunnel," *Niagara Falls Gazette*, Niagara Falls, New York, April 28, 1902, p. 1; "Keeping Close Watch," *Niagara Falls Gazette*, Niagara Falls, New York, May 3, 1902, p. 1.
8 A.J. Porter to Canadian Niagara, October 20, 1904, National Archives, Box 3; Rankine to Grier, July 7, 1904, National Archives, Box 3; Grier to Rankine, August 23-25, 1904, National Archives, Box 3.
9 This account is based on Dumych, *The Canadian Niagara Power Company*, pp. 43-45; "Red Coats Guarding Canada Power Work," *Cataract Journal*, Niagara Falls, New York, November 3, 1903, p. 1; "Strike Is Off: Plants Resume," *Cataract Journal*, Niagara Falls, New York, November 4, 1903, p. 1.
10 Van Cleve, *Transactions of the American Society of Civil Engineers*, pp. 243-244. He added that movement ceases "except in so far as changes of temperature may cause slight inward and outward movements, the maximum width of the pits being attained in May and the minimum width in November. The movements of the sixteen points in the wheel-pit of The Canadian Niagara Power Company have been measured at frequent intervals for 3 years, but the amount of seasonal movements varies from point to point so that no general law may be laid down as to the extent of the movement in a given period of time."
11 Fram, Mark, ed., *A Selective Guide to Industrial Archaeology in the Niagara Peninsula*. Toronto: Society for Industrial Archaeology, 1984, p. 70.
12 Grier to Rankine, August 3, 1903, National Archives of Canada, Box 3. Grier and Rankine often seemed to exchange ideas on design. In correspondence regarding their revised trademark, Grier acidly comments that "You must attribute the whole of its beauty to the ingenuity of the Government. Beyond question, it is now bad enough to pass muster." National Archives, Canadian Niagara papers, Grier to Rankine, September 8, 1903.
13 Based on material in an untitled trade catalogue for the Canada Crushed Stone Corporation Limited, Hamilton, Ontario, circa 1930. The catalogue is from the personal collection of John Burtniak.
14 "Brief Biographies of American Architects Who Died Between 1897 and 1947." The full entry reads: "Bell, Algernon, S. A retired architect, died December 5, 1936, at his home in New York City, aged seventy-five." The country house project is also mentioned in the Museum of The City of New York, www.mcny.org.
15 Smith to Grier, March 7, 1904, National Archives, Box 3.
16 Fram, *Selective Guide*, p. 70.
17 "The Royal Visit," *The People's Press*, Welland, Tuesday, October 15, 1901, p. 3.
18 "Royal Party Departs" and "Corner Stone Laid," *The Welland Tribune*, Friday, October 18, 1901, p. 5.
19 R.D. Sappington, Assistant Superintendent, Memorandum for A.D. Robb, November 1, 1921, National Archives, Box 22, File 2.
20 National Archives, Box 22, File 2.
21 Smith to Rankine, September 26, 1904, National Archives, Box 3.
22 Rankine to Smith, September 28, 1904, National Archives, Box 3.
23 Grier to Rankine, July 27-29, 1903, National Archives, Box 3.
24 National Archives collection, Correspondence, Box 7, 1903.

CHAPTER 4

Designing to Get Power to Customers Safely and Reliably

1 This account is based on *Information for Visitors*, an official guide to the generating plants of The Niagara Falls Power Company and the Canadian Niagara Power Company, which was first published jointly by both companies in 1907. Some sources give generation voltage as 12,000 and others as 11,000. The former is correct.

2 The description of the transmission system to Buffalo is based on four main sources in addition to the *Information for Visitors* noted above: Philip P. Barton, "Generating Plant of the Canadian Niagara Power Company," *The Canadian Electrical News*, June 1906, p. 155–156; H.W. Buck, "The Electrical Plant of the Canadian Niagara Power Company," *The Canadian Electrical News*, August 1906, p. 223; Cecil B. Smith, "Construction of Canadian Niagara Power Company's 100,000 H.P. Hydro Electric Plant at Niagara Falls, Ont." *The Canadian Engineer*, Vol. XII, No. 3, March 1905, p. 105, 110, 112; and A.H. Van Cleve, "The Hydro-Electric Development and Transmission Lines of the Canadian Niagara Power Company," *Transactions of the American Society of Civil Engineers*, Vol. 62, March 1909, p. 228–240.

3 This is based on watching and talking with cable splicers who worked for Ottawa Hydro in the early 1990s.

4 Trevor Kletz, a British chemical engineer who has done a great deal of accident investigation as well as design, is one of the world's leading advocate of inherently safe design.

5 Van Cleve gives the transmission line length from the CNP transformer station to Terminal House B in Buffalo as 16.09 miles. In the United States, the distance from the Niagara Falls Power Company transformer station to terminal was either 22.4 or 19.33 miles, depending on which route it took; extending the line to Terminal House B would have added another 2.5 or so miles of 22,000 volt lines. *Information for Visitors*, gives the Canadian distance as 16 miles to Terminal B and the American distance to Terminal A as 22.5. It shows the two terminals linked by underground cables.

6 Van Cleve speaks of 11,000-volt primary current and possible secondary currents at 22,000, 33,000, 38,000 or 57,300 volts. Buck's account mentions a primary of 12,000 and secondary at 24,000 volts, 36,000 volts, 41,500 volts or 62,500 volts. Buck rates the line to Buffalo at 24,000 volts.

7 The sixth of eleven generators came on line December 5, 1910.

8 The building later became a meter repair shop.

CHAPTER 5

Extending Power to Local Communities

1 *Many Voices: A Collective History of Greater Fort Erie*, Ridgeway: Fort Erie Museum Board, 1996, p. 36.

2 Raymond, Marguerite A., *Tales of the Old Town*, Ridgeway, Ontario: Fort Erie Historical Society Museum, 1990, pp. 141-143.

3 *Many Voices*, 1996, p. 41. There would be no bridge until the Peace Bridge opened in 1927, but one could cross the river by ferry or on an unusual passenger service on the International Railway Bridge called the "dummy car."

4 Alonzo Mather as quoted in *Many Voices*, 1996, p. 143.

5 When Canadian Niagara put up the high tower on the Fort Erie side of the river, Alonzo Mather asked for permission to put in an elevator. He wanted to charge tourists for a magnificent view. The tower had not been designed for tourists, and Canadian Niagara turned him down, because it was afraid of people falling or getting electrocuted. The view would be reserved for their own linemen, some of whom took photos while up there working and local adventurers, not all of whom survived.

6 For an introduction to natural gas in the Bertie Township area see Tripp, Allen W., and Tripp, Jean M., "Natural Gas: The Blue Flame" in *Many Voices*, 1966, pp. 328-333; Learn, H.L., "Natural Gas Development," in *Fort Erie Times-Review*, Fort Erie Centennial, 1857-1957, pp. 55-56; Glenny, Charles N., "Story of Gas," in *Fort Erie Times-Review*, Golden Jubilee Supplement, July 18, 1946, pp. 22, 30.

7 Based on Dumych, Daniel M., *Ninety Years of Light: How the Canadian Niagara Power Company Brought the Electric Light to Fort Erie in 1907*, Fort Erie: The Canadian Niagara Power Company, 1997, pp. 11-17.

8 Dumych, *Ninety Years of Light*, p. 1-5.

9 The Review Company, *Bridgeburg: The Gateway to Canada's Abundant Prosperity*. Bridgeburg, Ontario: The Review Company, Printers and Publishers, September 1913, p. 3 [unnumbered].

10 Paula Voell, "Check it out," *Buffalo News*, January 7, 1993.

11 "From a Hundredweight of Pi," *Fort Erie Times-Review*, Golden Jubilee Supplement, July 18, 1846, p. 16.

12 Discussion with Jane Davies, whose enthusiasm and knowledge have contributed greatly to this book.

13 In *The Philosophy of Railroads*, Thomas Coltrin Keefer, one of Canada's most eminent 19th-century engineers, talked of towns becoming Sleepy Hollows if they didn't get a railroad, but not just any railroad, one that was soundly engineered, well built, and had an economic base.

14 Disher, Robert M., *Bertie Town Hall Story 1874–1974*, Ridgeway: Robert M. Disher, 1974, p. 25.

15 Typescript of unidentified clipping from Crystal Beach historian Bill Kae and titled "Peg-Leg" by either Isabel Peacock or Kitty Collard.

16 Based on discussions with Bill Kae, who saw and made notes on an undated and unidentified newspaper clipping describing the tragic event.

17 Canadian Carborundum Co., Ltd., *The Romance of Carborundum*. Niagara Falls, Canada: Canadian Carborundum Co., Ltd., n.d., pp. 4-7.

18 The Carborundum Company, *Carborundum Brand Products for the Stone Working Trades*. Niagara Falls, N.Y., 1939, p. 4. The description is from the Niagara Falls, New York, plant but the processes and equipment were probably identical in the Niagara Falls, Ontario plant as the Chippawa plant is often referred to.

19 *Romance of Carborundum*, pp. 14-19.

20 The details of Perky's early life and career are far from clear. According to other versions of his biography, he was fairly successful early on. For example, in Sherman Zavitz, "One of the oldest Falls companies to expand facilities," *Niagara Falls Review*, Nov. 18, 2000, he went West from his home state of Ohio for the restorative air of Colorado and while studying nutrition, not visiting a hotel in Nebraska, came up with the idea of boiling whole unground wheat. John Curran, "Shredded Wheat's story brings Niagara Falls glory," *Niagara Falls [New York] Gazette*, January 11, 1987, has a more flattering picture. "Born in 1843, Henry D. Perky was a tall, charismatic Cleveland businessman who operated a mercantile business and later moved to Colorado and Nebraska, where he served for a period in the state Senate."

21 Some sources claim that the machine itself was developed in Watertown, New York; see, for example, Michael Beebe, "Falls, Shredded Wheat share centennial," *Niagara Falls [New York] Gazette*, February 14, 1992, p. B1.

22 Hamilton B. Mizer, "Shredded Wheat plant once leading tourist spot," newspaper clipping, identified by hand as August 17, 1980, probably *Niagara Falls [New York] Gazette*.

23 Daniel M. Dumych, "A Requiem for Shredded Wheat," unidentified newspaper clipping, Niagara Falls (Ontario) Public Library, vertical file "Nabisco." Dumych gives the figure of 844 windows in this account of how strikingly modern the Shredded Wheat plant in Niagara Falls, NY, was. For a longer discussion of the plant and its place in Niagara Falls and industrial history, see William Irwin, *The New Niagara*, pp. 181-203.

24 Claims such as this are impossible to document completely and often reflect local pride more than historical truth. There is a claim that the coffee break originated in Buffalo at the Barco and Boll company, now famous as the home of the Barcalounger chair.

25 "Triscuit is now on the market," *Niagara Falls Gazette*, Niagara Falls, New York, May 22, 1903, p. 1. This article clearly states that the conventional shredded wheat is "baked in a regular oven" and only Triscuits were baked in an electric oven. The same article concluded that "While electricity has been applied to baking before, this is the first time that it has been used extensively for baking for commercial purposes and the result is highly gratifying."

26 "Nabisco Shredded Wheat Celebrates Fifty Years," Niagara Falls (Ontario) Public Library, Petrie Collection, clipping dated September 17, 1954.

27 Clipping on "The Canadian Shredded Wheat Co." in Niagara Falls (Ontario) Public Library vertical files for "Nabisco."
28 Ibid.
29 Queen Victoria Niagara Falls Park, Annual Report for 1906, pp. 16-17.
30 A fourth company generating power in the Park, the International Railway Company, formerly the Niagara Falls Park and Railway Company, was not represented at the hearing. There was no need. Just as the hearing was closing, Mr. Ellis, one of the Park Commissioners, stated that the International Railway Company and the Park Commissioners had "come to an agreement precisely on the terms the Commission are now contending for," namely the increasing rental — peak power system. Queen Victoria Niagara Falls Park, Annual Report for 1908, p. 50.
31 Queen Victoria Niagara Falls Park, Annual Report for 1908, pp. 30-50. This story of the dispute is based almost entirely on the QVNFP Annual Reports up to the year 1917. The 1908 Report locates the meeting at the "Clifton Hotel" which is in Niagara Falls, NY wheras the Clifton House is in Niagara Falls, Ontario. Perhaps there was a reporting error.

Chapter 6

The First World War and the 1920s

1 Information on Lyall has been largely drawn from http://www.donlowconcrete.com/102.
2 When the Second World War broke out, he again did the patriotic thing. Now Major Lyall, he commanded the 3rd AA Division Workshop Company, Royal Army Ordnance Corps, and by 1940 was on active duty in North Africa as Lieutenant Colonel. He rose to Colonel and on November 28, 1941, died in his sleep. He is buried in the Halfaya Sollum Cemetery on the main coastal road from Mersa Matruh, only 11 miles from the Libyan border.
3 As quoted in *Many Voices: A Collective History of Greater Fort Erie*, Ridgeway: Fort Erie Museum Board, 1996, p. 288.
4 Queen Victoria Niagara Falls Park, Annual Report for 1915, p. 11.
5 Ronald L. Way, *Ontario's Niagara Parks: A History*, Niagara Falls, Ontario: Niagara Parks Commission, 1946, p. 116-17.
6 Queen Victoria Niagara Falls Park, Annual Reports, 1917, p.10. According the Annual Report for 1918, pp. 10 and 12, the Canadian Niagara Power Company, which was authorized to produce 100,000 horsepower, had a peak load of 103,351 horsepower for the half-year ending October 31, 1918. The Electrical Development Company, entitled to produce 125,000 horsepower, had an average generation of 134,010 horsepower for the six months ending July 31, 1918.
7 Adams, *Niagara Power: History of the Niagara Falls Power Company, 1886-1918*, vol. I, p. 223.
8 Memorandum for Mr. A.D. Robb, from R.D. Sappington, 1st November, 1921, Niagara Falls, Canada. This is one of the most complete summaries of where Canadian Niagara Power electricity went. In addition to summarizing power deliveries to Ontario Hydro, it also lists the power the company was allowed to export to the United States under the Act passed in 1907. National Archives, CNP Papers, Box 22, file 2.
9 Artificial abrasives and grinding machines are perhaps the least recognized of the late 19th and early 20th century technologies that revolutionized so much manufacturing by lowering costs and increasing output. For a good introduction to the this important industry, see Robert S. Woodbury, *History of the Grinding Machine. A Historical Study in Tools and Precision Production*. Cambridge, Massachusetts: M.I.T. Press, 1959, second printing, 1964.
10 *Ann Augusta Miller Diary*, Fort Erie Historical Society Museum, various dates.
11 Queen Victoria Niagara Falls Park Annual Reports, 1917, pp. 9 and 11; and 1918, p. 10.
12 The association represented the private electrical industry and worked to promote greater electrical knowledge, standardization, safety and an appropriate legal framework for the industry.
13 Canadian Electrical Association, *Proceedings of Annual Convention*, 1919, pp. 100-101.
14 "In the case of domestic electric appliances, there is, quite simply, very little 'pull' [consumer demand] and a great deal of 'push' [industry marketing] to be found. Try as one might to find a demand among middle class housewives looking for a solution to the shortage of servants, or among working class housewives looking to escape the drudgery of domestic labour, one finds instead, time and again, intensive advertising and novel marketing campaigns being used to create a demand for these new devices." Richard White, "Creating a Demand: The Origin of Domestic Electric Appliances," unpublished paper, 1989.
15 Canadian Electrical Association, *Proceedings of Annual Convention*, 1919, p. 111.
16 Canadian Electrical Association, *Proceedings of Annual Convention*, 1919, p. 104.
17 Advertisement in CNP papers, National Archives.
18 Westinghouse/Canadian Niagara Power Company ad, *Fort Erie Times*, Vol. 29, no. 6, December 7, 1923, p. 5.
19 "Cook by Wire," Canadian Niagara Power Co. advertisement, *Fort Erie Times*, Vol. 28, no. 3, November 17, 1922, p. 5 The same ad was repeated in November, December, and into early 1923.
20 Jane Busch, "Cooking Competition: Technology on the Domestic Market in the 1930s," *Technology and Culture*, vol. 24, January 1983, p. 240.
21 "Canadian Villages Reach High Kilowatt-Hour Average," p. 8 of two unidentified pages titled *Niagara Power Service*, no date other than October 9, 1929 for B.N. & E. [Buffalo, Niagara & Eastern Power Corporation] stock quotations." Found in CNP papers, Box 22, file 4.
22 Fort Erie Public Library, interviews, Mr. John Stickles, p. 3.
23 Fort Erie Public Library, interviews, Mrs. Shirley Beam, #094, p. 15.
24 Fort Erie Public Library, interviews, Mr. Sumner Beam, #072, p. 1-2, 6, 33.
25 Fort Erie Public Library, interviews, Mrs. Ruth Carver, #011, pp. 12, 15, 19.
26 A transcript of the diary of Mrs. Ann Augusta Miller is in the Fort Erie Historical Museum in Ridgeway, Ontario. Mrs. Miller was born December 19, 1848, and died October 25, 1932. The diary covers the period Feb. 17, 1870 to Sept. 30, 1932, but the author studied it only for the period 1905 to 1932.
27 This section is based on interviews with Norman Rouse and Jack Irish.
28 A.W. Spear, *The Peace Bridge 1927-1977*, as quoted in *Many Voices*, p. 145.

Chapter 7

The New Service Centre and the 1930s

1 Unidentified newspaper clipping titled "New Service Building" from Louis McDermott Reference Collection, Fort Erie Historical Museum, Ridgeway.
2 In addition to the documents cited, this chapter is based on interviews with Shirley Athoe, Frank and Sandra Flake, Audrey Troup, Leonard Heximer, Harvey and Shirley McAteer, and Sid Morris.
3 It was very much in the style of some of the Pan–Am buildings, a homage to that exceedingly influential exposition in praise of electricity.
4 The cards are still in the wooden file box they were originally stored in. They are owned by Frank and Sandra Flake to whom I am indebted for allowing me to use them.
5 The author's family got its first electric stove in the mid-1950s. The front left burner was the largest and it worked in either of two positions. It could be a surface burner or one could twist it and drop it down about 12 to 14 inches. In the lower position, one used a specially made aluminum pot with a lip that rested on the top of the stove to prevent it falling too low. The pot had two handles and a slightly domed lid with two steam vents. The compartment was well insulated and the side of the stove did not get uncomfortably hot. The author's mother liked it because it did not throw a lot of heat into the room and with five children in the house, she regarded it as safer because it could not tip and there was no hot surface to burn little fingers.
6 "The Most Remarkable Offer on Electric Ranges your light company has ever made," *Fort Erie Times* ad, vol. 33, no. 26, Friday, April 27, 1928, p. 5.

7 Electric companies bought out other electric companies at a rapid rate in the early 20th century. The Niagara Falls Power Company became part of the Buffalo, Niagara and Eastern Power Corporation, which was swallowed up by Niagara Hudson, which eventually became part of the Niagara Mohawk Power Corporation.
8 "Beach Store is Patronized," article in Niagara Power Service newsletter, October 1928.
9 When it closed at the end of 1949, complaints from Crystal Beach residents led to an arrangement whereby Crystal Beach residents could pay their bills at a local real estate store. At the same time, the company added a 24-hour, toll-free call-in service for problems with electrical service.
10 Canadian Niagara Power Company Board of Directors Minutes, November 12, 1936, National Archives.
11 Canadian Niagara Power Company Board of Directors Minutes, 1931, National Archives.
12 The rent paid to the Niagara Parks Commission dipped slightly in 1932, fell dramatically in 1933, bounced back in 1934, and returned to its usual level in 1936. See Niagara Parks Commission annual reports. This pattern is consistent with Ontario Hydro's revenues, which also fell to their lowest point in 1933, before returning to their pattern of annual growth. Merrill Denison, *The People's Power: The History of Ontario Hydro*, Toronto: McClelland and Stewart, 1960, p. 193.
13 "Provincial Hydro under heavy fire, a detriment to growth of the city," clipping from unidentified newspaper, March 21, 1934, photocopy in National Archives.
14 "Taking over of Canadian Niagara Power Company by Hydro would be disastrous to Stamford, is claim," *Niagara Falls Gazette*, April 18, 1936, clipping in CNP files, National Archives.
15 " 'Secret' clauses seem to be favorable to one side only," *Niagara Falls Evening Review*, February 28, 1935, clipping in CNP files, National Archives.
16 *Niagara Falls Times Review*, May 23, 1940, clipping in CNP files, National Archives, some words obscured on photocopy.
17 "Canadian Niagara Power is retained in Fort Erie," *Buffalo Evening News*, March 1, 1935, clipping in CNP files, National Archives.
18 "Why you should vote for the Canadian Niagara Power Company franchise bylaw," *Fort Erie Times-Review*, February 28, 1935, clipping in CNP files, National Archives.
19 Memo from J.E. Lawson to A.D. Robb, March 4, 1935, National Archives.
20 "Weir Construction at Niagara," Canadian Niagara Power Company pamphlet, c1937, National Archives, Box 99, File 36.
21 W. Bruce Leslie, "Through the Eyes of the Review," *Niagara Falls Evening Review*, January 28, 1938, clipping in CNP files, National Archives.
22 "People Flock Here to See Bridge Ruins," *Niagara Falls Evening Review*, January 29, 1938, clipping in CNP files, National Archives.
23 The minutes of April 28, 1921, mention the payment of $500.00 to "Mrs. Alex D. Robb (Miss F.C. Durdan) in recognition of her faithful services to the Company for 15 years."
24 Obituary from Niagara Falls, N.Y., newspaper, October 28, 1936, in CNP scrapbook, National Archives.
25 Announcement of New Employee Insurance Plan, February 1, 1933, pamphlet, National Archives.

Chapter 8
The War Years, 1939–45

1 In addition to the documents cited, this chapter is based on interviews with Don Kelly, Russell Sanderson, Sid Morris, and John Campsie.
2 Niagara Parks Commission, Annual Report for 1940, p. 5. This protection also included surge tanks.
3 Niagara Parks Commission, Annual Report for 1941.
4 Jackson et al., *The Mighty Niagara*, p. 311.
5 Both planes were made largely of made of fabric, wood and glue. In bitterly cold weather, particularly out on the Prairies, where many Canadian pilots and aircrew were trained, the glue tended to crystallize, and wings sometimes broke off during difficult manoeuvres. For more detail on the Commonwealth Air Training Plan and the Fleet Finch, see Ted Barris, *Behind the Glory*, Toronto: Macmillan, 1992.
6 *Canadian Niagara Power Company Gossip Monthly News Letter*, July 15, 1942. By the next issue, August 15, 1942, it had changed names *to The C.N.P. Co. Tattler*.
7 *The Tattler*, November 1942, December, 1942, February 1943.
8 *The Tattler*, September 1943.
9 *The Tattler*, December 1943.
10 *The Tattler*, June 1943, August 1943, September 1943.
11 *The Tattler*, January 1943. The Cornwall plant was affiliated with the American parent company.
12 *The Tattler*, September 1942.
13 *The Tattler*, February 1943.
14 *The Tattler*, March 1944.
15 *The Tattler*, May 1943.
16 *The Tattler*, January 1943, June 1943.
17 *The Tattler*, June 1943.
18 *The Tattler*, November 1942, January 1943, March 1944.
19 *The Tattler*, May 1943, July 1943.
20 *The Tattler*, July 1943, February 1943, August 1942, November 1942.
21 *The Tattler*, April 1944.
22 Letter from Paul Bloomfield, October 15, 1944.
23 Letter from Jack Delaney, June 25, 1943.
24 Letter from Jack Delaney, April 8, 1943.
25 *The Tattler*, August 1942 and February 1943.
26 *The Tattler*, March 1943, April 1943, May 1943. The June issue announced that Sid Morris had joined the Army and "this makes 15 of our boys in the services."
27 *The Tattler*, December 1943, March 1944, April 1944.
28 Letter from Edward Clarke, June 29, 1942.
29 Letter from Art Fretz, September 24, 1943.
30 Letter from Art Fretz, September 20, 1944.
31 Letter from Jimmy Bird, October 21, 1942. The June 1944 issue of *The Tattler* reported that "Mrs. Art Fretz has gone to join Art in Newfoundland."
32 Letter from J.D. Johnston, July 28, 1942.
33 Undated letter received from Scotty Sterling July 19, 1942, CNP papers, National Archives.
34 *The Canadian Power Company Gossip Monthly News Letter* of July 15, 1942, lists the six who were able to attend the buffet lunch given by Mr. Lawson: Scotty Sterling, Don Fraser, Omer Steele, Bud Reid, Jack Clark, and Jack Delaney.
35 Letter of July 14, 1942, to Corp. K.B. MacIntyre, R.C.A.F. Station, Yarmouth, N.S. National Archives. *The Canadian Niagara Power Company Gossip*, July 15, 1942 listed all 11 enlisted men.
36 These were found along with issues of *The Tattler*, CNP papers, National Archives.
37 *The Tattler*, July 15, 1942, November 1942, April 1943.
38 Letter from Scotty Sterling, July 1, 1943.
39 *The Tattler*, April 1944.
40 Letter from Jimmy Bird, January 1, 1945.
41 Letter from Joe Johnston, July 26, 1943.
42 Letter from Earl Snider, August 14, 1944.
43 Letter from Don Fraser, September 26, 1944.
44 Letter from Jack Delaney, July 23, 1942.
45 Letter from Jack Delaney, June 6, 1943.
46 Letter from Bud Reid, September 29, 1942.
47 Letter from Bud Reid, May 2, 1943.
48 Letter from Jimmy Bird, April 2, 1945.
49 Letter from Paul Bloomfield, Sunday, October 15, 1944.
50 "Thanksgiving Service marks local observance of V-E Day," *Fort Erie Times-Review*, May 10, 1945,
51 Ibid.
52 *Niagara Falls Evening Review*, May 11, 1945, p. 1.
53 *Fort Erie Times-Review*, June 14, 1945, p. 7; June 21, 1945, p. 3; July 5, 1945, p. 5; , July 19, 1945, p. 3; July 26, 1945, p. 3; August 2, 1945, p. 3.
54 "Local citizens hold service to mark peace," *Fort Erie Times-Review*, August 23, 1945, p. 1.
55 *Fort Erie Times-Review*, August 23, 1945, p. 10.

CHAPTER 9

Ice Problems in the Banana Belt

1 In addition to the documents cited, this chapter is based on interviews with Harry Saylor Sr., Sid Morris, and Walter Moroz.
2 CNP scrapbook, "Niagara Silent in the Grip of Ice," clipping from unidentified newspaper, April 12, 1909. National Archives.
3 CNP scrapbook, "The Test of Niagara Power. The remarkable Ice-Jam and its Lessons," *Financial Post*, April 17, 1909. National Archives.
4 CNP scrapbook, "The Test of Niagara Power" mentions an "average of thirty feet" and says "the ice was piled up forty and even fifty feet above the ordinary level." National Archives.
5 CNP scrapbook, "Niagara Silent in the Grip of Ice." National Archives.
6 Frank H. Quinn, Raymond A. Assel, Daniel W. Gaskill, "An Evaluation of the Impact of the Niagara River Ice Boom on the Air Temperature Regime at Buffalo, New York," *Journal of Applied Meteorology*, Volume 21, March 1982, p. 342.
7 Edward Dean Adams, *Niagara Power: History of the Niagara Falls Power Company, 1886-1918*, Niagara Falls, New York: Niagara Falls Power Company, 1927, vol. II, p. 145.
8 Adams, II, pp. 147-8.
9 Adams, II, p. 156.
10 Adams, II, pp. 153-154.
11 Adams, II, p. 157, also describes the placement and operation of the floating booms.
12 Adams, II, pp. 158-159.
13 Adams, II, p. 159.
14 Adams, II, p. 157.
15 Adams, II, p. 143.
16 CNP scrapbook, "Big Chippawa Project 'Delivers' In Ontario's Sudden Hour of Need," *Globe*, January 30, 1922. National Archives.
17 CNP scrapbook, "Great Chippawa Plant to be Opened Today," *Globe* December 28, 1921; "Drury Starts First Power at Chippawa," *Globe*, December 29, 1921; "Chippawa Canal is Opened, Develops its First Power," *Mail*, December 29, 1921. National Archives.
18 CNP papers, National Archives.
19 "Power is Affected by the Ice," newspaper clipping of January 10, 1942, Niagara Falls Public Library, vertical files, Ice.
20 "Power is Cut By Ice Jams," newspaper clipping of January 30, 1947, Niagara Falls Public Library, vertical files, Ice.
21 "Niagara's New Magic," *Ontario Hydro News*, October 1957, pp. 2-7, 22.
22 The Power Authority of the State of New York, The Hydro-Electric Power Commission of Ontario, *Brief on Proposed Ice Boom in Lake Erie at Entrance to Niagara River*, February 1926, p. 2. Mimeographed typescript with illustrations.
23 Panel on Niagara River Ice Boom Investigations, Water Science and Technology Board Commission on Physical Sciences, Mathematics, and Resources, National Research Council, *The Lake Erie-Niagara River Ice Boom: Operations and Impacts*. Washington, D.C. National Academy Press, 1983.

CHAPTER 10

Electric Meters and Meter Readers

1 In addition to the documents cited, this chapter is based on interviews with Dr. Jan Carr, Fred Barber, Jack and Ann McLeod, Leonard Heximer, and John Parisi.
2 R.C. Lanphier, *Electric Meter History and Progress*, Springfield, Illinois: Sangamo Electric Company, 1925, p. 5.
3 Lanphier, *Electric Meter History*, p. 72.
4 The gas industry, whether selling natural or manufactured gas, piped its product from one central source, and with the aid of gas meters, charged customers according to the amount they used. In England, electrical pioneer Sebastian Ziani de Ferranti independently foresaw the same model, as did Nikola Tesla in the United States.

CHAPTER 11

60-cycle Conversion — The Carnival and the Iceberg

1 In addition to the documents cited, this chapter is based on interviews with Robert Barnett, Walter "Babe" Douglas, Todd Docker, John Damato, Syd Freeman, Dave Renshaw, Bill Johnston, Gladys Johnston, and Moe Edwards.
2 At the time, various frequencies were used on the American side of the Niagara River, but 25-cycle was by far the commonest.
3 Edward Dean Adams, *Niagara Power: History of the Niagara Falls Power Company, 1886-1918*, Niagara Falls, New York: Niagara Falls Power Company, 1927, vol. II, p. 236.
4 Adams, *Niagara Power*, II, p. 238. Two alternations made a cycle. An alternation went from zero to maximum to zero through either positive or negative, whereas a cycle went through both negative and positive and back to the same starting point.
5 The term *polyphase* is sometimes used inconsistently. The Tesla system was two-phase but has often been called polyphase, as has three-phase. It is best not to use the term polyphase because it is too imprecise.
6 Henry J. Prout, *A Life of George Westinghouse*, quoted in Adams, *Niagara Power*, II, p. 193.
7 Robert D. Barnett, "The frequency that wouldn't die," *IEEE Spectrum*, December 1990, pp. 120-121.
8 For this aspect of the early history of electric power see Jill Jonnes, *Empires of Light: Edison, Tesla, Westinghouse and the Race to Electrify the World*, New York: Random House, 2003.
9 Letters to Canadian Niagara Power Co., from A.M.Cutt, July 9, 1958, and A.H. Anderson, November 17, 1958.
10 There was an earlier cut, perhaps several of them. An internal audit report from Ontario Hydro notes that "to accommodate a transformer change, a small special cut was made on Saturday, October 26, 1957, in advance of regular operations for Canadian Niagara" (A.M. Bizzell, Liaison Officer, Niagara Region, The Hydro-Electric Power Commission of Ontario, internal audit report). Probably this was work in a transformer substation that had 60-cycle equipment put in ready for the cutover, but the 25-cycle power was left connected until the important day. Another report notes that "a special cut was made in advance of November 26 [1957] to clear up a line for 60-cycle service. Under this schedule, 1,582 domestic customers were changed over for 60-cycle operation." (A.S. Robertson, Director of Frequency Standardization, Canadian Niagara Power Company Ltd., "Progress on Conversion Program," undated typescript report). That is a very large number of customers, and this notation is hard to explain.
11 Originally, the unit had been called a "niggerhead," but this unacceptable term was modified after Canadian Niagara linemen encountered black linemen during training in Buffalo.
12 A.M. Bizzell, letter of May 7, 1958, to Mr. A.S. Robertson, containing internal audit report for Nov. 21 to Dec. 31, 1957, Appendix D. National Archives.
13 Canadian Niagara Power Company, Limited, *Policies and Procedures relating to the Standardization of Frequency (25 to 60 cycle) Customer's Information, March 31, 1957* and Canadian Niagara Power Company, *Policies and Procedures, C.N.P. 2560-1, March 1, 1957*, p. 3
14 *Policies and Procedures, C.N.P. 2560-1, March 31, 1957*, p. 4.
15 Letter of C.A. Clancy, March 29, 1957, to Canadian Niagara customers, National Archives.
16 *Policies and Procedures...March 31, 1957*, p. 13.
17 This was the era when many electric player pianos were converted to conventional pianos that in turn sold for a higher price. Today, the rarer unconverted ones are generally valuable as antiques.
18 These items are drawn from the typescript "Cash-In-Lieu of Conversion, Canadian Niagara Power, Accounting Period January 2, 1958–January 29, 1958." National Archives.
19 A.M. Bizzell, The Hydro-Electric Power Commission of Ontario, "Canadian Niagara Power Company Limited Frequency Standardization Report as at September 1, 1958," unpublished typescript, National Archives.
20 Typescript report of A.S. Robertson to C.A. Clancy, June 27, 1958.

21 Based on Bizzell, September 1, 1958, and typescript report of A. S. Robertson to C.A. Clancy, February 19, 1959.
22 A.S. Robertson, "Progress of Conversion Program."
23 Appendixes C and D in Bizzell, September 1, 1958.

CHAPTER 12
In Praise of the Women Who Got the Bills Out

1 This chapter is based on interviews with Frank Flake, Shirley McAteer, Shirley Athoe, Annette Brinda and Mary McKinnon.

CHAPTER 13
The Great Ice Storm of 1976

1 *Niagara Falls Review*, March 1, 1976, p. 1.
2 In addition to the documents cited, this chapter relies on information from interviews with Bob Douglas, Babe Douglas, Blaine Desrosiers, Wendy Spry, Harvey McAteer, and Dave Durward.
3 "Rain and snow mix into a one-two punch," *Niagara Falls Review*, March 2, 1976, p. 1.
4 "Storm cuts F.E. power, shuts plants," *Niagara Falls Review*, March 2, 1976, p.1.
5 Canadian Niagara Power Company Limited, Sleet Storm, meeting minutes, March 15, 1976, p. 1.
6 "Storm cuts F.E. power, shuts plants," *Niagara Falls Review*, March 2, 1976, p.1. According to Canadian Niagara Power Company Limited, Sleet Storm, meeting minutes, March 15, 1976, p.1, "#5 line was back in service as were all feeders and stations by 0940 hours."
7 "Storm cuts F.E. power, shuts plants," *Niagara Falls Review*, March 2, 1976, p.1.
8 "Resume of storm damage," internal memo, March 17, 1976, National Archives.
9 Canadian Niagara Power Company Limited, Sleet Storm, meeting minutes, March 15, 1976, p. 1. It went out around midnight and was out of service until 1 p.m. Wednesday, March 3.
10 "Ice-covered Fort Erie shivers without power," *Niagara Falls Review*, March 3, 1976, p. 1.
11 "Ice-covered Fort Erie shivers without power," *Niagara Falls Review*, March 3, 1976, p. 1.
12 "The storm's gone, we're still here," *Fort Erie Times-Review*, March 10, 1976, p. 4.
13 "Resume of storm damage," internal memo, March 17, 1976, National Archives.
14 Letters from William Hunter to Charles Davidson and A. Highfield, both March 17, 1976, National Archives.
15 Canadian Niagara Power Company Limited, Sleet Storm, meeting minutes, March 15, 1976, p. 2. Letter from W. W. Hope, Superintendent of Operations, Water Division, The Regional Municipality of Niagara, to Canadian Niagara Power Company, March 18, 1976, National Archives.
16 Dave Schultz, "5,000 without power emergency continues," *Niagara Falls Review*, March 4, 1976, p. 1.
17 Dave Schultz, "Winds and floods pose new threat to Fort Erie area," *Niagara Falls Review*, March 5, 1976, p. 1.
18 "Long list of power users still waiting for hookups," *Niagara Falls Review*, March 8, 1976, p. 14.
19 Photograph caption, *Niagara Falls Review*, March 9, 1976, p. 11.
20 Clipping, Metro section of *Niagara Falls Review*, Thursday March 25, 1976, photo with main caption Fire + Smoke = Pollution, National Archives.
21 *Fort Erie Times Review*, March 10, 1976, p. 13.
22 "Five fireplace fires since storm hit," unidentified clipping, National Archives, Box 99.
23 "Merchants suffer losses as frozen foods go soft," *Niagara Falls Review*, March 5, 1976, p. 17.
24 "1,000 more homes get power," *Niagara Falls Review*, March 6, 1976, p. 15.
25 Dave Schultz, "5,000 without power emergency continues," *Niagara Falls Review*, March 4, 1976, p. 1.
26 Form titled "Canadian Niagara Power Company Limited Temporary Connection due to Storm Damage, March 2, 1976." National Archives, Box 99, File 52.
27 "Many still without power," *Niagara Falls Review*, March 9, 1976, p. 17.
28 "1,000 more homes get power," *Niagara Falls Review*, March 6, 1976, p. 15.
29 Dave Schultz, "Winds and floods pose new threat to Fort Erie area," *Niagara Falls Review*, March 5, 1976, p. 1.
30 "Power ready to flow to all homes—Hunter," *Niagara Falls Review*, March 11, 1976.
31 "Resume of storm damage," internal memo, March 17, 1976, Appendix A, National Archives.
32 Letter from William J. Hunter to Mr. Harold E. Brownhill, P.Eng., Niagara Falls Hydro, March 15, 1976. National Archives.
33 "Power line repairs still not complete," *Niagara Falls Review*, March 31, 1976.
34 "How efficient was it?" *Niagara Falls Review*, March 9, 1976, p. 11.
35 "Region will pick up work tab for its Fort Erie storm crews," *Fort Erie Review*, March 12, 1976.
36 "How efficient was it?" *Niagara Falls Review*, March 9, 1976, p. 11.
37 Canadian Niagara Power Company, internal memo from John E. Ferris to William J. Hunter, March 25, 1976, National Archives.
38 Canadian Niagara Power Company, "Costs Recorded to 1/5/76 Only Re: Storm Damage in Fort Erie, Ontario," internal document, National Archives. Niagara Mohawk billed Canadian Niagara for $155,353.63, but this included some of the costs incurred by Cleveland Illuminating. Bill from Niagara Mohawk in National Archives.
39 Canadian Niagara Power Company, internal memo from E.J. Morel to R.F. Torrey, J.M. Endries, W.J. Hunter, and A.J. Baratta, December 21, 1976, on "Ice Storm Loss." National Archives.
40 "Standard insurance doesn't cover some segments of damage," Niagara Falls Review, March 5, 1976, p. 11.
41 "EMO Head Says: Ontario deserted Fort Erie," *Niagara Falls Review*, March 4, 1976, p. 9.
42 "Region will pick up work tab for its Fort Erie storm crews" *Fort Erie Review*, Friday March 12, 1976.
43 "Scorecard on a Storm," National Archives and Fort Erie Historical Museum, Ridgeway, Ontario.

CHAPTER 14
Dangerous Work

1 This chapter is based on interviews with the following Canadian Niagara Power employees: Harry Saylor, Babe Douglas, Syd Freeman, Bob Douglas, Sid Morris, Hector Smith Sr., Hector Smith Jr., John Damato, Jim Nigh, Leonard Heximer, and Harvey MacAteer.
2 "Fatal Accident: Man Killed at Canadian Niagara Power Company's Work," *The Daily Record*, December 18, 1911, p. 1; "Power Plant Fatality: Jury on Melvin C. Brown Says Canadian Niagara Power is Negligent," *The Daily Record*, December 21, 1911, p. 1.
3 "Nearly killed by live wire," *Niagara Falls Daily Record*, September 21, 1903, Daniel Dumych papers.
4 Paul Harvey and Dave Claydon, "Woman saved near brink," *Niagara Falls Review*, May 30, 1983; Michael Clarkson and Doug Herod, "Woman saved from near certain death," St. Catharines *Standard*, May 30, 1983; Dave Claydon, "Unbelievable heroism," *Niagara Falls Review*, May 30, 1983; Paul Legall, " 'I thought: Oh God, I'm going to die," *Hamilton Spectator*, May 31, 1983.
5 Dave Claydon, "Rescuer and rescued recall brush with death above falls," *Niagara Falls Review*, May 30, 1984.
6 Sherry Vyverberg as told to Barbara Raymond, "Seconds from Death," undated article, possibly from *Reader's Digest*, National Archives, box 99, file 1.

Chapter 15
Linemen Then and Now

1 Those who started in the 1940s formed the second generation of linemen at the company. They started their careers when the previous generation (those who began their jobs when Canadian Niagara began generating power) retired in the 1930s.
2 This chapter is based on interviews with Harvey McAteer, Sid Morris, Syd Freeman, Harry Saylor Sr., Tom Trasmundi, Jim Nigh, Todd Docker, Babe Douglas, and Frank Flake.
3 In 1996, Canadian Niagara created a subsidiary that tested gloves and other safety equipment, but later sold off this company.

Chapter 16
A Rich Social Life

1 This chapter is based on interviews with Mary Genco-Brant, Maria McKinnon, Shirley Athoe, Wendy Spry, Frank and Sandra Flake, Annette Brinda, Harvey and Shirley McAteer, Hector Smith Sr., John Damato, Barry Bracken, Bob Douglas, John Fogg, David Gordon, Jack Irish, Donald Smith, and Dave and Cynthia Durward.

Chapter 17
Maintaining and Updating an Aging Power Plant

1 This chapter is based on interviews with Bob Douglas, Tom Trasmundi, Frank Flake, Doug Jacobi, Bill Daley, John Fogg, Harry Saylor Sr., Blaine Desrosiers, David Durward, Gino Visentin, Walter Lord Charest and Bill Harris.

Chapter 18
What's to Become of Us? Searching for the Future of Canadian Niagara

1 In addition to the documents cited, this chapter is based on interviews with Harry Macdonell, Donald Smith, Frank Flake, David Durward, Jimmy Fretz, Rick Heximer, David Gordon, Jim Nigh, Wendy Spry, Ray Gosine, Mardon Erbland, Bill Daley, and Craig Woodward.
2 It was a technological grudge match the likes of which would not be seen again until some of the software and media concentration battles of the 1980s and 1990s. Edison even condoned and implicitly promoted execution by electric chair as a means of demonstrating the dangers of AC. See "Edison vs. Westinghouse: The AC/DC War," in Hal Hellman, *Great Feuds in Technology*, New York: John Wiley and Sons, 2004, pp. 59-76.
3 J. Tait Davis, "Policy-Induced Distortions in Spatial Cost Surfaces: The Case of Ontario Hydro," *Canadian Geographer*, vol. XXVII, no. 1, Spring 1983, pp. 35-47. It is a revealing article that points out, among other things, how Ontario government policy has negated any historical economic advantage that industry had in the Niagara Falls region and all but killed the abrasives industry. It also shows the disparity between Farrell's ideas of fairness and the prevailing government attitude towards hydroelectric development.
4 Harry Macdonell remembers that when he was working for another American company, not Niagara Mohawk, he recovered several million dollars in expenses that had been improperly charged to the utility. But the company had already factored those expenses into rates and would have been happier had he not recovered the money. He compared this with his work in building the Churchill Falls plant, in which the agreement with Hydro-Québec stipulated that if the company exceeded cost estimates, it lost equity to Hydro-Québec.
5 The customers were in the Town of Prescott; the Village of Westport; the Village of Cardinal (now in the Township of Cardinal-Edwardsburg); the Village of Iroquois (now in the Township of South Dundas); the Village of Morrisburg (now in the Township of South Dundas); and the former village of Williamsburg in the Township of Williamsburg (now in the Township of South Dundas).
6 The customers were in the Town of Hanover; the Village of Neustadt (now part of the Township of West Grey); the Village of Clifford; the Town of Harriston and the Town of Palmerston (now part of the Town of Minto); the Town of Walkerton and Village of Elmwood (now part of the Municipality of Brockton); the former Village of Ripley and the former Village of Lucknow (now part of the Township of Huron-Kinloss); the former town of Kincardine (now known as Ward 1 of the Municipality of Kincardine); the former Town of Port Elgin and the former Town of Southampton (now part of the Town of Saugeen Shores); the former Village of Mildmay and the former Village of Teeswater (now part of the Municipality of South Bruce); and the former Town of Wingham (now part of the Township of North Huron, with the exclusion of the customer located at 200 Water Street).
7 George Robert Tregerthen Sawle, *Autobiography of G.R.T. Sawle. An average fellow with an average quota of good and bad luck.* Welland, Ontario, 1962. Printed at The Review Company of Fort Erie Ltd., p. 5.

Bibliography

While far from exhaustive, the following list of books and articles should serve as a starting point for other researchers. Naturally one would also want to consult some of the archival collections listed in the notes and acknowledgements. The papers of The Canadian Niagara Power Company are at Library and Archives Canada in Ottawa. When the author used them, they had not yet been organized, but they will be before they are open for general research purposes. It is a very rich collection.

Adams, Edward Dean, *Niagara Power: History of the Niagara Falls Power Company, 1886-1918:* Niagara Falls, N.Y.: privately printed for the Niagara Falls Power Company, 1927, on the Fiftieth Anniversary of its Foundation, in two volumes.

Anderson, George Stanley, *The Power of Niagara.* [Niagara Falls, N.Y.:] The Niagara Falls Power Company, 1923.

Ball, Norman, and John Vardalas, *Ferranti-Packard: Pioneers in Canadian Electrical Manufacturing.* Montreal and Kingston: McGill-Queen's University Press, 1994.

Barnett, Robert D., "The Frequency that Wouldn't Die," *IEEE Spectrum*, December 1990.

[Barton, Philip P.,] "Generating Plant of the Canadian Niagara Power Company," *The Canadian Electrical News*, June 1906, pp. 152-156.

Berton, Pierre, *Niagara: A History of the Falls.* Toronto: McClelland & Stewart Inc., 1992.

Buck, H. W., "The Electrical Plant of the Canadian Niagara Power Company," *The Canadian Electrical News*, August 1906, pp. 220-228.

Canadian Carborundum Co., Ltd., *The Romance of Carborundum.* Niagara Falls, Canada: Canadian Carborundum Co., Ltd., n.d.

"The Canadian Power Plant." *The Electrical World.* Vol. XXXIII, No.2, January 14, 1899, pp. 47-49.

Carborundum Company, *Carborundum Brand Products for the Stone Working Trades.* Niagara Falls, N.Y., 1939.

Cheney, Margaret, *Tesla: Man Out of Time.* New York: Dell Publishing, 1981.

Cheney, Margaret and Robert Uth, *Tesla: Master of Lightning.* New York: Metro Books, 2001.

Davis, J. Tait, "Policy-Induced Distortions in Spatial Cost Surfaces: The Case of Ontario Hydro," *Canadian Geographer*, vol. XXVII, no. 1, Spring 1983, pp. 35-47.

Disher, R.M., *Bertie Town Hall Story 1874–1974.* Privately printed, 1974.

Dow, Charles Mason, *Anthology and Bibliography of Niagara Falls.* Albany: State of New York, 1921, two volumes.

Dumych, Daniel M., *The Canadian Niagara Power Company: One Hundred Years.* Niagara Falls, Ontario: Canadian Niagara Power Company, 1992.

Dumych, Daniel M., *Ninety Years of Light: How the Canadian Niagara Power Company Brought the Electric Light to Fort Erie in 1907.* Fort Erie: The Canadian Niagara Power Company, 1997.

Dumych, Daniel M., *William Birch Rankine: One in Ten Thousand.* North Tonawanda, New York: Drumdow Press, 1991.

Fram, Mark, ed., *A Selective Guide to Industrial Archaeology in the Niagara Peninsula*. Toronto: Society for Industrial Archaeology, 1984.

Greenhill, Ralph, and Thomas D. Mahoney, *Niagara*. Toronto: University of Toronto Press, 1969.

The Harnessing of Niagara. London: The Louis Cassier Co., 1899, fourth edition.

Hellman, Hal, *Great Feuds in Technology*. New York: John Wiley and Sons, 2004.

Hulse, A.E., *The Story of Niagara Falls Power Company & William B. Rankine*. Southampton, N.Y.: The Yankee Peddler Book Company, printed at the Steamboat Press, 1965.

Imlay, L.E., "Some Special Features of the Canadian Niagara Power Company's Plant," *The Electric Journal*, Vol. XI, pp. 302-305.

Irwin, William, *The New Niagara. Tourism, Technology, and the Landscape of Niagara Falls 1776-1917*. University Park, Pennsylvania: The Pennsylvania State University Press, 1996.

Israel, Paul, *Edison: A Life of Adventure*. New York: John Wiley and Sons, 1998.

Jackson, John, John Burtniak, and Gregory Stein, *The Mighty Niagara: One River, Two Frontiers*. Amherst, New York: Prometheus Books, 2003.

Jonnes, Jill, *Empires of Light: Edison, Tesla, Westinghouse and the Race to Electrify the World*. New York: Random House, 2003.

Lanphier, R.C., *Electric Meter History and Progress*, Springfield, Illinois: Sangamo Electric Company.

Many Voices: A Collective History of Greater Fort Erie. Fort Erie: Fort Erie Museum Board, 1996.

Mavor, James, *Niagara in Politics: A Critical Account of the Ontario Hydro-Electric Commission*. New York: E.P. Dutton & Company, 1925.

National Research Council, *The Lake Erie–Niagara River Ice Boom: Operations and Impacts*. Washington, D.C. National Academy Press, 1983.

Negru, John, *The Electric Century: An Illustrated History of Electricity in Canada*. Montreal: Canadian Electrical Association, 1990.

The Niagara Falls Electrical Handbook: Being a Guide for Visitors from Abroad Attending the International Electrical Congress, St. Louis, Mo., September 1904. Niagara Falls: American Institute of Electrical Engineers, 1904.

Niagara Parks Commission (formerly Commissioners of the Queen Victoria Niagara Falls Park) *Annual Reports*. These are available as separately printed documents or as part of the Annual Sessional Papers for the Province of Ontario.

Quinn, Frank H., Raymond A. Assel, Daniel W. Gaskill, "An Evaluation of the Impact of the Niagara River Ice Boom on the Air Temperature Regime at Buffalo, New York," *Journal of Applied Meteorology*, Volume 21, March 1982.

Rankine, deLancey, compiler. *Memorabelia* [sic] *of William Birch Rankine of Niagara Falls, New York*. Niagara Falls, New York: Power City Press, 1926.

Raymond, Marguerite A., *Tales of the Old Town*. Ridgeway, Ontario: Fort Erie Historical Society Museum, 1990.

The Review Company, *Bridgeburg: The Gateway to Canada's Abundant Prosperity*. Bridgeburg, Ontario: The Review Company, Printers and Publishers, September 1913.

Sawle, George Robert Tregerthen, *Autobiography of G.R.T. Sawle*. Welland, Ontario: The Review Company of Fort Erie, 1962.

Seibel, George A., *The Niagara Portage Road. A History of the Portage on the West Bank of the Niagara River*. Niagara Falls, Ontario: The City of Niagara Falls, Canada, 1990.

Seibel, George A., editor, *Niagara — River of Fame*. Ridgeway, Ontario: Kiwanis Club of Stamford, Ontario, and City of Niagara Falls, Ontario, 1968

Seibel, George A., *Ontario's Niagara Parks: 100 Years 1885-1985. A History*. Niagara Falls, Ontario: The Niagara Parks Commission, 1985.

Smith, Cecil, B., "Construction of Canadian Niagara Power Company's 100,000 H.P. Hydro Electric Plant at Niagara Falls, Ont." *The Canadian Engineer*, Vol. XII, No. 3, March 1905, pp. 82-86, 109-112.

Smith, Cecil, B., "Construction of Canadian Niagara Power Company's 100,000 H.P. Hydro Electric Plant at Niagara Falls, Ont." *Transactions of the Canadian Society of Engineers*, Vol. XIX, 1905, pp 62-81.

Spieler, Cliff, and Tom Hewitt, *Niagara Power from Joncaire to Moses*. Lewiston, N.Y.: Niagara Power Publishers, 1960, revised edition.

Stetson, Frances Lynde, "The Use of the Niagara Water Power," *Cassier's Magazine*, Niagara Number, Vol. 8, No. 3, July 1895.

Van Cleve, A.H., "The Hydro-Electric Development and Transmission Lines of the Canadian Niagara Power Company," *Transactions of the American Society of Civil Engineers*, Vol. 62, March 1909, pp. 199-244.

Way, Ronald L., *Ontario's Niagara Parks: A History*. Niagara Parks Commission, 1946.

Acknowledgments

Acknowledgments are curious beasts. The ultra-short ones hint at anger over the fact that no one besides the author's mother, brother, and dog had any faith in the great work. The ultra-long ones seem to list everyone the author has met in his or her life. Surely not all of the 400 names represent significant contributors, but the mere act of listing might make them feel obliged to do so for the next book. And so we come to this author's dilemma: What to say when there really *were* so many people who were genuinely helpful?

Let me start by thanking those who, upon reading this list, will not find their names. For undoubtedly I will miss some, and to you I say, thank you most sincerely.

One of the other great mysteries of acknowledgments is why spouses and children are mentioned in the last-but-not-least section south of the legendary 400. I want to say right here that I owe my greatest debt to my wife Philippa for her patience, support, research assistance, companionship, editorial skills, and occasional impatience. Thank you very much and yes, we can now take a holiday. Did I mention this great idea I have for another book?

Many thanks to Mardon Erbland and Bill Daley for getting things rolling and to Bill for keeping them going. Mardon came to Canadian Niagara from Fortis to help a troubled company regain its confidence. He succeeded admirably and quickly realized that Canadian Niagara is a company whose history should not be allowed to disappear through the usual neglect of corporate records. When he saw that the records were too large to be handled locally, he authorized sending them to the National Archives and Library of Canada (Library and Archives Canada) in Ottawa and enlisted Peter Sindell to facilitate the process. Bill Daley, Erbland's successor as President and CEO, knew that it wasn't enough to preserve the records in an archives. He suggested that a book would help bring the company and its dedicated employees and retirees out of the shadows. Together we discovered that the history was far too big to fit into a single volume, but I am exceedingly grateful to have been privileged to include as much as I could into this one.

The number of people who were willing to be interviewed was one of many pleasant surprises during my research. Most, but not all, of the interviewees are employees or retirees. I have been honoured to be invited into your homes and offices, to talk with you at various social gatherings. I have particularly enjoyed the fourth-Monday-of-the-month breakfasts with retirees. You have taught me about much more than Canadian Niagara; you have taught me something about balance and taking time for social life. And let us hope that my list is pretty much complete when, in alphabetical order I say thank you for the interviews to: Shirley Athoe, Fred Barber, Bob Barnett, Barry Bracken, Annette Brinda, John Campsie, Jan Carr, Bill Daley, John Damato, Blaine Desrosiers, Todd Docker, Bob Douglas, Walter "Babe" Douglas, Cynthia and David Durward, Mardon Erbland, Frank and Sandra Flake, John Fogg, Syd Freeman, Jim Fretz, Mary Genco-Brant, David Gordon, Ray Gosine, Stanley Hall, Clayton Hesser, Leonard Heximer, Rick Heximer, Bill Hunter Jr., Jack Irish, Doug Jacobi, Gladys Johnston, Bill Kae, Don Kelly, Walter Lord Charest, Shirley and Harvey McAteer, Harry Macdonell, Maria McKinnon, John "Jack" and Ann McLeod, Herb Miracle, Walter Moroz, Walter "Sid" Morris, Jim Nigh, Fred

O'Brien, John Parisi, Dave Renshaw, Norm Rouse, Russell "Sandy" Sanderson, Harry Saylor Sr., Harry Saylor Jr., Don Smith, Hector Smith Sr., Mary Lou and Hector Smith Jr., Wendy Spry, Alan Stouffer, Tom Trasmundi, Lisa Trider, Audrey Troupe, Gino Visentin, and Craig Woodward.

I owe a special debt to the employees at the Rankine Generating Station. You were always there to open the door, show me around, answer my questions, share your pride and deep knowledge. Together, all of you helped me fall in love with Rankine, just as you have. Thanks to all of you, including Terry Bearss, Jim Docker, Walter Lord Charest, Denis Levesque, Steve Muir, Dave Steinschifter. And you too Frank, even though you were retired before I started on the job.

The lifeblood of historical research courses through the arteries of museums, archives and libraries. They stand ready to serve us when we get around to asking important questions such as who we are, where we came from, what we stand for, and what we have contributed.

During the course of my research I drew heavily on old institutional friends that have served me well before: the Metro Toronto Reference Library, the many libraries at the University of Toronto, and the National Archives of Canada and National Library of Canada, now merged as Library and Archives Canada, which hold the papers of the Canadian Niagara Power Company under the care of business archivist Steve Salmon. My thanks also to archivists Larry McNally and Andrew Rodger.

As my research carried me down the Queen Elizabeth Way, I visited other institutions where I had not done research before, but where I hope to do more research in future. The Special Collections and University Archives in the Gibson Library at Brock University in St. Catharines provided a quiet retreat with a wonderful view as well as rich research resources. Its head, Lynne Prunskus, and her assistant Edie Williams were unfailingly welcoming and helpful. On some trips I would find a pile of new material on the desk and Lynn or Edie would say, "You didn't request this, but it might be helpful," and it always was. And when I wandered down from the 10th floor to Library Copying and Printing Services on the 5th floor, Maureen McCart made electronic scans of illustrations with a speed that would leave most places in Toronto in the dust.

A little further down the QEW, Jane Davies, Museum Administrator, and Erin Wilson, Curator of Structural Heritage, introduced me to the surprising riches of the Fort Erie Historical Museum in Ridgeway. In addition to helping me find what I was looking for, both Jane and Erin anticipated what I might need but hadn't yet realized I needed, which was nice, and then found it for me, which was even nicer. Working with Jane and Erin once again reminded me of the importance of small regional and community museums. These institutions make a disproportionately large contribution to our historical understanding. Without them, we would be a much poorer nation.

On the Canadian side of the Niagara River, the Fort Erie Public Library graciously gave me access to transcripts of interviews conducted in the mid-1980s with elderly Fort Erie area residents. Thank you, Erin, for bringing these to my attention, they were every bit as good as you said they were. The Victoria Avenue branch of the Niagara Falls Public Library never failed to impress me. The collection is both rich and well-organized, and the librarians at the reference desk were always knowledgeable and helpful. I am sorry I don't know all of your names, but would like to thank Andrew Porteus, Manager of Adult Reference and Information Services, and Kathy

Simpson, Local History Librarian. It is so rewarding to visit a library that still appreciates and maintains vertical files. I am also indebted to Lundy's Lane Historical Museum.

The Canadian Niagara story is an international one, and so I often found myself on the American side of the Niagara River. The Buffalo and Erie County Historical Society Library and Museum, housed in the former New York State Pavilion, the only permanent building left from the Pan-American Exposition of 1901, is enough on its own to justify a trip there. The excellent collections and helpful staff lived up to the grand expectations created as soon as one enters the building. My thanks to Executive Director Dr. William H. Siener for his enthusiasm for my work, which he shared with his staff.

At the Lafayette Branch of the Buffalo and Erie County Public Library, Librarian Rob Alessi introduced me to the collections of the Grosvenor Room and provided help whenever I needed it. In Rob I have discovered an excellent Buffalo ambassador. He answered my questions, told me about special events, and even recommended some excellent places for lunch. He opened my eyes to the rich heritage of the city.

John Burtniak, my chief researcher, also helped introduce me to Buffalo. John made so many suggestions, introductions, and observations that I hardly know how to thank him. In that I am not alone. John is a scholar, author, collector and retired Brock University archivist who always has time to help others who are doing research. I shall miss our early morning phone calls and research jaunts, and his carefully worded corrections when what I wrote didn't quite hit the mark. I hope we can find another project that allows us to collaborate again.

During the research and writing, and my absences from the University of Waterloo, Wendy Stocker at the Centre for Society, Technology and Values was invaluable, as she always is. Her assistance ranged from shielding me from people when I needed time to think and write, to research and tracking down library materials, as well as transcribing tapes that could be anything from a one-or two-hour interview to something I thought of while driving down the highway and mumbled into my memo recorder. Wendy, you are a much-appreciated rarity and my thanks to you run very deep.

Ellen Garfield also transcribed some of my interview tapes and brought to the work the same good humour and lively intellect that I remember from the days when we were doing a radio series and distance education course on information technology.

Nor will I forget my summer research assistant, Alexandra Ishigaki, who, as an undergraduate at the University of Western Ontario, offered to work for free one summer. I had to insist on paying her, and she was worth every penny. As time went on, I had to impose other conditions, such as the time we were working in Ottawa on Canada Day and she insisted on helping me with research in the National Archives instead of joining in the festivities. At noon I said she was fired if she didn't get out and enjoy what the capital of Canada has to offer on July 1.

I owe a great deal to many Canadian Niagara employees for their unfailing patience, but most particularly to Mari Ann Rusnov, Wendy Spry and Mary Tibollo. If one couldn't figure out who or what I was talking about, one of the other two could. Mari Ann, Wendy and Mary are living proof that the extra X-chromosome gives women a lot of extra ability.

At National Grid USA, formerly Niagara Mohawk, I am indebted to Joe Santore, Archivist, and Dr. Robert F. Dischner, Director, Technical Training.

All historians follow in the steps of their predecessors, and I retraced many of the steps taken by the late Daniel Dumych, whose short books on the company helped me enormously. I am also grateful to his mother, Anne Dumych, who granted me access to Daniel's papers and became a personal friend.

And I could not end without thanking the people at The Boston Mills Press and I particularly appreciated the contributions of designer Gill Stead. Like small libraries and archives, small presses are essential to helping us understand who we are. My thanks also to Gill for introducing me to George Fischer, the photographer whose cover photo gave me a new angle on the Rankine Generating Station. Thanks very much, George. I hope we'll work together again.

Thank you all. I'll let you know when I start my next book.

And just when we thought we had everything finished, proofreader Stephen Jones helped Gill and me make the book even better. We appreciate you giving up weekends for *The Canadian Niagara Power Company Story.*

Index

"c" denotes caption text